Electronic Access Control

Electronic Access Control

Second Edition

Thomas L. Norman, CPP/PSP

Protection Partners International (PPI), Houston, TX, United States

Protection Partners International (PPI), Beirut, Lebanon

Butterworth-Heinemann
An imprint of Elsevier

Butterworth-Heinemann is an imprint of Elsevier
The Boulevard, Langford Lane, Kidlington, Oxford OX5 1GB, United Kingdom
50 Hampshire Street, 5th Floor, Cambridge, MA 02139, United States

Notices
Knowledge and best practice in this field are constantly changing. As new research and experience broaden our
understanding, changes in research methods, professional practices, or medical treatment may become necessary.

Practitioners and researchers must always rely on their own experience and knowledge in evaluating and using any
information, methods, compounds, or experiments described herein. In using such information or methods they
should be mindful of their own safety and the safety of others, including parties for whom they have a professional
responsibility.

To the fullest extent of the law, neither the Publisher nor the authors, contributors, or editors, assume any liability
for any injury and/or damage to persons or property as a matter of products liability, negligence or otherwise, or
from any use or operation of any methods, products, instructions, or ideas contained in the material herein.

British Library Cataloguing-in-Publication Data
A catalogue record for this book is available from the British Library

Library of Congress Cataloging-in-Publication Data
A catalog record for this book is available from the Library of Congress

ISBN: 978-0-12-805465-9

For Information on all Butterworth-Heinemann publications
visit our website at https://www.elsevier.com/books-and-journals

Working together
to grow libraries in
developing countries

www.elsevier.com • www.bookaid.org

Publisher: Candice Janco
Acquisition Editor: Candice Janco
Editorial Project Manager: Hilary Carr
Production Project Manager: Mohanapriyan Rajendran
Cover Designer: Mark Roger

Typeset by MPS Limited, Chennai, India

Contents

PART I   **THE BASICS**..1

CHAPTER 1   Introduction and Overview...3
Chapter Overview.. 3
Rules to Live By... 4
Who Should Read This Book?... 5
How Is Material Presented in This Book?.. 6
What Will You Learn, and How Will It Help Your Career?.............. 6
What Is in This Book? 8
    Part I: the basics 8
    Part II: how things work.. 9
    Part III: The things that make systems sing 12
Chapter Summary.. 15

CHAPTER 2   Foundational Security and Access Control Concepts...................21
Chapter Overview.. 21
Understanding Risk 22
    Types of organization assets 22
    Types of users...23
    Types of threat actors 24
    Understanding criticalities and consequences..........................26
    Understanding vulnerability 27
    Understanding probability 27
    What is risk?...28
Managing Risk... 28
    Methods of managing risk..28
    How security and access control programs help manage risk............29
    Security program elements...29
    The importance of a qualified risk analysis...............................30
    The importance of security policies and procedures 30
Types of Countermeasures 31
    Hi-tech 31
    Lo-tech..32
    No-tech..32
    Mixing approaches 33
    Layering security countermeasures...33
Access Control System Concepts.. 34
    Types of users...37
    Types of areas/groups...37
    User schedules 37

Portal programming 38
Credential programming 38
Group and schedule programming 38
Chapter Summary 38

CHAPTER 3 **How Electronic Access Control Systems Work**...**43**
Chapter Overview... 43
First, a Little History 44
The Basics.. 45
Authorized Users, User Groups, Access Zones, Schedules,
and Access Groups 46
Authorized users 46
User groups..46
Access zones..46
Schedules 47
Access groups 47
Portals 47
Types of portals 48
Credential readers..49
Electrified locks 49
Safety systems 49
Alarm monitoring 50
Request-to-exit sensors 51
Credentials and Credential Readers 51
Credential Authorization 52
Locks, Alarms, and Exit Devices.. 52
Electrified locks 52
Alarms..54
Exit Devices...54
Data, Data Retention, and Reports.. 55
Chapter Summary 56

PART II **HOW THINGS WORK** **59**

CHAPTER 4 **Access Control Credentials and Credential Readers** **61**
Chapter Overview... 61
Access Credentialing Concepts.. 62
Keypads... 63
Access Cards, Key Fobs, and Card Readers... 64
Wiegand Wire Cards 66
125 K Passive Proximity Cards.. 68
125 KHz (Low Frequency) Active Proximity Cards.............................. 68

13.56 MHz (High Frequency) Contactless Smart Cards 68
RFID Wireless Transmitter Systems 69
Multitechnology Cards 69
Mobile Phone Access Control 70
Capture Card Reader 70
Multitechnology Card Readers 70
Biometric Readers 70
Photo Identification 73
Chapter Summary 74

CHAPTER 5 Types of Access Controlled Portals 77
Chapter Overview 77
Portal Passage Concepts 78
  Card entry/free exit 78
  Card entry/card exit 78
  Tailgate detection 78
  Positive access control 80
  2-Man rule 80
  Schedules 81
  Antipassback 81
Pedestrian Portal Types 82
  Standard doors 82
  Automatic doors 83
  Revolving doors 85
  Turnstiles 87
  Man-traps 88
  Automated walls 89
Vehicle Portals 89
  Standard barrier gates 89
  Automated vehicle swing gates 91
  Automated sliding vehicle gates 91
  Automated roll-up vehicle gates 92
  High-security barrier gates 92
  Sally ports 94
Chapter Summary 94

CHAPTER 6 Life Safety and Exit Devices 99
Chapter Overview 99
Life Safety First 99
Security Versus Life Safety 101
Understanding National and Local Access Control
Codes and Standards 101
  NFPA 101 101

International building code 101
NFPA 72 102
More on these codes 102
UL 294 103
Life Safety and Locks 104
Life Safety and Exit Devices 107
Life Safety and Fire Alarm System Interfaces 109
Chapter Summary 111

CHAPTER 7   Door Types and Door Frames **115**
Chapter Overview 115
Basics About Doors and Security 115
Standard Single-Leaf and Double-Leaf Swinging Doors 117
Hollow metal doors 117
Solid core wood doors 118
Framed glass doors 119
Unframed glass doors 120
Total doors 121
Pivoting doors 123
Balanced doors 124
Door Frames and Mountings 125
Hollow metal—high-use and high-impact 125
Aluminum—medium-use and medium-impact 125
Wood—light-use and light-impact 126
Door mounting methods 126
Overhead Doors 126
Roll-up doors 126
Paneled overhead doors 126
Revolving Doors 126
Sliding Panel Doors 127
Bifold and Fourfold Doors 127
Chapter Summary 128

CHAPTER 8   Doors and Fire Ratings **131**
Chapter Overview 131
What Are Fire Ratings? 132
Basic fire egress concept 132
How should this be done? 132
Exceptions 132
Fire Penetration Ratings 133
Hose stream test 134
Door Assembly Ratings 135
The three-fourths rule 135
Doors with glass 135

Temperature rise doors...135
Louvers 135
Fire Door Frames and Hardware...136
Latching devices...136
Fire exit hardware..136
Pairs of Doors 137
Latching hardware 137
Inactive leaf on pair of doors 137
Double egress pairs..137
Astragals 138
Smoke and draft control 138
"Path of Egress" Doors 138
Electrified Locks and Fire Ratings 139
Additional References 140
Chapter Summary 140

CHAPTER 9   **Electrified Locks—Overview**..**145**
Chapter Overview..145
Why Electric Locks? 146
Types of Electrified Locks 148
How Electrified Locks Work 149
Electric strikes 150
Electrified mortise locks..151
Electrified panic hardware 152
Electrified cylinder locks 153
Magnetic locks..153
Electrified dead-bolts...154
Paddle-operated electromechanical dead-bolts 155
Lock Power Supplies...156
Electrified Lock Wiring Considerations 157
Voltage drop example 159
Electrified Lock Controls 159
Types of Locks Not Recommended..161
Chapter Summary 164

CHAPTER 10   **Free Egress Electrified Locks**...**169**
Chapter Overview..169
Types of Free Egress Locks 170
Electrified Mortise Locks 170
Mortise latch only—no lock..171
Mortise locks with no dead-bolt...171
Mortise locks with dead-bolts 172
Door frame considerations 174

Additional lock switch fittings 175
Door handing 175
Electrified "Panic" Hardware 176
Rim exit devices 176
Mortise lock exit devices 177
Surface-mounted vertical rod exit devices 177
Concealed vertical rod exit devices 178
Three-point latching exit device 179
Exit device functions 179
Electrical options 179
Popular double door applications 181
Electric Strikes 181
Switches available for electric strikes 184
Electrified Cylinder Locks 184
Self-Contained Access Control Locks 185
Chapter Summary 185

CHAPTER 11   Magnetic Locks 189
Chapter Overview 189
Standard Magnetic Locks 189
Standard magnetic lock applications 191
Magnetic Shear Locks 191
Magnetic shear lock applications 193
Magnetic Gate Locks 193
Cautions About Magnetic Locks 194
Egress assurance 195
Operational and maintenance warnings 196
Chapter Summary 198

CHAPTER 12   Electrified Dead-Bolt Locks 201
Chapter Overview 201
Surface-Mounted Electrified Dead-Bolt Locks 201
Concealed Direct-Throw Mortise Dead-Bolt Lock 202
Dead-Bolt Equipped Electrified Mortise Lock 203
Top-Latch Release Bolt 204
Electrified Dead-Bolt Gate Locks 205
Electrified dead-bolt lock safety provisions 205
Chapter Summary 206

CHAPTER 13   Specialty Electrified Locks 209
Chapter Overview 209
Electrified Dead-Bolt-Equipped Panic Hardware 210
Securitech Locks 210

Delayed Egress Locks 211
    Specialize school locks to protect against active shooters 212
Hi-Tower Locks...212
CRL-Blumcraft Panic Hardware 214
Chapter Summary 214

CHAPTER 14 Selecting the Right Lockset for a Door...............................217
Chapter Overview..217
Standard Application Rules..217
How to Select the Right Lock for Any Door 218
    Description of door...218
    Framed glass door..221
    Herculite lobby doors 222
    High-rise building stair-tower door...223
    Rear-exit door on warehouse with hi-value equipment...................225
    Office suite door..226
    Double-egress doors—hospital corridor 228
    Inswinging office door 229
    Revolving door—emergency egress side door 231
Chapter Summary 232

CHAPTER 15 Specialized Portal Control Devices and Applications 237
Chapter Overview..237
Specialized Portals for Pedestrians 238
    Automatic doors 238
    Man-traps..239
    Full-verification portals 241
    Electronic turnstiles 242
    Antitailgate alarm 245
Specialized Portals for Vehicles...246
    High-security barrier gates 246
    Sally ports 247
Chapter Summary 250

CHAPTER 16 Industry History That can Predict the Future............................255
Chapter Overview..255
A Little Background...256
First Generation 257
Second Generation..259
Third Generation...260
Fourth Generation...261
    Stalled progress...263
Fifth Generation..265

Avoiding Obsolescence 267
  Planned obsolescence 267
  Unplanned obsolescence..267
  What the future holds 268
Chapter Summary 269

CHAPTER 17   Access Control Panels and Networks 273
Chapter Overview... 273
Access Control Panel Attributes and Components.............................. 273
Communications Board 276
  Power supply and battery 277
  Central processing unit..277
  Erasable programmable read-only memory......................................278
  Random access memory..278
  Input/output interfaces..278
Access Control Panel Form Factors... 279
Access Control Panel Functions... 282
Access Control Panel Locations... 284
Local and Network Cabling 286
Networking Options.. 289
Redundancy and Reliability Factors 291
  Good wiring and installation...292
  Good design...292
  Good power 292
  Good data infrastructure 293
  Redundancy 294
Chapter Summary 294

CHAPTER 18   Access Control System Servers and Workstations....................299
Chapter Overview... 299
Server/Workstation Functions 300
  Store system configurations 300
  Store the system's historical event data...301
  Manage communications throughout the entire system 303
  Serve workstations with real-time data and reports 304
Decision Processes.. 305
System Scalability 306
Unscalable Systems 306
  Basic scalability...307
  Multisite systems 307
  System-wide card compatibility...307
  Enterprise-wide system...308
  Master host..308
  Super-host/subhost..308

Access Control System Networking 309
 The core network..309
 The server network 310
 The workstation network 310
 The access control panel network 311
 Integrated security system interfaces..311
 VLANs..312
 Multisite network interfaces...312
 Integration to the business information technology
 network 312
Legacy Access Control Systems 313
Chapter Summary 314

PART III **THE THINGS THAT MAKE SYSTEMS SING......................321**

CHAPTER 19 **Security System Integration** 323
 Chapter Overview... 323
 Why Security Systems Should Be Integrated 323
 Integration Concepts... 326
 Benefits of System Integration... 328
  Operational benefits...328
  Cost benefits 331
 Types of Integration 331
  Dry contact integration...332
  Wet contact integration 332
 Serial Data Integration.. 334
 TCP/IP Integration.. 334
 Database Integration 334
 System Integration Examples 334
  Basic system integration...335
  More advanced system integration.......................................335
  Advanced system integration 336
 Chapter Summary 343

CHAPTER 20 **Integrated Alarm System Devices** 347
 Chapter Overview... 347
 Alarm Concepts 347
  Detection and initiation 347
  Filtering and alarm states 348
  Communication and annunciation..350
  Assessment...351

Response 353
Evidence..353
Types of Alarm Sensors 354
Outer perimeter detection systems...354
Building perimeter detection systems 359
Interior volumetric sensors..361
Interior point detection systems 362
Intelligent video analytics sensors 367
Complex alarm sensing 368
Beyond Alarm Detection.. 369
Trend analysis..369
Vulnerability analysis 369
Alarm analysis 369
Chapter Summary 370

CHAPTER 21   Related Security Systems 375
Chapter Overview.. 375
Photo ID Systems 375
Visitor Management Systems.. 376
Security Video 377
Video history you need to know...377
Cameras and lenses 379
Lighting and light sources...380
Auto-white balance..382
Dynamic range...382
Display devices...382
Video recording devices 382
Video motion detectors 386
Video analytics 387
Video system interfaces...387
Security Communications... 388
Two-way radio..389
Telephones 389
Security intercoms and bullhorns..390
Public address systems 391
Nextel phones 391
Voice loggers...392
Smart phones and tablets..392
Consolidated communication systems 392
Security Architecture Models for Campuses and Remote Sites 393
Command, Control, and Communications Consoles............................ 394
Chapter Summary 395

CHAPTER 22   **The Merging of Physical and IT Security** **401**
Chapter Overview 401
There Is Only One Security Mission 402
IT Security and Physical Security Share the Same Mission 402
What Vulnerabilities Exist Between IT & Physical Security? 404
Sophisticated Threat Actors Are Exploiting Those Vulnerabilities 406
Learn How to Reduce and Mitigate Those Vulnerabilities 407
Chapter 21A—Chapter Summary 408
Chapter 21A Summary—The Merging of Physical and IT Security ... 409

CHAPTER 23   **Securing the Security System** **413**
Chapter Overview 413
Understand That the Organization Isn't Secure If the Security
System Isn't Secure 414
What Kinds of Threats Present a Problem to Securing the
System Data? 414
What Kinds of Vulnerabilities Can Exist in the Security
System Itself? 415
What Can We Do To Secure the Security System? 416
A 9 Point Plan for Securing the Security System 417
Chapter Summary 420
Chapter 21B—Chapter Summary 421

CHAPTER 24   **Related Building/Facility Systems and REAPS Systems** **425**
Chapter Overview 425
Building/Facility Systems 425
    Elevators 426
    Stairwell pressurization 427
    Lighting 428
Controlling and Automating Building Functions 429
    Direct action interfaces 429
    Proxy action interfaces 429
    Feedback interfaces 429
REAPS Systems 430
    Irrigation systems 430
    Deluge fire sprinkler control 431
    Acoustic weapons 431
    High-voltage weaponry 433
    Remotely operated weaponry 433
    Appropriateness 434
    Operationally 434
    Safety systems 435
Chapter Summary 436

CHAPTER 25   Cabling Considerations..**441**

Chapter Overview... 441
Cable Types 441
    Copper/fiber...442
    Cable voltage and power classes...442
    Wire gauge...443
    Insulation types..443
    Stranded versus solid core wires...444
    Cable colors 444
    Cable brands 445
Conduit or No Conduit... 446
    Why use conduit?...446
    Types of conduit...446
    Other wireways...448
    Indoor conduit applications...448
    Outdoor conduit applications 448
    When you can forget about conduit......................................448
    Conduit fill..449
    Conduit bends 451
    Conduit/cable fire protection...451
Cable Handling 451
    Cable handling nightmares..451
    Cable handling and system troubleshooting 451
Cable Dressing Practices 452
    What is cable dressing?...452
    Cable dressing nightmares..452
    Cable dressing and system troubleshooting 452
    The proper way to dress cables...453
    Cable cross-dressing 454
Cable Documentation 455
    What is cable documentation? 455
    Who cares about cable documentation?................................455
    When should cable documentation begin? 456
    What is the best way to document cabling? 456
    What is the best way to present cable documentation?..................456
Chapter Summary 457

CHAPTER 26   Environmental Considerations **463**

Chapter Overview... 463
Electronic Circuitry Sensitivities 463
Environmental Factors in System Failures 464
    Temperature extremes 464
    Humidity or condensation 465

Vibrations 466
Dirt 466
Insects, birds, snakes, and other creatures 467
Lighting (at access control system portals) 467
Securing the IP network 468
Access control in the cloud 469
Security-Systems As-A-Service 470
Chapter Summary 470

CHAPTER 27   **Access Control Design** **473**
Chapter Overview 473
Design Versus Installation Versus Maintenance
(The Knowledge Gap) 474
The Importance of Designing to Risk 474
The Importance of Designing for the Future 475
Design Elements 476
Drawings 476
Specifications 477
Interdiscipline coordination 478
Product selection 479
Project management 479
Client management 480
Designing Robust Portals—How Criminals Defeat
Common Locks, Doors, and Frames 481
Unlocking the door from the outside 482
Double glass door exploit 482
Defeating electrified panic hardware 483
Defeating door frames 484
Application Concepts 484
Robust design 485
Redundancy 486
Expandable and flexible 486
Easy to use 486
Sustainable 487
Implementing Design Ideas to Paper 488
Creating access control zones 488
Door types 488
Alarm devices 490
Racks, consoles, and panels 490
Conduits and boxes 490
Physical details 492
Riser diagrams 493
Single-line diagrams 493

Wiring diagrams 494
Schedules 495
System Installation 495
Project planning..495
Project schedule..496
Shop and field drawings...497
Product acquisition 497
Permits 498
Coordination with other trades...498
Access coordination..498
Preliminary checks and testing...499
Final works 499
System Commissioning 499
Completing Punch List Items..500
System Acceptance..500
Chapter Summary...501

CHAPTER 28   Access Control System Installation and Commissioning 509
Chapter Overview...509
Jobsite Considerations 510
Conduit Versus Open Cabling..511
Device Installation Considerations..511
The Importance of Documentation 512
Device Setup and Initial Testing..513
Alarm and Reader Device Database Setup...................................513
User Access Database Setup 513
Access Schedules and Areas 514
Chapter Summary...515

CHAPTER 29   System Management, Maintenance, and Repair 519
Chapter Overview...519
Management..519
Governance, Risk Management and Compliance.........................520
Database management 520
Merging databases of different systems on a common
corporate campus..521
Alarm management...522
Reports 522
Maintenance and Repair...522
System as-built drawings...522
Wire run sheets...524
System infant mortality 524
Maintenance versus repairs 525

Scheduled maintenance 525
Emergency repairs 525
Maintenance contracting options 526
Security System Integrity Monitoring..528
Chapter Summary 532

Index 537

The Basics

Chapter 1

Introduction and Overview

CHAPTER OBJECTIVES

1. Understand who should read this book and why
2. Understand what you can expect to learn
3. Understand how information is presented in the book
4. Understand how this book will improve your career

CHAPTER OVERVIEW

This book is about physical Access Control Systems. It does not cover Information Technology Access Control, which is another subject altogether.

Access control systems are electronic systems that allow authorized personnel to enter controlled, restricted, or secure spaces by presenting an access credential to a credential reader. Access control systems can be basic or highly complicated ranging across state and national borders and incorporating security monitoring elements and interfaces to other security systems and other building systems.

Security technicians, designers, and program managers who fully understand access control systems have a distinct advantage over those with only a passing knowledge. This book will give you that edge.

This book covers virtually every aspect of electronic alarm/access control systems and also includes insight into the problems you will face as you learn to install, maintain, or design them, including valuable information on how to overcome those challenges.

This book is designed to help you launch your career with alarm/access control systems well ahead of your contemporaries by equipping you with the knowledge that takes most people decades to assemble.

Electronic Access Control. DOI: http://dx.doi.org/10.1016/B978-0-12-805465-9.00001-4

New material for the 2nd Edition includes:

- Chapter 4: More information on Biometrics and Card Readers
- Chapter 13: Information on New Locks
- Chapter 17: New information on Networks and Network Options, Risks and Challenges
- Chapter 19: Information on a new technology to better handle multitenant hi-rise building access control systems
- Chapter 20: More information on integrating video and alarm functions into the Access Control System
- Chapter 24: Access Control Systems in the "Cloud" and Security Systems as a Service (SSaS)
- Chapter 25: Open Sourced Solutions
- Chapter 26: Card Data Management, Access Level Management, Cardholder Private Information and Improved Control
- Chapter 27: Security System Integrity Monitoring & Scheduled Maintenance Programs

RULES TO LIVE BY

While many security system designers today come out of Engineering Schools, I became a designer after first being a technician. I also spent a number of years learning the intricacies of Risk Management and Security Program Management. This background has allowed me to create three important rules that I follow when designing systems:

Rule #1: Design security systems based upon an understanding of the facility's risks and make sure the system can mitigate as many as possible.
Rule #2: Design the security system as a "Force Multiplier" to repeatedly expand the capabilities of the Security Force.
Rule #3: Always design security systems as if the person who will maintain it is a violent psychopath who knows where you live.

Rule #3 acknowledges that technicians have a very difficult job. If you are one, you already know this. If you are an aspiring designer, please remember all three of these rules. They will serve you well. If you design systems that embrace risk, that expand the capabilities of the security force, and that are easy to operate and easy to maintain, then you will be much loved by your clients, the installing contractors, and the maintenance technicians.

If you are a security system technician, it pays to learn more about risk and how to mitigate it. There is more to security than security systems.

In fact, I am well known for saying: "Electronics is the high priest of false security." It is understandable but unfortunate that very few security system technicians understand how organizations operate, why organizations assemble assets, what kinds of threat actors put those assets at constant risk, what threat scenarios the security program should be designed to mitigate, how to identify a very wide range of vulnerabilities, how to calculate threat/vulnerability probabilities, and indeed much about risk and risk management. There is much more to security than installing cameras, alarms, and card readers.

My background as an installation and maintenance technician has given me a deep appreciation of the difficulties in installing and maintaining security systems. Although today I design highly integrated security systems, security risk management programs, and antiterrorism technologies for extremely high-threat environments, I still draw daily on that early experience as a technician. Engineers can do plenty to make the installation and maintenance process much easier and to make designs resonate with the risk they manage.

WHO SHOULD READ THIS BOOK?

Alarm and access control systems can be complicated. Today they do much more than most engineers and technicians think. You will finish this book with a very high level of knowledge about alarm/access control systems. You will learn about security technologies that very few technicians, designers, and security program managers know anything about or indeed have ever heard of. After reading this book, you will rise to an expert level of knowledge.

This book should be read by:

- Security System Technicians
- Security System Engineers and Designers
- Risk Managers
- Security Program Directors and Managers
- Guard Company Managers and Supervisors
- Security Consultants
- Facility Managers
- Security Installation Project Managers
- Anyone new to Alarm and Access Control Systems

Each of these professionals will learn important things from this book that are not available from other common sources and that will further their careers.

HOW IS MATERIAL PRESENTED IN THIS BOOK?

This book is designed to teach concepts that may be unfamiliar to many if not most readers. It is a foundational book upon which your career may be built, especially if you are going into a technical career in the security industry. Whether you will be a technician, installer, or designer, this book includes information that is essential to your success.

As many of these concepts are highly detailed and have common threads that run through many variations of the technology, the information presented herein is presented in a similar way. You will see a design concept presented first in its simplest form. In another chapter you will see it presented again in a slightly different form, relating this time to something similar, but somehow different than the first form. You may see the design concept presented a third time in still another environment. Finally, you will see the design concept integrated later in the book into a much more complex assembly, often where it is used to integrate multiple systems or multiple concepts together into a higher, more complex concept.

This layering approach to concepts ensures that you will not be left in the dark at any time. As each layer is presented, one over the other, you will build brick by brick a structure of knowledge that is secure on a foundation of the simple concepts learned early in the book.

This books teaches in a way that is similar to how a user may use a web forum to enhance their knowledge on a subject by hearing a concept being described by various people talking about the same idea from many different points of view and each time talking about different applications of the concept. Each time a concept is presented, it is shown in a slightly different way, enlightening your understanding of the idea each time, but from a different angle.

WHAT WILL YOU LEARN, AND HOW WILL IT HELP YOUR CAREER?

Security technicians will develop a profound understanding of how alarm/access control systems work, which will demystify previous complexities and make them stone-cold experts among their peers. Reading this book carefully will instill a depth of knowledge that few people in the industry have, making you an instant expert amongst your peers and much more valuable to your employer.

Security system designers and engineers will learn details about how alarm/access control systems work and interface the building architecture that will keep them from making alarmingly common design errors. Designers will learn about rare and unusual technologies and how to think "outside the box" to create designs that are powerful and economical and extremely capable.

Risk Managers will learn how alarm/access control systems can reduce operating costs by minimizing the number of security personnel on staff and making the most of the security staff. Alarm/access control systems can be a "Force Multiplier" when they are designed and installed correctly, and as such can save hundreds of thousands annually in operating costs as compared to a guard staff. Risk Managers will also learn secrets for identifying and managing risk of which few Risk Managers are aware.

Guard Company Managers and Supervisors will learn how to use alarm/access control systems to their fullest to amplify the reach of their staff and mount energetic responses to alarms in a timely manner. The knowledge gained from this book will also help you keep the account in-house at services bid time and be competitive when the client is looking toward a technology solution. You will learn how to integrate guards and technology for the best value for your clients and therefore keep clients rather than losing them to technology.

Security Consultants will gain a depth of understanding about this technology that few other books can offer (one exception may be from my book, *Integrated Security Systems Design*). This book can keep consultants and designers from making common design and integration errors that I see frequently from otherwise knowledgeable consultants. It will also give them insight into high-efficiency system design that can reduce the time to design while improving the level of detail in the design documents.

Facility Managers and Security Installation Project Managers who are given responsibility for security systems will learn how to make the most of tight budgets and get more performance while reducing capital and operating costs. They will also learn how to avoid common mistakes made by facility managers that seem at first to save money but ultimately cost the organization big dollars. Facility Managers will also learn secrets to making sure that a large security project does not go badly. Learning the horror stories of others is the best prevention against learning a horror story firsthand.

Anyone new to alarm/access control systems will learn a depth of understanding that few in the industry possess. And knowledge is always good for your career.

WHAT IS IN THIS BOOK?

This book is comprised of three parts:

Part I: The Basics
Part II: How Things Work
Part III: The Things That Make Systems Sing

Part I: the basics

This section includes Chapters 1–3.

Chapter 2

Chapter 2 of this book introduces foundational security and access control concepts. In it there is a section entitled Understanding Risk that includes the types of assets organizations have to protect and how all of those assets relate to the mission of the organization. It also includes information about the types of users in each kind of facility, the types of threat actors, understanding criticalities and consequences, understanding vulnerabilities and probabilities or likelihood of security events, and finally understanding risk itself.

Chapter 2, Foundational security and access control concepts, also includes the section Managing Risk, which outlines methods for managing risk, how security programs help organizations manage risk, security program elements, the importance of basing all security program elements on a qualified risk analysis, and the importance of basing all security program elements on a sound set of security policies and procedures.

Chapter 2, Foundational security and access control concepts, is also comprised of a comprehensive discussion of types of security countermeasures, including the three basic types: Hi-Tech, Lo-Tech, and No-Tech. This chapter includes a discussion on why it is important to mix all three to make a complete security program. Chapter 2, Foundational security and access control concepts, also contains a brief introduction to access control system concepts.

Chapter 3

Chapter 3 covers how access control systems work. This chapter begins with sections on authorized/unauthorized users; types of security portals; types of devices at access control portals; a discussion about the various types of credential readers; how credential authorization works; information about locks, alarms, and exit devices; a section on alarm/access control system data and data retention and reports.

Part II: how things work

Chapter 4

Chapter 4 delves in depth into the subject of access control credentials and credential readers. It includes a section on access credentialing concepts, access cards, biometrics, and photo ID systems.

Chapter 5

Chapter 5, Types of access controlled portals, introduces information about types of access control system portals. A portal is the way through which a person must pass and an access control system portal is a portal controlled by access control system devices. Here you learn all about pedestrian and vehicle portals. Sorry, no wormholes through space— maybe in my next book.

Chapter 6

Chapter 6, Life safety and exit devices, is one of the most important chapters in the book. It covers Life Safety and Exit Devices. It teaches why life safety is the primary concern in security and access control systems and the challenges between life safety and security that confront the designer and installer. There is a section on understanding National and Local Life Safety and Access Control Codes, a section on Life Safety and Locks, and Life Safety and Exit Devices.

Chapter 7

Chapter 7 covers door types and door frames. There is an introduction about doors and security, information about the types of door construction and how they affect system performance and security, a section on door frames and why this is essential to understanding how to make security systems work reliably, and on other types of doors including roll-up doors and revolving doors.

Chapter 8

Chapter 8, Doors and fire ratings, extends the information on doors seen in Chapter 7, Door types and door frames. It includes information about door fire ratings, how door assemblies are given fire ratings, fire penetration ratings, "Path of Egress" doors, and Electrified Locks and Fire Ratings. For designers and installers, this chapter can literally save the company. A number of firms have been forced out of business after designing or installing access control systems that violated fire codes and ratings.

Chapter 9

Chapter 9, Electrified locks—overview, begins a multichapter section on Electrified Locks. This book includes one of the most comprehensive sections on electrified locks ever put into print. When you finish these chapters, you may be an electrified lock expert. Chapter 9, Electrified locks—overview, includes foundational information on types of electrified locks, how they work, lock power supplies, electrified lock control systems, and electrified lock wiring considerations. It also includes information about standard applications and types of electrified locks that are not recommended for use in security systems. (This is essential to save your career and possibly even more important to know than the right locks to use.)

Chapter 10

Chapter 10, Free egress electrified locks, talks about what should be the most common lock type—"Free Egress Locks." These locks require no special knowledge to exit, and do not rely on any electronic controls to disengage. Free egress locks should be the basic locks in your arsenal. They include electrified mortise locks and "panic" hardware.

Chapter 11

Chapter 11, Magnetic locks, covers another common type of lock (actually, it is the most common, but should be the second-most common). These are magnetic locks, the "workhorse" of most security integrators. This chapter discusses the various types of magnetic locks and when and where each should and should not be used. You career and your employer's business risk may depend on this knowledge.

Chapter 12

Chapter 12 covers electronic dead-bolt locks. Although rarely used, these locks have an important place in your knowledge base. Types discussed will include Electrified Mortise Dead-Bolts, Hybrid Mortise Locks with Dead-Bolts, and Electrified Frame-Mounted Dead-Bolt Locks. This chapter also has a section that discusses the dangers of these locks and what you should be certain of before specifying, designing, or installing plans that use them.

Chapter 13

Chapter 13, Specialty electrified locks, lends insight into a very interesting and valuable set of specialized Electrified Locks. These include locks that can solve specialized problems common in high-threat and high-esthetic environments. This is the kind of knowledge that will set you above your peers.

Chapter 14

Chapter 14, Selecting the right lockset for a door, covers standard door and lock combinations. It discusses standard application rules and a very valuable tool, the Lock/Door/Frame Types List. This is a single source for understanding how to apply all types of locks to all types of frames and doors. It can save you hours of research on many projects.

Chapter 15

Chapter 15 delves into specialized portal control devices and applications. These are the head-scratchers that can cost too much time in respect to their budget in a project. They include for pedestrians—Automatic Doors, Man-Traps, and Electronic Turnstiles; and for Vehicles—Sally Ports and High-security Vehicle Gates.

Chapter 16

Chapter 16, Industry history that can predict the future, may be one of the most important chapters in the book. It covers the history of alarm/access control systems. Altogether, there have been five generations of alarm and access control systems, and there are still many fourth generation systems on the market and just a few fifth generation systems. Understanding the history of these systems will give you deep insight into why things work the way they do (and why some systems are so maddening to work with). After reading this chapter, you will be able to recognize instantly whether a proposed system is emerging technology or a system that will be abandoned soon to the trash heap of history.

Chapter 17

Chapter 17 covers access control panels and networks. This chapter continues the information in Chapter 16, Industry history that can predict the future, and will expand your knowledge about how alarm and access control systems communicate. Believe me, thousands of hours are wasted annually on troubleshooting systems because some designer, engineer, or installer did not know what is in this chapter. In it you learn about alarm and access control system panel functions, typical locations and form factors and the problems you can incur or avoid with that knowledge, various types of data communications cabling schemes and what works and what doesn't, local and network cabling methods, and extremely valuable insight into redundancy and reliability factors.

Chapter 18

Chapter 18 reviews access control system servers and workstations. Most modern alarm/access control systems use some combination of servers and workstations. On small systems, these are often combined into a single computer; nonetheless the functions of both remain. Understanding how servers and workstations function and perform their roles in small, medium, large, and global systems can help you meet any challenge. This chapter covers decision processes, system scalability, access control system networking, and more information about legacy alarm/access control systems.

Part III: The things that make systems sing

Chapter 19

Chapter 19 introduces security system integration. This is the process of making multiple building systems "talk" to each other to perform advanced system functions. This chapter covers integration concepts, types of integration, and the benefits of security system integration. Building Facility Managers will be especially interested in this chapter as it shows how to make a building more efficient and reduce operating costs.

Chapter 20

Chapter 20 expands on Chapter 19, Security system integration, and covers integrated alarm system devices. This chapter goes in depth into alarm concepts including Alarm Initiation, Alarm Filtering, Alarm Communication, Alarm Assessment and Response, Evidence Gathering, and Trend Analysis, as well as using alarm systems to perform daily Vulnerability Analysis. It also includes a section on types of alarm sensors including Point Sensors, Volumetric Sensors, Glass Break Detectors, Intelligent Analytics Sensors, and information on both indoor and outdoor applications as well as complex alarm sensing. The chapter ends with a discussion on application rules for both indoor and outdoor alarm systems.

Chapter 21

In Chapter 21, Related security systems, you learn all about various types of video and communications systems and how to integrate the security system to each. Chapter 21, Related security systems, expands further on Chapter 19, Security system integration, with the introduction of how to integrate alarm/access control systems with other security systems including Analog and Digital Video Systems, Security Communications Systems, Security Architecture Models for Campuses and Remote Sites, and Security Command and Control Centers. This chapter delves into the history of video systems and how they became what they are today, including cameras and lenses, lighting and dynamic range, display devices, video recording devices (analog and digital), video motion detectors, video analytics, and video system interfaces. The section on communications systems includes a discussion of 2-way Radios, Telephones, Security and Building/Elevator Intercoms, Public Address Systems, Nextel™ Phones, Smart Phones and Tablet Computers, and Public Address Systems. Also covered are Consolidated Communications Systems, which combine all of these systems into a single manageable system.

Chapter 22

Chapter 22, The merging of physical and IT security, expands on Chapters 19−22 and covers Related Building and Facility Systems. The first part of this chapter covers Elevators, Stairwell Pressurization, Lighting, Automated Barriers, and Specialized Applications. The second part describes Controlling and Automating Building Functions including Direct Action Interfaces, Proxy Action Interfaces, and Feedback Interfaces. This chapter can make your life easy as you approach difficult and challenging building interfaces.

Chapter 23

Chapter 23, Securing the security system, covers the apparently boring but necessary subject of Security System Cabling. This is another subject where what you don't know can bite you big time. This chapter covers security cable types, the subject of conduit and conduit types, and when and when not to use each. There is valuable information on cable handling (this is another subject wherein lack of expertise has cost installers hundreds of thousands of dollars and reputation, and has cost many installers their careers). The chapter also covers cable documentation, why it is so important that you can't ignore it, and how to document cables easily in the field and on paper.

Chapter 24

Chapter 24, Related building/facility systems—and REAPS systems, covers Security System Environmental Considerations. Electronic circuitry sensitivities and environmental factors that can cause costly maintenance and repairs are discussed. Understanding this is the key to keeping maintenance costs low and owner satisfaction high. Also discussed are temperature extremes; the effects of humidity and condensation; vibrations; dirt; insects, birds, snakes, and other creatures; and the effect of lighting at access control system portals.

Chapter 25

Chapter 25, Cabling considerations, gets into the subject of how to design alarm and access control systems. Let me be clear about one thing, a security technician is a security designer in the same way that an automobile mechanic is an automobile designer. That is, although the technician may think he understands design, he does not. There is much more to designing security systems than understanding product data sheets and knowing how to install and maintain the products. Just because an auto mechanic can tell you which car gets the best fuel economy, which has the most power, or which is the least costly to maintain, it does not mean that he has any insight into how to design the system. Knowing how to hook up a system is not the same as knowing how to design one. This will be a new idea for many technicians, but in this chapter you can learn why security systems that are designed by installers and technicians disappoint owners and often do not work; and about design elements; how to design robust portals; how criminals defeat common locks, doors, and frames; technology application concepts; how to implement designs to paper; system installation design considerations (how to design for the ease of installers and maintenance technicians); and all about system commissioning and acceptance testing.

Chapter 26

Chapter 26, Environmental considerations, is all about Access Control System Installation and Commissioning. This chapter covers Jobsite Considerations, Conduit versus Open Cabling, Device Installation Considerations, Device Setup, and the importance of Documentation to make installation and maintenance easy and how it helps to keep costs low now and in the future. Device Setup and Initial Testing, Alarm and Access Control System Database Setup, and Access Control Schedules and Groups are also covered.

Chapter 27

Chapter 27, Access control system management, covers System Management, Maintenance, and Repair. This chapter reviews how to keep maintenance costs low and engender system longevity. Also discussed is System Management including Database Management, merging databases of different systems on a common corporate campus, and Alarm Management and Reports. You can also learn about Maintenance and Repair including how to create System As-Built Documents, Wire Run Sheets, why some systems suffer from system infant mortality, the difference between Maintenance and Repairs, and how knowing that difference can save tens of thousands of dollars a year on a system. How to set up Schedule Maintenance Programs, how to handle Emergency Repairs and Maintenance Contracting Options including In-house Technicians, Extended Warranties, Annual Maintenance Agreements, and On-call Maintenance and Repairs, are also discussed.

So, let's get on to the basics and up with your career.

CHAPTER SUMMARY

1. Who should read this book, what will you learn, and how will it advance your career?
    a. Security System Technicians to gain a deep knowledge about installing and maintaining Alarm/Access Control Systems.
    b. Security System Engineers and Designers to get a thorough understanding of Alarm/Access Control Systems, system design principles, a cost-effective way to achieve effective designs with greater detail, and how to avoid common mistakes in installations.
    c. Risk Managers to learn more about how to understand, estimate, and mitigate security risks of all types and how to tell if the alarm/access control system as designed will be effective in managing risk.
    d. Security Program Directors and Managers to learn how to use the alarm/access control system as a "Force Multiplier" to expand the capabilities of the Security Staff, how to use it to generate metrics that help prove the effectiveness of the Security Program, and how to use it to generate metrics that can improve the Security Program's performance year after year.
    e. Guard Company Managers and Supervisors to learn how to work with instead of against security technology and how to use that knowledge to secure a long-term relationship with their client.
    f. Security Consultants to learn about advanced technology applications, how to improve their designs, how to make them easier to install and maintain, and how to avoid costly design mistakes.
    g. Facility Managers to learn how to avoid costly installation mistakes and how to maintain an alarm/access control system

cost-effectively and for the greatest dependability and sustainability.

   h. Security Installation Project Managers to learn how to avoid costly installation mistakes and how to ensure that the system is installed to the highest standards.

   i. Anyone new to alarm/access control systems to learn a depth of understanding that few in the industry possess.

2. What can you expect to learn?

   a. Security System Technicians will gain a deep knowledge about installing and maintaining alarm/access control systems.

   b. Security System Engineers and Designers will receive a thorough understanding about alarm/access control systems, system design principles, a cost-effective way to achieve effective designs with greater detail, and how to avoid common mistakes in installations.

   c. Risk Managers will learn more about how to understand, estimate, and mitigate security risks of all types and how to tell if the alarm/access control system as designed will be effective in managing risk.

   d. Security Program Directors and Managers will learn how to use the alarm/access control system as a "Force Multiplier" to expand the capabilities of the Security Staff, how to use it to generate metrics that help prove the effectiveness of the security program, and how to use it to generate metrics that can improve the security program's performance year after year.

   e. Guard Company Managers and Supervisors will learn how to work with instead of against security technology and how to use that knowledge to secure a long-term relationship with their client.

   f. Security Consultants will learn about advanced technology applications, how to improve their designs, how to make them easier to install and maintain, and how to avoid costly design mistakes.

   g. Facility Managers will learn how to avoid costly installation mistakes and how to maintain an alarm/access control system cost-effectively and for the greatest dependability and sustainability.

   h. Security Installation Project Managers will learn how to avoid costly installation mistakes and how to ensure that the system is installed to the highest standards.

   i. Anyone new to alarm/access control systems will learn a depth of understanding that few in the industry possess.

Q&A

1. *What are Access Control Systems?*
   a. Electronic systems that allow anyone to enter everywhere.
   b. Electronic systems that allow anyone to enter controlled, restricted, or secure spaces.
   c. Electronic systems that allow authorized personnel to enter controlled, restricted, or secure spaces by presenting an access credential to a credential reader.
   d. Electronic systems that allow unauthorized personnel to enter controlled, restricted, or secure spaces by presenting any credential.
2. *How do you design Security Systems?*
   a. Understand the facility's risks.
   b. Design the Security Systems as a "Force Multiplier."
   c. Design it as if the guy who will maintain it is a violent psychopath who knows where you live.
   d. All of the above.
3. *How can Engineers contribute to Security Systems?*
   a. They can do a great deal to make the installation and maintenance process much easier and to make designs resonate with the risk they manage.
   b. Design the Security Systems as a "Force Multiplier."
   c. Design it as if the guy who will maintain it is a violent psychopath who knows where you live.
   d. Learn how to use alarm/access control systems to their fullest to amplify the reach of their staff and mount energetic responses to alarms in a timely manner.
4. *What can Security System Designers and Engineers learn from Security Systems?*
   a. They can do a great deal to make the installation and maintenance process much easier and to make designs resonate to the risk they manage.
   b. Reduce operating costs.
   c. Learn details about how alarm/access control systems work and interface to the building architecture that will keep them from making alarmingly common design errors; designers will learn about rare and unusual technologies and how to think "outside the box."
   d. Learn how to use alarm/access control systems to their fullest to amplify the reach of their staff and mount energetic responses to alarms in a timely manner.
5. *What can Risk Managers learn from Security Systems?*
   a. They can do a great deal to make the installation and maintenance process much easier and to make designs resonate to the risk they manage.
   b. Reduce operating costs by minimizing the number of security personnel on staff and making the most of the security staff.

    c. Learn details about how alarm/access control systems work and interface with the building architecture that will keep them from making alarmingly common design errors; designers will learn about rare and unusual technologies and how to think "outside the box."

    d. Learn how to use alarm/access control systems to their fullest to amplify the reach of their staff and mount energetic responses to alarms in a timely manner.

6. *What can Guard Company Managers and Supervisors learn from Security Systems?*

    a. They can do a great deal to make the installation and maintenance process much easier and to make designs resonate to the risk they manage.

    b. Reduce operating costs by minimizing the number of security personnel on staff and making the most of the security staff.

    c. They will learn details about how alarm/access control systems work and interface with the building architecture that will keep them from making alarmingly common design errors; designers will learn about rare and unusual technologies and how to think "outside the box."

    d. Learn how to use Alarm/Access Control Systems to their fullest to amplify the reach of their staff and mount energetic responses to alarms in a timely manner.

7. *What can Security Consultants learn from this book?*

    a. They can do a great deal to make the installation and maintenance process much easier and to make designs resonate to the risk they manage.

    b. Reduce operating costs by minimizing the number of security personnel on staff and making the most of the security staff.

    c. Learn about advanced technology applications, how to improve their designs, how to make them easier to install and maintain, and how to avoid costly design mistakes.

    d. Learn how to use alarm/access control systems to their fullest to amplify the reach of their staff and mount energetic responses to alarms in a timely manner.

8. *What can Facility Managers and Security Installation Project Managers learn from this book?*

    a. They can do a great deal to make the installation and maintenance process much easier and to make designs resonate to the risk they manage.

    b. Reduce operating costs by minimizing the number of security personnel on staff and making the most of the security staff.

    c. Learn how to make the most of tight budgets and get more performance while reducing capital and operating costs.

    d. Learn how to use alarm/access control systems to their fullest to amplify the reach of their staff and mount energetic responses to alarms in a timely manner.

9. *Can anyone new learn about Security Systems?*
   a. That person has to know how to reduce capital and operating costs and mitigate risks.
   b. That person must have technical security background.
   c. It is hard to learn if anyone new needs to enter the field.
   d. Anyone new to alarm/access control systems will learn a depth of understanding that few in the industry possess.

Answers: (1) c; (2) d; (3) a; (4) c; (5) b; (6) d; (7) c; (8) c; (9) d.

Foundational Security and Access Control Concepts

CHAPTER OBJECTIVES

1. Understand risk and how to manage risk
2. Understand types of countermeasures
3. Understand access control system principles

CHAPTER OVERVIEW

Access control systems are electronic systems that facilitate automated approval for authorized personnel to enter through a security portal without the need for a security officer to review and validate the authorization of the person entering the portal, typically by using a credential to present to the system to verify their authorization. A security portal is a door or passageway that creates an entry point in a security boundary. Common security portals include standard doors (such as shown in Fig. 2.1 using an HID Global Corporation reader), vestibules, revolving doors, and vehicle entry barriers.

Access control systems are an important part of an overall security program that is designed to deter and reduce both criminal behavior and violations of an organization's security policies. But it is important to remember that it is only a part.

First, it is important to understand that access control is not an element of security; it is a concession that security programs make to daily operational necessities. Perfect security involves perfect access control. By that, I mean that in a perfect security environment, not one person can enter who is not absolutely known without question to be an ardent supporter of the security portion of the overall mission of the organization. In a real organization, this virtually never happens. Access control systems are an automated method to allow "presumed" friendlies to enter controlled, restricted, and secured areas of a facility with only minimal vetting at the

Electronic Access Control. DOI: http://dx.doi.org/10.1016/B978-0-12-805465-9.00002-6

■ **FIGURE 2.1** Access control portal. *Photo by HID Global Corporation.*

access control portal. Indeed, access control portals are doorways through a security perimeter in which the entrants are "assumed" to be friendly, due to their status as an employee, contractor, or softly vetted visitor.

Understanding this, a new light is shed on the role of access control systems. They are a vulnerability that exists right in the heart of the security system. As such, one can now understand why one must understand Risk, Managing Risk, and Types of Countermeasures to understand how to properly utilize access control systems. One must also completely understand access control system principles so as not to create unseen vulnerabilities in the heart of the Security Program.

UNDERSTANDING RISK
Types of organization assets

Every organization begins with a mission, which may be healthcare, mercantile, financial, education, or something else. Each organization establishes programs in support of its mission. If healthcare, it may establish a hospital, outpatient clinics, and a pharmacy. If financial, it may establish a retail bank, investment bank, or stock brokerage. Each type of mission produces different types of programs in support of the mission.

Once the mission and programs are determined, the organization will set about acquiring assets to support the programs. There are only four types of assets (Fig. 2.2) common to all types of organizations, missions, and programs:

■ People
■ Property

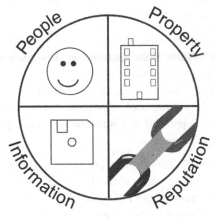

Asset classes

■ **FIGURE 2.2** Asset classes.

- Proprietary Information
- Business Reputation

People include employees, contractors, vendors, and visitors. Property may include real estate, fixtures, furnishings, and equipment. Proprietary information is information that is unique to the organization and necessary for it to function and is not common to other organizations. This may include information such as the formula for Coca-Cola, vital records, customer lists, and accounting records. Finally, the organization will establish and try to grow its business reputation. Indeed, its future is tied intrinsically to the success of this effort, for any organization that loses its business reputation is going to suffer badly, almost always in ways that fundamentally affect its ability to carry out its core mission. Every organization of every type has these four types of assets, and they are what the security program is designed to protect.

The organization's assets become the targets of those who oppose the organization's mission.

Types of users

Each organization has three types of users (people who use the facility). In the first category are people who intrinsically share and support the mission of the organization. In the second category are people who oppose the mission of the organization. In the third category are people who sometimes work in support of the mission and sometimes work in opposition to the mission of the organization.

The smallest number of users are those who either totally support or totally oppose the mission of the organization. Those who totally support the mission typically include shareholders, founders, and owners. Those who oppose typically include competitors, predatory criminals, and so forth. By far the largest segment among these is the third—those who sometimes work in support of and sometimes work in opposition to the mission of the organization.

This typically includes almost all employees, contractors, vendors, and visitors. As long as the mission of the organization aligns (in that moment) with the needs of the individual, then the individual supports the organization's mission. Where there is no alignment, the individual does not support the mission.

It is this vast third group that comprises most of the people who occupy the organization's facilities. Security programs must effectively allow the normal operation of the organization and deal effectively with those who oppose (however temporarily) the mission's organization.

The purpose of the security program is to help ensure that all activity within the organization's facilities supports the organization's mission to the greatest extent possible, and works in opposition to the mission to the smallest extent possible.

Types of threat actors

A threat is any person who can create a manmade danger to the organization's security. A hazard is a natural or manmade equivalent of a threat where there is no intent to do harm, but harm nonetheless can result. Hazards include storms, earthquakes, fires, and safety risks. Threats include bombings, weapons usage, theft, diversion of assets, and undermining the organization's Business Reputation.

There exist only four types of security threat actors (Fig. 2.3):

- Terrorists
- Violent Criminals
- Economic Criminals
- Petty Criminals

Terrorists are at the extreme end of the spectrum and have the greatest capacity to harm life and property. Although the probability of a terrorist event is very low, the consequences are high. Terrorists do not care about

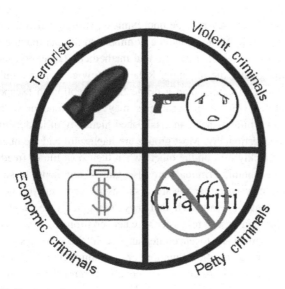

■ **FIGURE 2.3** Threat actors.

being quiet, quite the opposite. They often do not care about escaping with their lives. They only care about inflicting the maximum amount of death, injury, and damage possible at the least financial cost to themselves. Their only real concern is about not succeeding in their attack. Terrorism typically involves substantial planning; acquisition of weapons, chemicals, or explosives; and conducting surveillance. Sometimes there is a practice run and then the terrorists go operational.

Violent criminals want to inflict control, injury, and/or death on their victims. If they care at all about economic benefit, then their violence is seen as a means to control their victims. Those with a tendency toward violence often prefer confrontation to a quiet economic crime. Violent criminals may be controlled or emotionally erratic. They may incorporate planning or spontaneously combust into violence without apparent reason or in response to an innocent comment or minor insult. Once violence ensues, all participants tend toward emotional responses. As emotional responses are not conducive to carrying out a plan, plans often go awry and the situation escalates. Escape may or may not be a part of the violent criminal's plan, and his actions may or may not allow for escape.

Economic criminals are after money or goods. They do not want to be caught, and they place this goal above all others. Economic criminals may be strategic (planning, acquisition of tools and weapons, conduct surveillance, planning their escape, perhaps a practice run and then they go operational), or they may be opportunistic, striking when the opportunity presents itself. Many economic criminals do not put much effort into planning. Only a few are strategic and methodical. Generally, the higher the risk, the greater the planning. Relatively few economic criminals are strategic enough to fully plan a crime from inception to escape in any methodical way. Economic criminals may work alone or in a group. Better group criminals have an established hierarchy of leadership that is followed unquestionably. Most groups are haphazard and members of the group will quickly turn on the others when their own life or freedom is at risk. Some economic criminals are also violent criminals. In such cases their "lean" may be either toward a successful economic crime or their violent side may prevail such that they will quickly abandon the economic crime and resort to violence when confronted. Many economic criminals do not want to encounter anyone in the commission of their crime. However, a significant number of economic criminals intentionally encounter victims and often use violence to control their victims to get them to comply with their directions. In such cases, some intend to leave their victims alive and others wish to leave no witnesses.

Petty criminals are individuals who commit petty crimes such as graffiti, drunken disturbances, loud behavior, and prostitution. Petty criminals reduce the attractiveness of the environment to normal behavior and create an environment that encourages unacceptable behavior.

It is important to understand how each type of potential threat actor can affect the facility to be protected. One must develop scenarios for each type of threat actor and evaluate vulnerabilities and develop countermeasures based on such scenarios. Effective security managers understand threat scenarios and run them through their heads effortlessly. It is a good skill to develop as a form of very practical and effective game theory.

Understanding criticalities and consequences

Not all assets are equal. Some are more critical to the operation of the facility than others. Entire factory operations have stopped because of the failure of a 25-cent part. Such assets are called critical assets. A critical asset is any asset that is critical to the flow of operations or down-line operations. Factors affecting criticality include the importance of its role in the operation of the organization and whether or not it is deployed in a nonredundant manner and also how difficult or time-consuming it is to repair or replace the asset to get the operation going again. Criticality assessment is the skill of assessing how critical assets are to the mission of the organization. Similarly, the loss of some assets has a higher consequence than others. The loss of any human is devastating, but the loss of the person with the necessary skills to maintain the organization's viability is a greater loss and therefore more consequential. The loss of a critical asset is more consequential than the loss of a noncritical asset.

The loss of some assets may present a higher consequence to the organization or to the community than others; e.g., an accident in the manufacture or storage of methyl isocyanate gas represents a very high consequence not only to the organization but also to the community where it might occur (In the world's worst industrial catastrophe, a leak of methyl isocyanate gas and other chemicals at the Union Carbide plant at Bhopal, Madhya Pradesh, India, resulted in an immediate death of 3787 people and ultimately claimed over 15,000 lives.). Many organizations have fallen due to the loss of their business reputation. Most of those organizations did not consider their business reputation to be either critical or of significant consequence before the event. Consequence analysis is the skill of determining which assets have the greatest potential for consequence.

Security programs should be designed to protect those assets with the greatest consequence first.

Understanding vulnerability

A vulnerability is an attribute of a facility that a threat actor can exploit to carry out a threat action. Vulnerabilities may include opportunities for entry, for access to assets, for concealment of activities or staging of stolen items prior to taking from the facility, for diversion of attention, for taking control of an area, or for escape. Vulnerabilities may be physical or procedural. Vulnerabilities exist in every facility and every security system and cannot be completely eliminated; however, it is imperative to mitigate vulnerabilities that could be easily exploited.

Vulnerabilities can be mitigated through physical, procedural, or electronic measures, or a combination of these. Vulnerability equals opportunity.

Understanding probability

Probability is the likelihood of an event occurring. Probability is a key factor in determining risk, and equally it is a key factor in determining how best to mitigate risk. Probability can be affected by the number of employees, contractors, vendors, and visitors; by the frequency of occurrence of an action such as an access control event; by the number of items affected; and by the environment.

Probability and consequence often go hand in hand. High-probability, low-consequence events present relatively minor risk whereas low-probability, high-consequence events may present very high risk.

Factors affecting probability include:

- Interest level including the value of assets to potential threat actors
- Opportunities available to exploit
- Number of potential threat actors who might be interested in carrying out a threat scenario
- Effectiveness of existing security countermeasures (what is the risk to potential threat actors?)

What is risk?

Risk is the potential for a threat action to occur. Risk factors include probability, vulnerability, and consequences. Risk is the final calculation taking into concern the potential threat actors, their potential interest, and available vulnerabilities to exploit. Risk works both ways. An organization can be at risk of theft of an asset and the thief can be at risk of not succeeding or being discovered and prosecuted.

A key factor in the risk formula is consequence. The higher the possible consequences, the higher the risk. Thus, low-probability, high-consequence threat scenarios rank higher on risk than high-probability, low-consequence threat scenarios. Although terrorism is low in probability, its consequences make it a higher risk.

See the author's earlier book *Risk Analysis and Security Countermeasure Selection* for a complete discussion on risk.

MANAGING RISK
Methods of managing risk

Risk can be managed by reducing the potential number of threat actors who have access to a given asset, reducing the vulnerabilities related to the asset, or reducing the potential consequences of the asset being attacked. Risk may also be reduced by increasing the risk to the potential threat actor. These may be achieved by modifying the physical, electronic, or procedural environment.

For example, by placing an asset in a more physically protected environment, risk is reduced. By requiring greater scrutiny of persons entering the area of the asset to be certain that they do not pose a threat, risk is reduced. By placing video cameras along the pathway to and from the asset and around the asset and recording those cameras, risk is reduced because it creates a higher risk to potential threat actors and provides evidence for investigations and prosecution if the asset is taken or damaged.

How security and access control programs help manage risk

Organizations develop Security Programs to manage risk in an organized and methodical way. By assigning a manager who is competent in security matters and also a competent personnel manager, the organization can organize the security process into a business process that supports the goal of protecting the assets of the organization while not interfering with the natural business of the organization.

Effective security programs use a layered approach to protecting the organization's assets. The most critical and consequential assets are the most protected, and a threat actor must engage multiple layers of physical, procedural, and electronic security measures to acquire such assets.

A good security manager will develop a security program that:

- Maintains focus on the mission of the organization at all times
- Supports the natural activities of the organization's programs and the business culture
- Is aware of all of the assets of the organization and has prioritized the assets by criticality and consequences
- Has analyzed risk comprehensively
- Has developed a set of security policies and procedures to address the risks in conformance with the business culture
- Utilizes a layered approach to security countermeasures
- Engages local, state, and federal law enforcement appropriately
- Trains the security force, employees, and visitors as needed to achieve compliance with the organization's security policies
- Continuously evaluates its own effectiveness and continuously improves the effectiveness of the program

Security program elements

A good security program will include all of the following elements:

- The program will be based on a comprehensive Risk Analysis
- The program will be based on a set of Security Policies and Procedures (Security Plan) that addresses all of the elements of the Risk Analysis

The program will include:

- An Access Control element
- Physical Security elements
- Intrusion Detection elements
- Security Communications
- Coordination and Response elements
- Evidence Gathering and Analysis elements
- Investigation capabilities
- Continuous Program Effectiveness Analysis and Improvement

The importance of a qualified risk analysis

It is surprising how many security programs are developed and run without a comprehensive risk analysis. Any security program that is not based on a comprehensive risk analysis will have unknown vulnerabilities. Unknown vulnerabilities present unknown risk that can be easily exploited by a thoughtful threat actor.

Security programs that are not based on a proper risk analysis are virtually certain to cost more and be less effective than programs based on a proper risk analysis.

Running a Security Program without a proper Risk Analysis is analogous to running a business without an accounting system or audit. The business will go broke, and the security program will fall victim to security exploits.

The importance of security policies and procedures

It is equally surprising how many security programs are run haphazardly without a formal set of security policies and procedures. Security policies and procedures are the foundation documents of a good security program. Without policies and procedures, the program will necessarily be haphazard and disorganized. Without policies, every response is spontaneous and based solely on the judgment of the individual security officer or manager. There is no continuity. Such programs are a wellspring of litigation opportunities, because there is no consistency and the judgment of individual security officers is rarely based on criminal law or company policy.

Further, the lack of security policies and procedures is a vulnerability due to the uneven approach of patrols, investigations (if any), and enforcement. In such cases the program generally swings toward accommodating employees and visitors to the detriment of the security of the organization, because individuals' complaints against individual security officers cannot be refuted.

Such programs often depend heavily on electronic systems to "enforce" security, and such systems are developed to address the most urgent security issues without regard to dozens or hundreds of other vulnerabilities that remain unaddressed. Thus money spent on electronics is money wasted.

No security program can succeed without proper security policies and procedures.

TYPES OF COUNTERMEASURES

There are three types of security countermeasures: hi-tech, lo-tech, and no-tech. These three must be used in combination to create a layered and effective security program. No single security countermeasure is effective against all threat scenarios. Access control systems, in particular, are a type of "controlled vulnerability" as each access control portal is a hole in the security boundary.

Hi-tech

Hi-tech systems include all electronic systems. Typically these include alarm/access control systems, video systems, communications systems, integrated security systems, specialized detection systems, and computerized systems.

Hi-tech systems serve to automate repetitive functions, monitor continuously without error, and report to and facilitate communication and coordinated response by security staff.

Increasingly, hi-tech systems are used to handle vast amounts of information that could never be handled cost-effectively by humans. Examples of these include intelligent video analytics that analyze a video scene evaluating human activity for unwanted behaviors and integrated systems that use multiple credential methods (such as a card reader and facial recognition), automated weapons screening, and package X-ray screening to allow entry to a high security area via an automated portal.

Although there is an increasing focus on hi-tech systems in security programs, they should be the last element considered. Lo-tech and no-tech elements form the basis for an effective security program and hi-tech elements amplify the capabilities of the lo-tech and no-tech elements. No matter how elegant an integrated electronic security system may be, they cannot substitute for a solid security program based upon physical (lo-tech) and procedural (no-tech) countermeasures.

Lo-tech

Lo-tech elements are physical security elements that are usually among the most cost-effective security measures any organization can employ. Lo-tech elements include such things as Locks and Barriers, Lighting, Fencing, Signage, Other Physical Barriers, and Crime Prevention Through Environmental Design (CPTED) measures. CPTED is an architectural discipline that involves creating architectural spaces that encourage appropriate behavior and discourage inappropriate behavior. It comprises three basic elements:

- Territorial Reinforcement
- Natural Surveillance
- Natural Access Control

Lo-tech elements are effective because in most cases they represent a single one-time investment that works daily without fail. Lo-tech elements can be used to funnel people to controlled access points, alert visitors to security policies, prevent entry to unauthorized people, provide a deterrent, and stop intruders cold with unexpected barriers.

No-tech

No-tech elements are those security elements that have no technology:

- A Comprehensive Risk Analysis
- Policies and Procedures
- Security Guard Programs
- Security Awareness and Training
- Law Enforcement Liaison
- Investigations
- Dogs and other nontechnology related programs

No-tech elements are among the most effective due primarily to their active nature. No-tech elements are the parts of the security program that users notice most. Every interaction with a Security Officer is somehow memorable; the same cannot be said for every interaction with a card reader. Users are more likely to criticize or complain to or about their interactions with security officers than their interactions with technology.

Mixing approaches

Successful security programs mix hi-tech, lo-tech, and no-tech counter-measures to achieve an effective and cost-effective program. Each element has its strengths and weaknesses, and each element has its own effect on the organization's business culture. By mixing hi-tech, lo-tech, and no-tech approaches, a security program is more likely to achieve repeatable desired results.

Layering security countermeasures

Layering means that an asset is protected by multiple layers of counter-measures from the outside in. To get to a valuable asset such as the information technology server room, a threat actor must encounter a number of security elements, each one a risk to his success. The more layers, the less likely it is that a threat action will succeed.

Layers should use mixed technology approaches. At the outside, the threat actor encounters the outer fence, then lights, then perimeter sensors, then cameras, then the building boundary, then door/window detection, then more cameras, then motion alarms, then the department perimeter with its alarms and access control and cameras, and finally, the controls in the server room. The threat actor will encounter all of these countermeasures on the way out as well, further reducing the likelihood of a successful exit and unnoticed escape (Fig. 2.4).

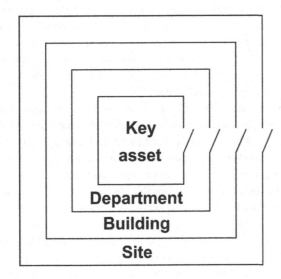

■ **FIGURE 2.4** Layered countermeasures.

ACCESS CONTROL SYSTEM CONCEPTS

Access control is all around us. From the lock on your front door or car door to the Personal Identification Number (PIN) for your Automated Teller Machine (ATM) card, we encounter access control systems every day. Even with the club doorman at the velvet rope looking down his nose at eager young club-goers, access control is everywhere. Access control systems are designed to help ensure that only authorized or qualified people are allowed to enter an exclusive space.

Electronic access control systems utilize a computer, credential, credential reader, and door lock to control access electronically. Elements at the door also include an alarm point to tell if the door has opened without authorization and a request-to-exit sensor to allow the door to unlock for those inside to exit without setting off an alarm. The access control portal is coupled with a wall or fence to prevent going around the portal.

Credentials: Access control systems rely on credentialed users. Credentials may be something you have, something you know, or something about you that is unique. Something you have might be an access card, something you know might be a keypad code, and something about you that is unique might be your fingerprint.
Credential Readers: The system combines the credential with a credential reader (card reader, keypad, or biometric reader) to send a code to the computer where it is compared to a database of codes. If a match is found, the computer orders the portal to unlock the door and allow your entry (accompanied by a green light on the credential reader). It also makes an entry into a data log to record the entry. If no match is found, it rejects your entry attempt (buzzer at the door accompanied by a red light on the credential reader) and it enters an entry into the log recording the improper attempt including the card number or improper keypad code.
Who, Where, and When: Electronic access control systems focus on who, where, and when, and access is granted according to a "Who, Where, and When" formula. Each authorized user (Who) is valid for given portals (Where) and on a certain authorized schedule (When)— who is allowed to enter, where can they enter, and when are they allowed to enter or exit. Historically, this was handled using keys and mechanical locks. Authorized users were given a key for each lock, and the keys were good at any time. This resulted in many keys and when a key was lost, the lock had to be rekeyed and all key holders had to be issued a new key. This was not practical for large facilities and did not provide for time-sensitive access control. With electronic

access control systems, each user is given only one key (like an access card) and the access control system database records where and when the card is valid. If the user loses his card, the card is voided in the database and a new card is issued. No other cardholders are affected, and no locks must be rekeyed.

Two-Factor Authorization: Additional security can be ensured by using two-factor authorization. In this case, access to a critical area such as a server room or vault requires the presentation of two credentials in a short period of time, typically one from an employee and one from a manager. Another form of two-factor authorization is created by requiring the user to present two factors such as a card and a keycode, or a card and a biometric credential such as a fingerprint. Threats have evolved such that two factor authentication should be offered for all logins. Nothing less is safe today.

Employee and Visitor Access Control: Cards are typically issued to employees and worn on their clothes by means of a clip or lanyard. Access control systems in facilities that receive visitors can be set up with a special visitor card to allow for contractors, vendors, and visitors to have access to certain designated areas. Typically, the visitor access control system operates as a subsystem of the access control system and works with either a temporary paper access card or with a conventional access card specially marked to show that the cardholder is a visitor.

Photo Identification: Most electronic access control systems utilize access cards that bear the photo and name of the cardholder. This aids in identification of users and helps ensure that the cardholder is indeed the valid user of that specific card.

Alarms: Most access control systems incorporate an alarm monitoring element. This allows intrusion detection to the facility or zones within the facility. The access control system will also record the event into a log file. Many if not most security portals utilize an alarm sensor to detect if the portal has been forced open or left (propped) open. The software for a portal is intrinsically capable of managing the security portal alarm. Those alarms are also bypassed (ignored) automatically when the door is opened legally to enter or to exit.

Events: The concept of events is central to the real power of advanced alarm/access control systems. An event is an automatic action that can be taken by the system in response to a stimulus. Examples of events are as simple as unlocking a door in response to a valid card or as complex as turning on after-hours path lights from the parking garage to a worker's office space while simultaneously starting up the air conditioning system for his office and warming up the coffee pot. Advanced systems allow for complex events that take multiple conditions and logic into consideration; e.g., changing what a guard will see on a workstation in response to the time of day and the Department of Homeland Security threat level.

Video System Integration: In response to an alarm or an unauthorized access attempt (and many other kinds of events), the alarm/access control system can cause the security video system to display one or more video cameras that are related to the event. This can help security staff to quickly assess the nature and seriousness of the security incident and to coordinate an appropriate response.

Security Intercom System Integration: Security Intercoms can be interfaced with the alarm/access control system to perform two important functions:

❑ Make the intercom call an event in the system's database
❑ Cause the access control system to display and record appropriate video cameras in response to an Intercom call

Guard Tours: Access control systems can also be used to ensure that security staff is conducting a scheduled guard tour of the facility several times each night; many access control systems can be programmed to require the tour guard to use access card readers as the checkpoints for their guard tour. These are sequenced and timed to ensure that each guard tour stop is checked in an appropriate time window.

Antipassback: As with many things, there is always somebody who tries to cheat the system. This is also true for access control systems. One common method is a technique called "Passback." This occurs when a valid access card holder uses his card to allow entry to one or more unauthorized persons who are with him. This happens often at parking access control systems when a valid card holder uses his card to allow entry to the employee parking garage for his own card, but he has a visitor in the car behind him who is not allowed to park in the employee garage. As the parking gate rises, the valid card holder leaves his card on top of the card reader pedestal and then he drives into the garage. The visitor enters the gate area, rolls down his window, and picks up the card that has been left on top of the card reader pedestal and presents the card to the parking garage card reader. The access control system sees that the card is valid and opens the gate again for the same person who has already entered. Access control systems with an antipassback feature require that each card leave the garage before they can reenter.

Roll Call Reports, Mustering, and Demustering: A roll call report is a function that can display all users in (or out) of a certain alarm zone at a given time (manually or automatically). The mustering feature is used to gather users in a safe area during an emergency and to account for their presence. When the emergency is over, users may be required to return to their work area. A roll call report can then verify that all users are back at their work areas.

Occupancy Restrictions and N-Man Rules: Organizations can limit access to an access zone to a designated minimum or maximum number of users. This can help ensure that fire code occupancy rules are obeyed. The N-Man Rule ensures that a minimum number of people are present before entry is allowed, such as ensuring that four bank employees are present before opening the external employee entry door in order to deter robberies.

Escorted Visitors: Some systems can enforce the Escorted Visitor Rule, which ensures that all escorted visitors must be associated with a valid user before entering a given access zone.

Types of users

Typical access control system users include management, employees, contractors, vendors, and visitors. Each of these is given access to areas as they are needed and for times as may be appropriate.

Types of areas/groups

Most access control systems define access portals according to a logical grouping. This might include grouping by departments, buildings, or functions (janitorial employees). Thus, a janitorial employee might have access to all floors and the service lift of the building to which they are assigned. Certain users may be authorized for all areas.

User schedules

Most access credentials are authorized on a schedule. The schedule may correlate to the user's working hours or some other reasonable program. Certain employees (management, maintenance, etc.) may be authorized for all hours.

Schedules may also include holiday and special events schedules. Most systems accommodate 7-day schedules and are programmed around the annual calendar. Some systems are also flexible enough to handle special schedules, e.g., offshore oil rigs may work on a 10-day on/10-day off schedule.

Portal programming

Typically each portal is assigned to an access group, such as a department or an elevator bank.

Credential programming

Typically each user's credential is assigned to a user group. This simplifies programming so that each user does not have to be assigned to individual doors and schedules.

Group and schedule programming

Access portals and users can be assigned to Access Portal Groups or Access User Groups, the combination of which allows for simplified system programming. Otherwise, every user would have to be programmed for access to each individual portal on individual schedules.

CHAPTER SUMMARY

1. Every organization begins with a mission and develops programs to support its mission. These programs acquire assets to support the programs, and those assets become targets of users who oppose the mission of the organization.
2. Every organization has only four types of assets: people, property, proprietary information, and the organization's business reputation.
3. There are three types of users of every facility: those who support the organization's mission, those who oppose the mission, and those who sometimes support and sometimes work in opposition to the organization's mission.
4. The purpose of the security program is to help ensure that all activity within the organization's facilities is in support of the organization's mission to the greatest extent possible, and works in opposition to the mission to the smallest extent possible.
5. Potential threat actors include terrorists, violent criminals, economic criminals, and petty criminals.
6. Some assets are more critical than others to the operation of the organization and the loss of some assets presents a higher potential for consequences to the organization and the community than others.
7. Vulnerabilities are any condition that can be exploited by a threat actor to carry out a threat action.
8. Probability is the likelihood of a security event occurring.

9. Risk is the potential for a threat action to occur. Risk factors include probability, vulnerability, and consequences.
10. Risk can be managed by reducing the potential number of threat actors who have access to a given asset, reducing the vulnerabilities related to the asset, or reducing the potential consequences of the asset being attacked.
11. Effective security programs use a layered approach to protecting the organization's assets.
12. Any security program that is not based on a comprehensive risk analysis will have unknown vulnerabilities.
13. Security policies and procedures are the foundation documents of a good security program. Without policies and procedures, the program will be haphazard and disorganized.
14. There are three types of security countermeasures: hi-tech, lo-tech, and no-tech.
15. Hi-tech countermeasures are electronic systems such as alarm/access control, CCTV, and communications systems.
16. Lo-tech countermeasures include locks, lighting, security landscaping, and CPTED.
17. No-tech countermeasures include risk analysis, policies and procedures, guards, dogs, security awareness programs, and law enforcement liaison.
18. Security countermeasures should be layered for complete protection. Layering means that an asset is protected by multiple layers of countermeasures from the outside in.
19. Electronic access control systems utilize a computer, credential, credential reader, and door lock to control access electronically.
20. Access credentials may be something you have, something you know, or something about you that is unique.
21. Credential readers read credentials and send identity data to an access controller that verifies the data against a database of authorized users.
22. Access is granted according to a "Who, Where, and When" formula. Each authorized user (Who) is valid for given portals (Where) and on a certain authorized schedule (When).
23. Threats have evolved such that two factor authentication should be offered for all logins. Nothing less is safe today..
24. Larger systems often use both employee and visitor access control systems.
25. Users include employees, contractors, vendors, and visitors.
26. Most access control systems define access portals according to a logical grouping such as departments, floors, and elevator banks.

27. Most access credentials are authorized on a schedule. Users can also be assigned to a group that includes a schedule.
28. Many portals are also part of a group, such as doors in a department.

Q&A

1. *What is access control?*
   a. An element of security.
   b. A concession that security programs make to daily operational necessities.
   c. An organization's security program.
   d. All of the above.
2. *The four types of assets in organizations are:*
   a. Employees, Equipment, Proprietary Information, Business Reputation.
   b. Employees, Property, Accounting Records and Customer Lists, Business Reputation.
   c. People, Property, Accounting Records and Customer Lists, Business Reputation.
   d. People, Property, Proprietary Information, Business Reputation.
3. *Types of users who use the facility are:*
   a. People who intrinsically share and support the mission of the organization.
   b. People who oppose the mission of the organization.
   c. People who sometimes work in support of the mission and sometimes work in opposition to the mission of the organization.
   d. All of the above.
4. *Who are the terrorists?*
   a. Those who incorporate planning or spontaneously combust into violence without apparent reason, or in response to an innocent comment or minor insult.
   b. Those who care about inflicting the maximum amount of death, injury, and damage possible at the least financial cost to themselves.
   c. Those who are after money or goods.
   d. Those who reduce the attractiveness of the environment to normal behavior and create an environment that encourages unacceptable behavior.
5. *Who are the violent criminals?*
   a. Those who incorporate planning or spontaneously combust into violence without apparent reason, or in response to an innocent comment or minor insult.
   b. Those who care about inflicting the maximum amount of death, injury, and damage possible at the least financial cost to themselves.
   c. Those who are after money or goods.
   d. Those who reduce the attractiveness of the environment to normal behavior and create an environment that encourages unacceptable behavior.

6. *Who are the economic criminals?*
   a. Those who incorporate planning or spontaneously combust into violence without apparent reason, or in response to an innocent comment or minor insult.
   b. Those who care about inflicting the maximum amount of death, injury, and damage possible at the least financial cost to themselves.
   c. Those who are after money or goods.
   d. Those who reduce the attractiveness of the environment to normal behavior and create an environment that encourages unacceptable behavior.

7. *Who are the petty criminals?*
   a. Those who incorporate planning or spontaneously combust into violence without apparent reason, or in response to an innocent comment or minor insult.
   b. Those who care about inflicting the maximum amount of death, injury, and damage possible at the least financial cost to themselves.
   c. Those who are after money or goods.
   d. Those who reduce the attractiveness of the environment to normal behavior and create an environment that encourages unacceptable behavior.

8. *A critical asset is:*
   a. Any asset that presents a lower consequence to the organization or to the community than others.
   b. The loss of any person with the necessary skills.
   c. Any asset that is not critical to the flow of operations or down-line operations.
   d. Any asset with medium importance of its role in the operation of the organization.

9. *What is vulnerability?*
   a. An attribute of a facility that a threat actor can exploit to carry out a threat action.
   b. Any asset that is critical to the flow of operations or down-line operations.
   c. An asset that can be completely eliminated.
   d. All of the above.

10. *Factors affecting probability include:*
    a. Interest level including the value of assets to potential threat actors.
    b. Opportunities available to exploit.
    c. Number of potential threat actors who might be interested in carrying out a threat scenario.
    d. All of the above.

11. *Risk can be managed by:*
    a. Reducing the potential number of threat actors who have access to a given asset.
    b. Reducing the vulnerabilities related to the asset or reducing the potential consequences of the asset being attacked.
    c. Increasing the risk to the potential threat actor.

      d. All of the above.
12. *Security countermeasures come in these types:*
      a. Hi-tech, lo-tech.
      b. Hi-tech, no-tech.
      c. Hi-tech, lo-tech, no-tech.
      d. Hi-tech, me-tech, no-tech.

Answers: (1) b; (2) a; (3) d; (4) b; (5) a; (6) c; (7) d; (8) b; (9) a; (10) d; (11) d; (12) c.

Chapter 3

How Electronic Access Control Systems Work

CHAPTER OBJECTIVES

1. Understand about access control system users
2. Understand access control system portals
3. Understand credential readers
4. Introduction to locks, alarms, and exit devices
5. Learn about access zones and schedules
6. Learn about data, data retention, and reports
7. Introduction to database maintenance
8. Learn about access control system architecture

CHAPTER OVERVIEW

This chapter is about electronic access control systems concepts and how they work. It is important to read this chapter very carefully and be certain that you understand everything in it. Go over the Chapter Objectives and the Chapter Summary. Everything else you read in this book is based on the material in this chapter.

Electronic access control systems comprise electronic elements, physical elements, operational elements, information technology elements, and logical elements to create a complete working system that facilitates rapid and reliable access to authorized users in a facility at minimum long-term cost to the organization.

This book is written in a hierarchical fashion, i.e., the concepts are related first and then expanded in greater detail later in the book. Accordingly, you will see repetition throughout the book, but that repetition is designed to instill learning in a layered fashion; laying down the

Electronic Access Control. DOI: http://dx.doi.org/10.1016/B978-0-12-805465-9.00003-8

foundation and then building layer-by-layer until the understanding is a complete structure in your head.

Access control elements include:

- Users
- Access Portals
- Credentials and Credential Readers
- Credential Authorization Process
- Locks, Alarms, and Exit Devices
- Access Zones and Schedules
- Access Control System Database
- Communications Infrastructure
- Access Control Policies and Procedures

FIRST, A LITTLE HISTORY

While electronic access control systems have only been around for about 50 years, the need for Access control has been around a lot longer.

First it is essential to understand how access control needs were met prior to the use of electronic access control systems. Good access control programs have always included all of the following elements:

Basic Access Control Policies:

- All areas under the purview of the organization will be organized logically into access areas (includes many portals that are logically related together such as all of the doors in a department).
- Each organization department or unit will determine where its employees need access. All organizational departments and units will be organized into access groups (includes the access areas that that department or unit's employees will need access to and the schedule for which the group may have access to an access area).
- Individual organization employees will be assigned to one or more departmental access groups.
- Each employee will receive an access credential (have a unique number to look up on an authorized user list).
- Each employee may use their access credential to acquire access to a portal within an authorized access group during the authorized schedule for that access group.
    Use:
- Authorized users approach an access portal (door, gate, etc.) and present their access credential to a credential reader (in the old days, this was a guard).

- The credential reader verifies the credential against a database (in the old days, this was a daily authorized list) of authorized credential holders.
- The credential reader then verifies the holder against the photo on the credential (usually a card).

    Contractors and visitors:

- Similar policies will be developed to handle contractors and visitors.
- Typically a department will notify the front desk of a pending visit ahead of time.
- Contractors may be given their own cards or such cards may be held at the security reception desk.

    Audit:

- All access control records should be audited regularly to ensure that policies are applied properly.

In the days before electronic access control systems all of these policies were carried out manually by a staff of trained security officers. Electronic access control systems embed all of those functions (except possibly visual confirmation of the photo) into electronics.

THE BASICS

Electronic access control systems are digital networks that control access to security portals. A security portal is an entry into or out of a security boundary. Most electronic access control systems also function as an intrusion alarm system. From this point forward we will assume that the systems we are discussing have an alarm system element. Electronic access control systems are comprised of field equipment (sensors and controlled devices), decision modules, a communications network, one or more databases, and one or more human interface terminals (computer workstations).

What are not so obvious are the "soft" elements of the Access Control System. These include the users, policies and procedures, the management and reporting structure, and the use of the system to enhance continuing evaluation of the overall security program.

The most obvious elements of an electronic access control system are the field elements: access control system portals (for pedestrians or vehicles), alarm sensors, and any controlled devices such as roll-up doors and lights. Pedestrian access control portals typically include a door, a credential reader, an electronically controlled door lock, a door position switch, and a request-to-exit sensor.

These devices connect to an access control panel, which grants access authorizations based upon comparing the credential presented at the door against a database of authorized credentials. The access control panel communicates to a server via a proprietary or TCP/IP computer network. The server maintains one or more databases including the master database of authorized users, equipment configuration records, access control groups, and schedules. It also includes access control events (requests/authorizations/denials) and alarm events. The server is operated by one or more workstations that are used for system configurations, interactive access and alarm notifications, and reports.

The entire system should be operated according to an established access control policy.

AUTHORIZED USERS, USER GROUPS, ACCESS ZONES, SCHEDULES, AND ACCESS GROUPS

Authorized users

Just as you give keys to your home's front door out carefully, access to access control portals is granted only to authorized users. User authorizations are granted based upon need. Users may be authorized because they are employees, regular contractors or vendors, or temporarily legitimate visitors.

Each access control system utilizes a type of credential that authorized users can use to submit to the access control system as evidence that they are authorized. The system will analyze the credential and verify that it is valid. The system then allows the user to pass through the portal.

User groups

User programming can be simplified by putting users of a common type into user groups. Thus, all employees might be in the employee group, janitors in the janitorial group, and managers in the manager's group.

Access zones

In most systems, a group of logically related security portals may be grouped together to form an access zone, which might include:

- Building public perimeter doors
- All doors within a department
- Freight elevators
- Public elevators
- The entire ninth floor

The use of access zones simplifies access control permissions. Instead of giving a single user authorization to every individual door to which he/she needs access, the security department can just grant access to the access zones to which this user needs to go. Thus instead of programming a user into 35 doors, he/she can be programmed into only four access zones.

Schedules

In most cases, users do not need, nor should they have access to, all authorized doors at all times. Accordingly, most users' access privileges are assigned to schedules, which might include:

- 24/7/365
- Daytime shift
- Evening shift
- Late night shift
- Weekends
- Holidays
- Special event

Access groups

So guess what? If access zones simplify user programming and schedules simplify it even more, wouldn't it be a good idea to keep up this simplification trend? You bet it would. That is how access groups came to be.

An access group is a combination of user groups, access zones, and schedules. In this way, a large number of users can be programmed for access to a logical group of portals on a particular schedule or schedules, such as office hours plus weekends.

PORTALS

The idea of an access control portal is central to the entire concept of access control systems. An access control portal is a passageway through which a person or vehicle must pass from one space into a more controlled or restricted space and in which only authorized persons are allowed. The discussion on portals in this chapter is introductory. Portals are covered in greater detail in Chapter 5, Types of access controlled portals.

Types of portals

There are two basic types of access control portals: those for pedestrians and those for vehicles. Each type has many variations (Fig. 3.1).

Virtually every access control portal has the following five common elements:

- A lockable, operable barricade
- An identity verification method or device
- A locking mechanism
- An alarm-sensing device
- A request-to-exit sensor

From the most common single-leaf door to the most complex vehicle security checkpoint, all have these elements in common.

Common Portals: The most common type of pedestrian portal is a single- or double-leaf door. This is a common door with a Credential Reader, Electrified Lock, Door Position Switch (DPS; alarm-sensing device), and some type of Request-to-Exit sensor (push button, motion detector, panic bar, etc.). Other common types of portals include revolving doors, automatic doors, and Man-Traps. A Man-Trap is a vestibule with a door leading into and out of the vestibule with no exit in between. Typical man-traps require a credential to enter and a credential to exit. The primary purpose of a man-trap is to ensure that no unauthorized user can pass through the portal while the door is open. Other common portals include elevator lobbies, elevators, and automatic doors.

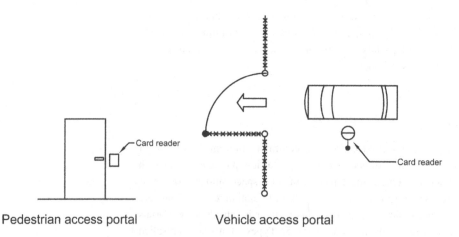

Pedestrian access portal   Vehicle access portal

■ **FIGURE 3.1** Access control portals.

Tailgating: The most common security problem related to electronic access control system portals is "tailgating." This is when one or more people follow an authorized user through an access portal after it has been opened by the authorized user. In common usage, an authorized user presents their credential and opens the door. As they walk through, an unauthorized person catches the closing door and enters behind the authorized user. This is a serious problem with electronic access control systems and it is one that security program managers have to address. We will deal with this in more detail later in the book.

Credential readers

The operational entry barrier to every access control portal is its credential reader. Authorized users have a valid credential and may enter and unauthorized users do not have a valid credential and thus cannot enter.

The three types of credential readers include card readers, keypads, and biometric readers. We will cover these in great detail in Chapter 4, Access control credentials and credential readers.

Electrified locks

The two fundamental components at an access-controlled door are the credential reader and the electrified lock. One authorizes entry and the other allows it to occur. There are two basic types of electrified locks from a safety standpoint: (1) Fail Safe and (2) Fail Secure. Fail safe locks can be opened for exiting when power is not on and fail secure locks cannot be opened for exiting when power is off.

Some locks by their very nature are fail safe, such as magnetic locks and electrified panic hardware. As a magnetic lock requires electricity to hold secure, when power is lost the lock unlocks. Electrified panic hardware uses the principle of "Free Mechanical Egress"; i.e., regardless of the state of the lock, when one pushes on the panic bar, the door opens. Other locks can be found in both types, such as electrified mortise locks, electrified cylinder locks, and electrified dead bolts.

Safety systems

No matter what else, it is imperative that people inside a locked area are able to exit in case of an emergency. This is a "Life-Safety" function. Electronic access control systems *must* be designed and installed to place life safety at the top of the list in priorities. There are *very few* circumstances in which life safety takes a back seat to security. Having designed access control systems for over 35 years and having designed some of the most secure facilities in the world, I can tell you that there is almost never a case in which a life safety provision is not paramount in the design of an access control portal.

One key safety system is the interface between the electrified locks and the building fire alarm system. Wherever there is a fire alarm in a building, there must be a fire alarm interface to unlock all doors in the building in response to fire detection.

There has been a great deal of discussion on this (A quick Google search on "access control fire alarm interface" will uncover many competing points of view on how and in what manner fire alarms should be interfaced to access control systems; and indeed some opinions that they should not be interfaced.). The discussion falls into three camps: (1) unlock all doors in the building so that everyone can exit quickly without relying on the access control system in any way; (2) unlock only doors related to the floor of the fire so that the rest of the building remains secure; and (3) allow the access control system to do its job so that all doors remain secure.

There is, however, no discussion on this among fire authorities. Every fire authority insists that all doors are unlocked (but should remain latched) in a fire emergency. This is not only for the quick exiting of occupants but also the quick entrance of fire department responders. Fire officials want doors to remain latched (but not locked) in order to prevent the movement of smoke and fire throughout the facility. Chapter 8, Doors and fire ratings, deals with locks and fire ratings.

Alarm monitoring

Because access control portals are entries through a security boundary, it is a likely point of illegal entry. It is essential that those monitoring the boundary for illegal intrusions be alerted if a person attempts to use the access control portal improperly.

In its most basic form, this will include a door position switch (DPS) that monitors the closure of the door. When the door is opened without authorization from the credential reader, an alarm is sent to the alarm-monitoring software.

Request-to-exit sensors

Now we understand how authorized users enter through an access control system portal, but they must also exit through it. This involves several things:

- Unlocking the door to allow exiting
- Bypassing the DPS so that no alarm occurs when the door is opened. The passage must be interpreted as an authorized exit, not an unauthorized entry
- Logging the exit in the database as an authorized exit

CREDENTIALS AND CREDENTIAL READERS

A credential is an evidence of authority, status, right, or entitlement to privileges (from dictionary.com). As we said earlier, authorized users utilize credentials to submit to the system so that it can make an authorization decision. The discussion on credentials and credential readers in this chapter is introductory. A complete discussion on access credentials and credential readers is covered in Chapter 4, Access control credentials and credential readers.

Before electronic access control systems, procedural methods using coded cards, passwords, and codes were used to gain authorized entry to secure areas in high-security military and military industrial complexes, mints, and other very high-security facilities. At the Manhattan Project, where the first atomic bombs were designed and built, security was handed over to the Army because it was their conviction that it was the organization best prepared during wartime to enforce a foolproof system of security (National Counterintelligence Center—*Counter-Intelligence in World War II*, Chapter 1. Manhattan Project physical security included a technical fence separating the industrial project from the housing areas, patrols, and access controlled gates using officers and credentialed employees. This was supplemented by a comprehensive operational security program including plant inspections and technical and undercover investigations. Employees would present their credential each day to an officer at a gate who would check the credential and the employee's identification papers against a list of authorized employees and their credential information. Check and countercheck.

In modern electronic access control systems, users present an electronic credential to an electronic credential reader. Credentials may be something you have (an access card), something you know (a keypad code), or something unique to you (fingerprint, voice, eyeball iris, pattern, etc.). Credential readers may be card readers, keypads, or biometric readers.

CREDENTIAL AUTHORIZATION

The Credential reader is an electronic device that reads one or more types of access credentials and converts the credential information into a coded data stream that it passes to the access control panel for interpretation. From there, all action is controlled by the access control panel.

The access control panel contains a connection to the credential reader, lock (or portal activating device), alarm point, request-to-exit sensor, and data communications to a server through some type of network. It also contains a small computer that communicates with the server and with the access portal(s) connected to it.

The access control panel is the principal device making access control decisions. It does this by receiving a coded data stream from the credential reader (corresponding to the credential presented at the reader) and comparing that with a database that it has received from the server. It performs a look-up comparison and then issues an access decision (grant/deny access). If access is granted, it deactivates the lock by triggering an output control point, bypasses the alarm (in software), and (typically) logs the event into an access control event database. This event is also sent to the server to a master dataset.

LOCKS, ALARMS, AND EXIT DEVICES

An access control portal requires some method to lock the portal or it would not be much of a deterrent against unauthorized access. Almost all portals also have an alarm point of some kind to detect unauthorized intrusion attempts. Finally, almost all portals also have a request-to-exit sensor so that persons inside the secured space can exit freely.

The discussion on electrified locks, alarms, and exit devices in this chapter is introductory. Electrified locks are discussed in detail in Chapters 9 through 13, Alarms are discussed in detail in Chapter 20, Integrated alarm system devices, and exit devices are discussed in detail in Chapter 6, Life safety and exit devices.

Electrified locks

There are four basic types of electrified locks:

- Electrified Mortise and Cylinder Locks
- Magnetic Locks

- Electrified Panic Hardware
- Electric Strikes

There are two common variations of electrified locks: fail safe and fail secure. Fail safe locks unlock when power is removed and Fail secure locks remain locked when power is removed. This decision has life-safety and code-compliance implications, so be certain to use the correct lock. A complete discussion of this is presented in Chapter 6, Life safety and exit devices.

Electrified mortise and cylinder locks are simply electrified versions of conventional mechanical door locks. They use an electric solenoid within the lock to engage a small pin that keeps the lock from opening. Electrified mortise locks can be very robust and secure locks because they are well fitted into the door and their strength relies on the strength of the door. Electrified cylinder locks are to be avoided for anything but very light access control. They are not a security device because they cannot stand up to even moderate force. Electrified mortise locks can also be ordered with an integral DPS, lock monitoring switch, and integral request-to-exit sensor within the handle itself.

Perhaps the most common type of electrified lock is the magnetic lock. Magnetic locks work by putting electricity through an electromagnet attached to the door frame that holds tightly to a steel plate attached to the door. All magnetic locks are fail safe and come in one of two varieties: plate locks and shear locks. Plate locks place the steel plate on the vertical surface of the door. The magnet is attached to the door frame with the electromagnetic exposed vertically. The steel plate and electromagnet come together as the door closes and they touch each other. Shear locks place the steel plate in a recessed area on the top of the door and the electromagnet is placed in the top of the door frame pointing down. As the door closes the plate takes its position under the electromagnet. When the electromagnet energizes, the plate is drawn up to the magnet, locking the door. When power is removed, the plate falls back into the recess in the top of the door.

Electrified panic hardware is conventional or specialized panic hardware (a push bar on the door) that is electrified to perform a remotely controlled locking function. Often the panic bar also contains a request-to-exit switch that activates when the panic bar is pushed. Electrified panic hardware is available in a rim-mounted type (surface mounted) or mortise-lock type, or it may lock using vertical bars that latch into strike pockets at the top, bottom, or both top and bottom of the door. Vertical bars may be surface mounted or concealed inside the door.

Electric strikes are remotely operable latchbars that replace a conventional fixed strike faceplate in a door frame. The door lock for that door is typically a conventional mechanical cylinder or mortise lock that latches into the latchbar of the electric strike. The electric strike typically has a beveled or angled surface that allows the door lock latchbolt to close into a keeper space in the electric strike pocket, behind the electric strike latchbar. When the power state of the electric strike is changed (turned off or on from its static state), the latchbar is released and allows the door to open without retracting the latchbolt (the lock swings freely out of the electric strike latch pocket when the strike is unlocked). Electric strikes are available in either fail secure or fail safe and in either AC or DC versions. The AC versions make a buzzing sound when unlocked, notifying the person at the door that the door can be opened.

Alarms

Most Portals are equipped with an alarm point to let the security staff know if the door is being forced opened by an unauthorized user, without the assistance of the access control system, or if it is being held open after a legitimate opening. This is typically in the form of a magnetic DPS at the top of the door. A magnetic switch in or on the door frame is held close to a door magnet in or at the top of the door. When the door opens the magnet moves away from the switch and allows the switch to change state. Magnetic switches may be normally open or normally closed.

Other types of alarm devices exist that are used in conjunction with access control systems. These will be discussed in detail in Chapter 20, Integrated alarm system devices.

Exit Devices

Since the access portal on the boundary of a secure space is typically locked, those inside the secure space have a need to exit by unlocking and opening the access portal. This can be done by one of two methods: Request-to-exit sensors or free mechanical egress devices.

Request-to-exit sensors are electronic sensors (usually a motion sensor or a push button) that signal the access control system to unlock the door and to bypass the alarm when the door is opened for an authorized user to exit.

Free mechanical egress devices are electrified locks that function mechanically to allow exiting no matter what the condition of the access control system. Examples include electrified mortise locks and electrified panic hardware. In both cases, the user only has to operate the lock as any other normal mechanical lock and open the door to exit. Typically these types of locks are also coupled up with request-to-exit sensors, which are used only to notify the access control system to bypass the door alarm because an authorized exit is occurring. The request-to-exit sensor may be a motion detector or a switch within the lock.

Chapter 6, Life safety and exit devices, covers exit devices and life-safety principles in detail.

DATA, DATA RETENTION, AND REPORTS

Electronic access control systems use stored data to operate and analyze security patterns.

From all of the information previously mentioned, it is clear to see that electronic access control systems manage a wide variety of data including:

- Access Portals
- Alarm Points
- Output Controls
- Schedules
- Users
- User Groups
- Access Zones
- Access Groups
- Access Control Request/Granting History
- Event Programming
- Event History

Depending on the brand and model of electronic access control system, these and other datasets may be kept in the server and access control panels. In every case, the server will hold a master alarm/access control database to which access control panels link and report their event history.

These data will be held for a determinate time period or indefinitely (depending on hard disk space). These data not only provide the foundational information from which all access control and event decisions are made and executed, but they are also the source of system reports. In most cases the data will be stored internally in the alarm/access control system server(s). However, if the system is part of a larger system that includes a digital video archiving system, a choice may be made to store the alarm/access control system data on a storage area network along with the digital video system data.

Reports can be run on system configurations, system status, and system events.

Chapter 17 contains detailed information on access control panels and networks, and Chapter 18 includes expanded information on access control system servers and workstations including more about data, data retention, and reports.

CHAPTER SUMMARY

1. Good access control programs have always included:
   a. Basic access control policies
   b. Usage policies
   c. Provisions for contractors and vendors
   d. An audit method
2. Access control systems are digital networks that control access to security portals.
3. Most Access control systems also function as intrusion alarm systems.
4. Systems must be supported by "soft" elements including the users, policies and procedures, the management and reporting structure, and the use of the system to enhance continuing evaluation of the overall security program.
5. The hardware elements include field elements, access control panels, servers, and workstations.
6. Authorized users may be employees, regular contractors or vendors, or visitors.
7. Users may be combined into a user group according to their common access needs.
8. Access portals may be combined into an access zone.
9. Access is normally granted according to a schedule.
10. Access groups can be assembled from user groups, access zones, and schedules.
11. Virtually every access control portal has the following five common elements:
    a. A lockable, operable barricade
    b. An identity verification method or device
    c. A locking mechanism
    d. An alarm-sensing device
    e. A request-to-exit sensor

12. Credential readers may read something you know (keycode), something you have (access card), or something that is unique to you (biometric).
13. Credentials are authorized in the access control panel.
14. There are four common kinds of electric locks:
    a. Electrified mortise and cylinder locks
    b. Magnetic locks
    c. Electrified panic hardware
    d. Electric strikes
15. Most portals are equipped with an alarm sensor to let security staff know if the door is being forced opened by an unauthorized user, or held open after a legitimate opening.
16. Request-to-exit sensors signal the access control panel that the door is being opened legitimately for exiting, and not being forced open.
17. Electronic access control systems use stored data to operate and analyze security patterns.
18. Data may be held in the server(s) and/or in the access control panels.
19. Reports can be run on system configurations, system status, and system events.

Q&A

1. *What do electronic access control systems comprise?*
    a. Electronic elements
    b. Physical elements
    c. Operational and logical elements
    d. All of the above
2. *What does the Server maintain?*
    a. Electronic access control panels
    b. Tcp/ip computer networks
    c. Pedestrian access control portals
    d. Master database of authorized users, equipment configuration records, access control groups, and schedules, access control, and alarm events
3. *Access Group is:*
    a. The combination of user groups, access zones, and schedules
    b. The combination of four access zones
    c. The combination of employee and manager groups
    d. A logical group of portals
4. *What is Tailgating?*
    a. The act of one or more people following an authorized user through an access portal after it has been opened by the authorized user
    b. Having a valid credential and entering into the access zones

    c. Granting access authorization to employees, regular contractors or vendors, or temporarily to legitimate visitors

    d. All of the above

5. *The Request-to-Exit Sensor involves:*

    a. Unlock the door to allow exiting

    b. Bypass the DPS so that no alarm occurs when the door is opened

    c. Log the exit in the database as an authorized exit

    d. All of the above

6. *What are electronic credential readers?*

    a. Card readers

    b. Keypads

    c. Biometric readers

    d. All of the above

7. *Electrified locks include:*

    a. Electrified mortise and panic locks

    b. Electrified magnetic hardware

    c. Electrified mortise and cylinder locks, magnetic locks, electrified panic hardware, electric strikes

    d. Magnetic strikes

Answers: (1) d; (2) d; (3) a; (4) a; (5) d; (6) d; (7) c.

Part

II

How Things Work

Chapter 4

Access Control Credentials and Credential Readers

CHAPTER OBJECTIVES

1. Learn the basics in the chapter overview
2. Learn more about access credentialing concepts
3. Learn all about access cards
4. Why and when to use keycodes
5. Learn all about biometrics
6. Learn photo ID concepts
7. Pass a quiz on access control credentials and credential readers

CHAPTER OVERVIEW

The idea of the access credential and credential reader and a comparison database of authorized users is the centerpiece of the concept of access control systems. These elements are essential to any type of access control system from the most sophisticated global enterprise-wide integrated electronic security system to the most humble procedural system.

Access credentials can take many forms, the most common being access cards, keycodes and biometric attributes (fingerprint, etc.). Common electronic access credential readers include card readers, keypads and biometric readers. Common card types include magnetic stripe cards, wiegand cards, and proximity and contactless smart cards. Common card readers include insertion and swipe readers for magnetic stripe and wiegand cards and proximity type readers for proximity and contactless cards. There is a separate type of biometric reader for each type of biometric credential.

Access cards are commonly coupled with photo identification, usually imprinted on the front of the access card, to help ensure that the bearer is the authorized card holder.

Electronic Access Control. DOI: http://dx.doi.org/10.1016/B978-0-12-805465-9.00004-X

ACCESS CREDENTIALING CONCEPTS

As discussed in Chapter 3, How electronic access control systems work, the most fundamental concept of access control is the idea of the access control credential. A credential is a method of proof of identity and evidence of authority, status, rights, and entitlement to privileges. In its original form, a credential was used by messengers in times of war and by anyone who needed access to the king or general. This practice was first recorded in Egyptian history over 3000 years ago. The purpose of the credential is to provide evidence of authority to a gatekeeper whose purpose it was to challenge unauthorized people seeking entry to the king's court. It did not matter whether or not the gatekeeper knew the person seeking entry; the credential gave that person the authority to enter.

Access credentials in modern history (preelectronics version) came in the form of a laminated identification card, usually with the bearer's name and photo and some emblem that identified the facility for which it was to be used. Along with that there was usually a number (to verify that the card was still valid), sometimes an expiration date, and very often some color codes (often two bars—top and bottom), where the top bar defined the areas and the bottom bar defined the times allowed.

Early electronic access control systems used a variety of different card technologies including Magnetic stripe (almost everybody), Barium Ferrite (Cardkey), Hollerith (Ving), Rare-earth (DKS-Australia), a very early form of Proximity technology (Schlage), and Wiegand wire cards (Cardkey). More recent card technologies utilize 125 KHz Proximity, MiFare, and contactless Smart Card RFID. It is rare today to see any other technologies.

Keypads are still in use, although less common today than in the past. Another common type of credential is a biometric. Biometric credentials may include fingerprint, iris, retina, voice, handwriting, hand geometry, and blood vessel patterns.

Although the use of plain, unprinted, or printed access cards is common for relatively minor security buildings, most governmental and larger corporate facilities utilize a photo ID card similar to the one described in controlled high-security facilities. At a minimum, the photo ID card typically includes a photo of the authorized bearer, their name, usually an icon representing the facility, and an access card number.

KEYPADS

The most basic types of ID readers are keypads (what you know). Basic keypads are simple 12-digit keypads that contain the numbers 0–9 and a * and # sign. The most desirable attributes of keypads are that they are simple to use and they are just dirt-cheap. The most undesirable attribute of keypads is that it is relatively easy for a bystander to read the code as it is being entered and suddenly, Wah-Lah! YOU have been duplicated in the access control database (i.e., now two people know your code so now no one is sure if the person who used the code is really you). Also, the pizza guy always knows a code since there is always some fool who will give out his code for things as important to the organization's mission as pizza. This sort of defeats the whole purpose of access control, since now management has no idea who has the codes. Although there are shrouds for keypads, they are cumbersome and never seem to be well accepted; and the pizza guy still knows the code.

Two other variants are the so-called "ashtray" keypad, which conceals the code quite nicely, and the Hirsch Keypad, which works really, really well. The ashtray keypad is somewhat rare now, but puts the 0–9 keys on five rocker switches just inside the lip of the reader where prying eyes cannot see.

The Hirsch Keypad (Fig. 4.1) displays its numbers behind a flexible, transparent cover using seven-segment LED modules. Then, just to confuse the guy across the parking lot with binoculars, it scrambles the

■ **FIGURE 4.1** Hirsch keypad. *Image courtesy of Hirsch Electronics.*

position of the numbers so that they almost never show up in the same location on the keypad twice. This ensures that even though Binocular Guy can see the pattern of button pressing, it will be useless since that pattern does not repeat itself very often, if ever. We have also found that in many organizations, there is something about the hi-tech nature of the Hirsch Keypad systems that seems to make its users more observant of the need not to give out the code to unauthorized people. Sorry, pizza guy.

ACCESS CARDS, KEY FOBS, AND CARD READERS

One step up the scale of sophistication from keypads are ID cards and card readers. Access control cards come in several variants, and there are a number of different card reader types to match both the card type and the environment; there will be more on that later in the chapter. Common access card types include:

- Magnetic stripe
- Wiegand wire
- Passive proximity
- Active proximity
- Smart cards (both touch and touchless types)

Increasingly rare types include:

- Barcode
- Barium ferrite
- Hollerith
- Rare-earth magnet

Magnetic stripe cards have a magnetic band (similar to magnetic tape) laminated to the back of the card. These were invented by the banking industry to serve automated teller machines (ATMs). Typically there are two to three bands that are magnetized on the card. The card can contain a code (used for access control identification), the person's name, and other useful data. Usually in access control systems, only the ID code is encoded. Magnetic stripe cards come in two types: high and low coercivity (how much magnetic energy is charged into the magnetic stripe). Bank cards are low coercivity (300 Oersted) and most early access control cards were high coercivity (2750 or 4000 Oersted). However, as clients began to complain that their bank cards failed to work after being in a wallet next to their access card, many manufacturers switched to low coercivity for access cards as well.

Desirable attributes of magnetic stripe cards are that they are easy to use and inexpensive. Undesirable attributes are that they are easy to duplicate and thus not suitable for use in any secure facility.

Magnetic stripe readers are available in two versions: insertion and swipe readers. Insertion readers, as the name suggests, require insertion of the card. This is common on ATM machines. The other type is the swipe reader. It has a slot through which the card is swiped. A read-head is present in the reader that reads the card's ID code as it is swiped through. Magnetic stripe readers were a common type of access control reader, but they have been largely replaced by proximity readers.

Barcode Cards: Barcode cards use any of several barcode schemes, the most common of which is a conventional series of lines of varying thicknesses. Barcodes are available in visible and infrared types. The visible type looks similar to the UBC barcode on food articles. Infrared barcodes (Fig. 4.2) are invisible to the naked eye, but can be read by a barcode reader that is sensitive to infrared light. The problem is that either type can be easily read and thus duplicated, so barcodes are also not suitable for secure environments. Barcode readers were also available in both insertion and swipe versions, and the swipe version was the most common.

Barium Ferrite Cards: These are based on a magnetic material similar to that used in magnetic signs and refrigerator magnets. A pattern of ones and zeros are arranged inside the card and because the material is essentially a permanent magnet, it is very robust. Barium ferrite card readers can be insertion or swipe type in the form of a stainless steel plate placed within a beveled surface. For the latter type, the user simply touches the card to the stainless steel surface and the card is read. Swipe and insertion barium ferrite cards and keys are almost nonexistent today, relegated only to legacy systems. The stainless steel touch panel is still common in some locales.

Hollerith: The code in Hollerith cards is based on a series of punched holes. The most common kind of Hollerith card is used in hotel locks. Some Hollerith cards are configured so their hole patterns are obscured by an infrared transparent material. One brand of Hollerith is configured into a brass key (Fig. 4.3). Hollerith cards are not common in secure facilities. The most common application of Hollerith cards is in the hotel industry.

Rare-Earth Magnets: An extremely rare type of access credential is the Rare-earth key. The Rare-earth magnets are set in a pattern of 4 wide by 8 long and each can be positioned so that north is pointing left or right, making a pattern of ones and zeros. Such keys are very

■ **FIGURE 4.2** Barcode card.

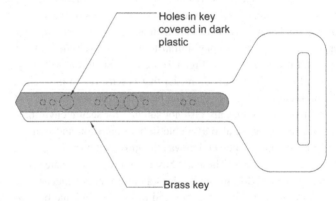

■ **FIGURE 4.3** Hollerith key.

difficult to duplicate and are suitable for high-security facilities, although their cost is high since each key must be handmade. Rare-earth cards were always unusual, but in some parts of the world, notably Australia, Rare-earth keys were pretty common. All were of the insertion type.

WIEGAND WIRE CARDS

After 40 years of research, the Wiegand effect was discovered by John R. Wiegand. The Wiegand effect is a way to cause the magnetic fields of specially processed, small-diameter ferromagnetic wire to suddenly reverse, generating a sharp uniform voltage pulse. Sensors based on the proprietary, patented Wiegand effect require only a few simple components to produce Wiegand pulses. These sensors consist of a short length of Wiegand wire, a sensing coil, and alternating fields, generally

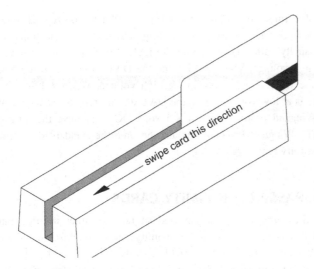

swipe card this direction

■ **FIGURE 4.4** Wiegand wire card and reader.

derived from small permanent magnets. (Dave Dlugos, Manager, Marketing Services, HID Corporation)

Wiegand wires are twisted into an "E" shape with two windings.

Wiegand Wire Cards (Fig. 4.4) are access cards that use a series of short lengths of Wiegand wires embedded in it. These encode the identity of the card based upon the pattern of the embedded wires. The identity code of the wire is comprised of a combination of presence (wires present) and absence (wires absent) along the length of the card. A second set of wires provides a clock track against which the first set is compared. The card is read by swiping it through a slot in a Wiegand card reader that has a fixed magnetic field and a sensor coil. As each wire passes through the magnetic field its magnetic state flips, indicating a 1, and this is sensed by the coil. Where there is no wire, the number is registered as a 0. The Wiegand Protocol is a standard bit-code (typically 26 bits) comprising one parity bit, 8 bits of facility code, 16 bits of ID code, and a trailing bit for a total of 26 bits. The Wiegand protocol can achieve many permutations, including many specialized bit patterns, providing a unique card for each individual facility.

Wiegand wire cards were extremely common for many years and presented the state of the art for nonproximity cards. Most Wiegand card readers were the swipe type.

The Wiegand Protocol provided for a significant advancement when it was introduced in access control systems in 1975, because it allowed for long cable runs from card readers (up to 500 ft). The typical Wiegand interface uses three wires, a common ground, and two data transmission wires usually called DATA0 and DATA1. When no data are sent, both are at high voltage. When a 0 is sent, the DATA0 wire is at low voltage while the DATA1 wire remains at high voltage. When a 1 is sent, the DATA1 is at low voltage while the DATA0 wire remains at high voltage. Most Wiegand protocols use a +5 volts DC to achieve the long cable runs. The Wiegand Protocol has also become the standard data protocol for proximity type cards.

125 K PASSIVE PROXIMITY CARDS

All of the cards we have discussed so far are either inserted into or swiped through a card reader. Proximity cards are different. They are Radio Frequency Identification (RFID) cards that are read by placing the card near (in proximity with) a proximity card reader. Proximity card is the generic name for two types of these cards. It can refer to the older (125 KHz) or the newer (13.56 MHz) contactless RFID Cards.

Proximity cards are tiny data transmitters that use no battery or energy storage device to operate. Instead, they do so by using resonant energy transfer. Proximity cards have three parts: a coil (antenna), a capacitor, and an integrated circuit (IC), which bears the ID code. The antenna of the card receives a radio frequency (RF) signal from the nearby card reader, which excites a coil and capacitor that are connected in parallel. The card reader presents an RF field that excites the coil (the antenna) and charges the capacitor. This energizes and powers the IC, which discharges its data code through the coil back to the card reader. The card reader transmits the card's code using the Wiegand Protocol (see the section Wiegand wire cards).

125 KHZ (LOW FREQUENCY) ACTIVE PROXIMITY CARDS

Active cards are typically used in long-range reader applications, most notably on toll roads and for vehicle or container tracking. Such cards have a battery inside that powers a more powerful card ID transmitter.

13.56 MHZ (HIGH FREQUENCY) CONTACTLESS SMART CARDS

125 KHz Proximity cards were a great advance over previous cards. 13.56 MHz tags were created in an effort to lower card costs and provide

additional functions. In 13.56 MHz cards, the coil does not need to be made of hard copper wrappings. The coil can actually be printed ink on a paper-like substrate that has an EEPROM added to it. During the mid- to late-1990s 13.56 MHz was the technology that many researchers thought could address very high tag usage applications such as library books, laundry identification, and access control. Typical tag costs run from 50 cents to $1.00. 13.56 MHz access cards often utilize so-called Smart Card technology, which enables read/write storage of data up to 64 MB per card. This also allows the cards to be used in cash management applications. 13.56 MHz cards can also be encrypted to reduce the possibility of their signal being intercepted and hacked.

RFID WIRELESS TRANSMITTER SYSTEMS

RFID Wireless Transmitter Systems are common in vehicle access control. These are usually configured similar to a garage door opener, but unlike a standard garage door opener, each individual unit has its own unique code it transmits to a receiver that converts this code to a card number using the Wiegand Protocol. The output of the receiver is connected to a card reader input on an access control panel, and the access control system operates the vehicle gate exactly as though a standard card were presented.

RFID Wireless Transmitter Systems are ideal where there is a high traffic area and a potential for a large queue to extend into a major road due to delays in presenting normal cards and the time it takes for the gate to open and reclose for each vehicle. It is common to see RFID Wireless Transmitter Systems used in high-end residential compounds.

MULTITECHNOLOGY CARDS

As organizations grow, it is common for some employees to travel to multiple offices and facilities where different card technologies may be used. There are three solutions for this problem. One solution is to have the traveling employees carry a different card for each facility they visit. Another is to convert the entire organization's access control system to a single card standard, which can be expensive. Finally, technology can come to the rescue by creating a card that contains codes that are readable by two or more access control systems. Multitechnology cards can include Magnetic Stripe, Wiegand, 125 KHz Proximity, and/or 13.56 MHz Contactless Smart Cards all in one card, such as the one by HID Global Corporation.

MOBILE PHONE ACCESS CONTROL

Organizations are moving towards allowing users to utilize their mobile phone as an access control credential, replacing access cards altogether. This will require the use of a new type of credential reader which will read the Near Field Communication (NFC) device near the door to communicate with the NFC device within the smart phone. A user would place their phone near the reader. The phone will communicate its unique ID to the reader, which has been registered in the access control system.

CAPTURE CARD READER

A class of card readers called "Capture" Card Readers can be used to capture and hold the access control card while a task is underway. These can be used to allow a computer workstation user to use their card to login to their workstation, and the card reader automatically logs the user out when they remove their card from the reader to leave for the day, or just go to visit another employee or to use the toilet. The Capture Card Reader is also useful for Security Console personnel to assure that all actions taken at their workstation are in fact by the actual user, and that a user does not walk away without logging out, allowing another user to grant access to portals under the first user's login credentials.

MULTITECHNOLOGY CARD READERS

Like Multitechnology Cards, some organizations need to be able to read the cards of several facilities where different types of cards exist. One of the most common types of Multitechnology Card Readers serves 125 KHz Proximity and 13.56 MHz Contactless Smart Cards. It is also common to see Proximity or Contactless plus a keypad all in one reader. In the past other Multitechnology card readers existed, but they are rare today.

BIOMETRIC READERS

Any device that reads the identity of a person by comparing some attribute of their physiological being or behavioral traits against a sample database is called a Biometric Reader. Biometric Readers come in many types, the most common including Facial Recognition, Hand Geometry (Fig. 4.5), and Fingerprint and Iris Recognition readers (Figs. 4.6 and 4.7).

■ **FIGURE 4.5** Hand geometry reader. *Photograph provided courtesy of Schlage Lock Company LLC.*

■ **FIGURE 4.6** Fingerprint reader. *Image courtesy of Hirsch Electronics.*

In high-security applications, conventional card readers are often coupled with Biometric Readers to assure that the person holding the credential (card) is the person who the card belongs to. This "Two-Factor" authentication helps assure strict access control.

Examples of physiological traits include fingerprint, face recognition, DNA, iris recognition, retinal scan, palm print, and blood vessel.

■ **FIGURE 4.7** Iris reader.

Examples of behavioral methods include, but are not limited to, walking gait, voice print, and typing rhythm.

Biometric traits can distinguish unique attributes about the individual that are permanent and collectable in a rapid, reliable fashion. The trait must also present a high barrier to circumvention. There are two basic types of Biometric analyses used in Access Control Systems: Verification and Identification.

Verification readers usually work in conjunction with a card reader or keypad wherein the user presents the first credential (card or keycode), thus claiming to be a given authorized user in the access control database. Then the biometric reader captures a biometric sample and compares the result of the capture against a previous sample drawn from the record belonging to the person that the card or keycode claimed them to be. If the samples match, the identity is verified. Verification readers are a "One-to-One" match, which is very simple technologically.

Identification readers use a "One-to-Many" comparison and do not require the user to present a card or keycode first. Upon presenting the biometric credential (fingerprint, iris, etc.) to the reader, the access control system tries to look up a match for the credential presented by the unknown user. When it finds a match, it opens the portal and records the access to that authorized user.

All biometric systems require authorized users to enroll in the database, usually by presenting several samples until the system recognizes the commonality of the biometric attributes taken. This is then associated with a given authorized user's name and other information. The process of enrollment also processes the biometric credential image so those attributes needed for future comparison are properly stored in the biometric database template.

PHOTO IDENTIFICATION

Access cards grant access and identification cards provide visual evidence that the bearer is authorized to be in the area. Identification badges can have many visual attributes. These may include a photo of the bearer, a logo of the organization (not necessarily a wise thing), the bearer's name, a color scheme that may identify areas where the person is authorized, and sometimes a color or code may designate if the bearer is a contractor or vendor.

To help verify the authenticity of the card, it is common to laminate a holographic overlay, which provides a visual indication that the card has not been tampered with. Some organizations use a separate access card and identification card, but most have combined the two functions together into a single credential. These are printed on a Photo ID card printer.

CHAPTER SUMMARY

1. The idea of the Access credential and credential reader and a comparison database of authorized users is the centerpiece of the concept of access control systems.
2. All access control systems, whether electronic or procedural, use these same elements.
3. Early Electronic access control systems used a variety of different card technologies including Magnetic Stripe, Barcode, Barium Ferrite, Hollerith, Rare-Earth, a very early form of Proximity technology, and Wiegand Wire Cards. More recent card technologies utilize 125 KHz Proximity, MiFare, and 13.56 MHz Contactless Smart Cards.
4. Keypads are also still in use, although they are less common.
5. The other type of credential and reader is the biometric system, which compares a physical or behavioral attribute against a previously taken sample.

6. Access cards in most advanced systems are also printed with information unique to the user and the facility using a Photo Identification System.
7. Multitechnology cards and card readers allow organizations to service people from various facilities without having to issue multiple cards to each user.
8. Any device that reads the identity of a person by comparing some attribute of their physiological being or behavioral traits against a sample database is called a biometric reader.
9. Biometric traits can distinguish unique attributes about the individual that are permanent and collectable and that can be collected in a rapid, reliable fashion.
10. Verification readers typically use a card reader and a biometric reader creating a one-to-one comparison wherein the biometric sample is compared to only the record matching the card or keycode.
11. Identification readers use a "one-to-many" comparison and do not require the user to present a card or keycode first.
12. All biometric systems require authorized users to enroll in the database.
13. Photo identification cards add a photograph and other visually identifying information to the access card so that users can be certain that the card used actually belongs to that user.

Q&A

1. *In its original form, a credential was used by:*
   a. Carrier Pigeons
   b. Messengers
   c. All the King's Men
   d. Servants
2. *Early electronic access control systems used different card technologies including:*
   a. Magnetic Stripe
   b. Barium Ferrite
   c. Hollerith
   d. All of the above
3. *Early electronic access control systems used different card technologies including:*
   a. Rare-Earth Magnets
   b. Wiegand
   c. Early form of Proximity
   d. All of the above

4. *Early electronic access control systems used different card technologies including:*
   a. 125 KHz Proximity
   b. MiFare
   c. Contactless Smart Card
   d. None of the above
5. *Keypads are the most sophisticated types of ID readers.*
   a. True
   b. False
6. *Access control cards come in several variants and there are a number of different card reader types to match both the card type and the environment.*
   a. True
   b. False
7. *Magnetic Stripe Cards*
   a. Use any of several conventional series of lines of varying thicknesses
   b. Are set in a pattern of 4 wide by 8 long and each can be positioned so that north is pointing left or right, making a pattern of ones and zeros
   c. Have a magnetic band laminated to the back of the card
   d. Are access cards that use a series of short lengths of Wiegand wires embedded in them
8. *Wiegand Wire Cards*
   a. Use any of several conventional series of lines of varying thicknesses
   b. Are set in a pattern of 4 wide by 8 long and each can be positioned so that north is pointing left or right, making a pattern of ones and zeros
   c. Have a magnetic band laminated to the back of the card
   d. Are access cards that use a series of short lengths of Wiegand wires embedded in them
9. *Barcode Cards*
   a. Use any of several conventional series of lines of varying thicknesses
   b. Are set in a pattern of 4 wide by 8 long and each can be positioned so that north is pointing left or right, making a pattern of ones and zeros
   c. Have a magnetic band laminated to the back of the card
   d. Are access cards that use a series of short lengths of Wiegand wires embedded in them
10. *Rare-Earth Magnets*
   a. Use any of several conventional series of lines of varying thicknesses
   b. Are set in a pattern of 4 wide by 8 long and each can be positioned so that north is pointing left or right, making a pattern of ones and zeros
   c. Have a magnetic band laminated to the back of the card
   d. Are access cards that use a series of short lengths of Wiegand wires embedded in them
11. *Proximity cards have three parts: a coil (antenna), a capacitor, and an Integrated Circuit (IC), which bears the ID code.*
   a. True
   b. False

12. *Biometric Readers*
    a. Make you able to read the cards of several facilities where different types of cards exist
    b. Are any device that reads the identity of a person by comparing some attribute of their physiological being or behavioral traits against a sample database
    c. Include a photo of the bearer, a logo of the organization, the bearer's name, a color scheme that may identify areas where the person is authorized, and sometimes a color or code may designate if the bearer is a contractor or vendor
    d. All of the above
13. *Identification readers usually work in conjunction with a card reader or keypad wherein the user presents the first credential (card or keycode), thus claiming to be a given authorized user in the access control database.*
    a. True
    b. False
14. *Identification readers use a "one-to-many" comparison and do not require the user to present a card or keycode first.*
    a. True
    b. False

Answers: (1) b; (2) d; (3) d; (4) d; (5) b; (6) a; (7) c; (8) d; (9) a; (10) b; (11) a; (12) b; (13) b; (14) a.

5

Types of Access Controlled Portals

CHAPTER OBJECTIVES

1. Learn the basics in the chapter overview
2. Portal passage concepts
3. Understand the most commonly used type—pedestrian portals

   a. Standard doors
   b. Automatic doors
   c. Revolving doors
   d. Turnstiles
   e. Man-traps
   f. Automated walls

4. Learn about the other common type—vehicle portals

   a. Standard barrier gates
   b. High-security barrier gates
   c. Sally ports

5. Pass a quiz on types of access controlled portals

CHAPTER OVERVIEW

An Access Portal is a passageway through which a person must pass in order to go from one access zone to another. When an access control portal is confronted, one knows that they are moving from one access area into another. Depending on the security configuration of the portal, access authorization may be required to enter, to leave, or to enter and leave through the portal. In most cases, access can be granted to a single individual, but for some higher security zones, access may require the presence of two or more authorized people. Access portals can be configured to work on a schedule so that access is free during some hours and requires authorization during others, or may be limited to certain individuals during certain hours.

Electronic Access Control. DOI: http://dx.doi.org/10.1016/B978-0-12-805465-9.00005-1

Pedestrian portals include standard doors, automatic doors, revolving doors, turnstiles, Man-Traps, and automated walls. Vehicle portals may include standard barrier gates, high-security barrier gates, and sally ports.

PORTAL PASSAGE CONCEPTS

Card entry/free exit

When a person confronts an access controlled portal they must show authorization to pass. The most common form of portal design is one designed to allow entry only to authorized users but also to allow anyone inside the access zone to exit freely. This is normally accommodated by using a card reader, keypad, or biometric reader (or some combination of these) to authorize the user to enter. Exit is possible without being an authorized user; i.e., any visitor or nonauthorized person who was escorted inside by an authorized user is free to exit at any time and without any impediment. A typical portal has a locking system that must be unlocked in order to exit. Various Request-to-Exit sensors may be used, including an Exit Push Button, Panic Exit Touch-Bar, or a motion detector over the door (Fig. 5.1).

Card entry/card exit

Another common portal configuration uses a Credential Reader on both sides of the door. This variation helps ensure that people are accounted for as they enter and leave an area. It is common to see this type of portal in areas where either financial instruments or proprietary information is controlled, such as in a cash-counting room or archive room. When using this type of portal, it is imperative to ensure that the door can be opened in an emergency without requiring a credential, which might not be immediately available. It is common to interface the locking mechanism on card in/out portals with the fire-alarm system and also mount an emergency door release next to the exit reader so that occupants can exit the door in a fire or other type of emergency.

Tailgate detection

An aggravating problem related to access control portals is the issue of "tailgating" (Fig. 5.2). Tailgating is the practice of an unauthorized user following an authorized user through an access controlled door by catching the door before it fully closes and sneaking in behind the authorized user.

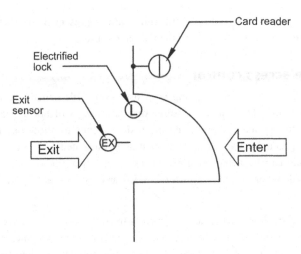

■ **FIGURE 5.1** Card entry/free exit.

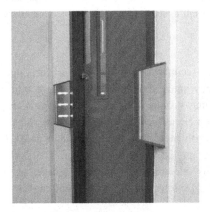

■ **FIGURE 5.2** Tailgate detector. *Image courtesy of Smarter Security Systems.*

This can be reduced by using a tailgate detection system that places a pair of infrared beams across the width of the door and allows for one passage through in the direction of the authorized entry. The beams comprise a set of infrared emitters and receivers that are stacked vertically to prevent an unauthorized person from crawling under or jumping over the beams; better systems also stack the beams horizontally so that the system can detect the direction of travel. These systems allow one person to pass freely but sound a local alarm at the door if a second person also tries to enter behind the authorized user. If a visitor is to be escorted into the space by an authorized user, the valid card is presented first for the visitor who enters separately from the authorized user.

Positive access control

The best way to prevent tailgating is to use a principle called "Positive Access Control." This method ensures that tailgaters will not abuse the access control system by providing a physical means to allow one user through a portal for each time a credential is presented to a credential reader. This can be achieved using an electronic turnstile (Fig. 5.3), a revolving door or a man-trap (for pedestrians), or a sally port (for vehicles).

Such applications are often needed in high-traffic areas where a significant number of unauthorized people could enter. In such applications, an electronic turnstile is a good method to maintain high-traffic flow (up to one person per second with some models) while still ensuring one passage per credential. It is also important in such applications to use high-speed credentials such as a proximity card or contactless smart card.

2-Man rule

Where a very high-security zone exists, such as for a nuclear missile silo or gold or currency vault in a central bank where billions of dollars in gold or currency may be stored, or where credit cards are being manufactured, or integrated circuit central processing unit chips are in bulk storage (which are worth more than their weight in gold), it is common to see a 2-Man Rule applied to the access door. The 2-Man Rule requires two people, usually one employee and one manager or two managers, to present their authorized credentials to a single card or biometric reader

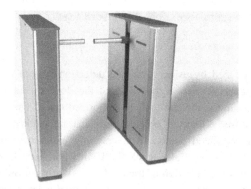

■ **FIGURE 5.3** Electronic turnstile. *Image courtesy of Smarter Security Systems.*

within a certain minimum time (such as within 30 seconds). The 2-Man Rule ensures that no single person ever has access to the secured area alone, thus providing a higher level of security by ensuring that one person will always be watching another.

Schedules

Organizations do not operate a single shift 24 hours/day, 365 days/year. They typically operate a day shift, evening shift, and night shift. Holidays, weekends, and special events also change the normal routine of every organization. Some organizations, notably offshore oil rigs, do not operate on a 7-day calendar schedule at all, but rather on a 10-day on/10-day off or similar schedule frequently with 12-hour instead of 8-hour shifts. The electronic access control system must adapt and accommodate all of these diverse schedules.

A normal system is set up with a number of different schedules, all working together:

- 24/7/365 (24 hours/day, 7 days/week, 365 days/year—all hours)
- Daytime working hours
- Evening working hours
- Nighttime working hours
- Weekend hours
- Holidays
- Special events (unique schedule for each event)

Antipassback

As discussed in Chapter 2, Foundational security and access control concepts, antipassback schemes are designed to prevent one authorized user from allowing his card to be used by a second unauthorized person. Antipassback schemes set up an antipassback zone that is bounded by access control portals. The access control portals are included in the antipassback zone so that an authorized user must enter the zone and exit it again (from the same or another portal in the antipassback zone) before his card can be used again to enter. The same card cannot be used twice in a row to enter. Once in, once out, then in again. This way, the authorized user who passes his card back to an unauthorized visitor will find that the card will not allow the visitor to enter because the card has not been used to exit.

Antipassback Portals must be configured as Card In/Card Out to keep track of each individual passage by each individual authorized user.

Antipassback works best with positive access control portals, where an authorized user cannot exit with a group, forgetting to use his card to go out, thus leaving him outside with no way back in.

PEDESTRIAN PORTAL TYPES
Standard doors

The standard door is the basic unit of an electronic access control system. Simple to understand, commonplace, and inexpensive, wherever there is a room or department to secure in most cases you can bet that it already has a standard door enclosing it. Standard doors are intuitive (everyone knows how to use one) and can allow for the passage of one or more users at a time.

A typical standard door access control portal comprises a door (single- or double-leaf) with hinges and door handle hardware; an electrified lock; a card reader, keypad, or biometric reader; one or two door position sensors to tell the system if the door is open or closed; and a request-to-exit sensor. These devices report to an access control panel that may be located near the door or centrally in a utility room.

Typically the request-to-exit sensor is not a credential reader, so the system does not know which authorized user has left the secured space, only that someone has left.

The most obvious problem with standard doors is that they allow any number of people to enter, once opened. Only one authorized user is required to open the door and there is no way for the system (thus the organization) to know if other persons entering were authorized or not. Accordingly, standard doors should not be used for high-security areas unless there are other sound procedures to ensure that each person entering is authorized.

Additionally, this deficiency makes tracking of authorized users difficult. The system cannot know for certain where any given authorized user is located. It knows that a given authorized user has entered, but cannot know that they have left and would not know if they have entered into the same space again, or perhaps another space by entering with a group when the door was opened by a single authorized user.

Standard doors can be configured for Card In/Free Out or Card In/Card Out and can be configured with a Tailgate Detector to deter the entry of multiple people on a single card; but a single door cannot be configured for positive access control. A single door can be used as part of an

■ **FIGURE 5.4** Standard type door with access control showing a card reader by HID Global Corporation. *Photo by HID Global Corporation.*

Antipassback Zone if it is equipped for Card In/Card Out and there is good access control procedural discipline to ensure that two people do not leave the zone under use of a single card (Fig. 5.4).

Automatic doors

In high-traffic areas, or where materials must be moved, it is common to use an automatic door. Automatic doors are used in hospitals where patients must be moved in to or out of an Intensive Care Unit, or for materials handling areas where a worker must move a cart of materials to a secure area for printing, such as in a credit card printing shop.

Automatic doors should not be used as a portal to a high-security zone unless there are very good procedural controls, because it is too easy for an unauthorized person to enter behind an authorized person.

The most common type of automatic door is the one- or two-leaf swing door. This is a standard single- or double-leaf door that has an automatic door operator attached to it. Before automating a standard door, it is important to ensure that the frame, hinges, and door are robust enough to stand up to the torque applied by the automatic operator at the top of the door. Remember, this action is applied to an area of the door that does not usually get pressure applied to it. Typically a door is made to be opened at the handle, not at the top. Opening a door from the top, midway between hinge and strike edge, applies considerable pressure that the door was not designed to accept. Accordingly, the door must be sturdy enough to withstand such action many times a day for many years. Lesser doors and doors that are in poor maintenance can fail when an automatic operator is installed.

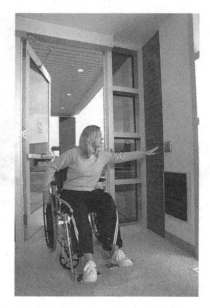

■ **FIGURE 5.5** Automatic swing door. *Image courtesy of Harton Automatics.*

The operation of an automatic swing door can be a surprise to a person who does not expect it (Fig. 5.5). It is advisable to paint or mark "door sweep zones" onto the floor to show where the door will open and as an alert to stay out of that area so that the door does not swing into the path of a person who is standing or walking there. This is especially important when an automatic door will open into a hallway where people are passing by.

Another common type of automatic door is the sliding door (Fig. 5.6). Automatic sliding doors are in many ways superior to automatic swing doors because there is no need to be concerned about the door swinging into the path of a person walking nearby. Sliding doors may be manufactured from steel or glass (I have even once seen a wooden door, but that is rare). Sliding doors have the distinct advantage of taking very little space in the room so that furniture or equipment may be located nearby.

Automatic sliding doors may be configured either as single-leaf where one door slides to the side or double-leaf where two doors slide to the side and then close to the center of the opening. Most automatic sliding door operators provide very little resistance to pulling the door open. This is especially true of double-sliding automatic doors using frameless glass panels. It is all too easy to place one's fingers between the glass

■ **FIGURE 5.6** Automatic sliding door. *Image courtesy of Harton Automatics.*

panels and peel the doors open. All automatic sliding doors used in access control applications should be equipped with magnetic locks to ensure that the door cannot be pried open by an unauthorized user. However, provisions must be made for opening in an emergency such as a fire, since one cannot simply push the door open unless the door is especially manufactured to have a swing partition hanging from a sliding frame. This is somewhat rare but is an option from some manufacturers.

Another common type of automatic door is the bifold or fourfold door. A bifold door is a double-leaf door in which one leaf is hinged to the frame at one side and both leaves are hinged where they come together in the center. The unhinged edge of the final leaf is set in a track at the top of the door so that the entire assembly may be pushed to the side of the leaf that is hinged to the frame. Thus, the bifold door folds in half to the side of the opening. Bifold doors are typically configured as half-width doors placed across a single door opening. A fourfold door is simply two bifold doors placed across a double-width opening (Fig. 5.7).

Bi- and fourfold doors, like sliding doors, open close to the frame and do not take up much space like a swing door does. Bi- and fourfold doors open and close quickly, which is their greatest asset. Bifold doors open twice as fast as a standard single-leaf sliding door across the same frame. A fourfold door opens four times as fast as a single-leaf sliding door across a double frame.

Revolving doors

Where space is available and speed of throughput is not such an issue, revolving doors offer a good solution (Fig. 5.8). Use of revolving doors

■ **FIGURE 5.7** Automatic fourfold door. *Image courtesy of Harton Automatics.*

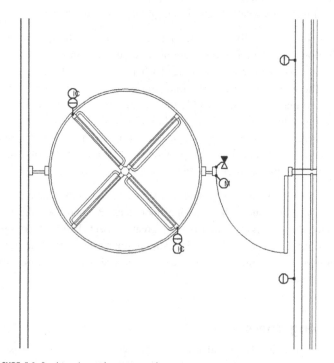

■ **FIGURE 5.8** Revolving door with access control.

for positive access control ensures that no person or package can be easily passed through the portal. Use of a revolving door for positive access control requires the addition of an adjacent emergency exit door activated by the fire-alarm system and equipped with an emergency door release that is similar to a fire-alarm pull station but marked Emergency Door Release. These are usually colored blue to ensure that no one assumes it is a fire-alarm pull station.

Like the turnstiles, a card reader is placed on each side of the door and allows a 180-degree turn to pass a single individual. Designs vary. The simplest versions allow two or more tightly crammed people to wedge their way through on a single card entry in the same direction (in or out). Better designs ensure that only one person is occupying the revolving door. All types typically have provisions (usually a floor sensor) to ensure that two people are not using the door at the same time in opposite directions.

Where the area is extremely sensitive, a man-trap can be used. A man-trap is a series of two doors configured so that only one can open at a given time. This way, an authorized user can enter the outer door that closes behind him/her and when both doors are again closed, the user can present his/her credential to the second internal reader, which then opens the inner door. The operation is reversed to exit. In many cases only an exit push button is used to release the outer door for exiting.

Turnstiles

Turnstiles have been around in mechanical form for about a 100 years. We have seen so-called tripod turnstiles (these have three arms that pivot) used in sports arenas, subways, and amusement parks for many years. In corporate lobbies, turnstiles are often configured as a pair of paddles or glass wings enclosed in a confined pathway. Electronic turnstiles come in a variety of types. Those most often used for positive access control are either a set of "paddles" or "glass wings" that physically prevent passage without use of an authorized credential (Fig. 5.9).

The simplest type of electronic turnstiles uses only electronic beams and a circuit to sound an alarm if an unauthorized user tries to pass. This type is of no use for positive access control as it presents no physical barrier. Even the paddle and glass wing types should be used under observation of a security officer to ensure that offenders do not jump over or duck under the paddles or jump between turnstiles using glass wings. All of these actions have occurred in actual installations. For observed public areas such as main lobbies, turnstiles are a good choice because of their

■ **FIGURE 5.9** Access control turnstile. *Image courtesy of Smarter Security Systems.*

speed of throughput and safety provisions. Electronic paddle turnstiles are also often equipped with break-away arms so that crowds of people can exit quickly in an emergency.

Man-traps

A man-trap is a series of two doors on each end of a vestibule where only one door can be opened at any time. Entry is made by sequencing through the two doors: first the outer door opens and then closes, and then the inner door opens, then closes. Exiting is done in the opposite sequence. The doors are electrically locked and interfaced such that only one door may open at a time. This ensures that no unauthorized person can enter behind an authorized user by closing the outer door before the inner door can open so that the authorized user can be certain that they are not being followed (Fig. 5.10). Man-traps utilize two different types of access control credentials, one for each door. The outer door may utilize a card reader and the inner door may utilize a biometric or keypad credential. Man-traps to high-security spaces typically utilize all three means (something you have (an access card), something you know, (a code), and something that you are (biometrics)).

Man-traps have serious life-safety implications that must be addressed. This typically includes a fire-alarm system interface and emergency pull-stations on the secure side of both doors that override the man-trap security function. Typically, pulling the inner emergency door release unlocks both doors, and pulling the outer door emergency lock release unlocks only the outer door. It is imperative to seek a variance with local code authorities before using a man-trap to ensure that all local life-safety codes and regulations apply. Note: more and more municipalities do not

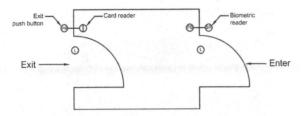

■ **FIGURE 5.10** Typical man-trap.

allow any type of man-trap, so it is especially important to check before designing or building one.

Automated walls

A rare but useful type of access control portal is the automated wall. Automated walls are not used for conventional access control; they are used to prevent the spread of fire throughout a building and to prevent a specific type of attack on certain types of facilities. Automated walls are high-security elements designed to protect people during a so-called "moving shooter" attack, such as happened in Mumbai in December 2008.

An automated wall is placed between areas of the facility, making certain that there is a ready fire exit on both sides of the wall. Placement is critical to comply with fire codes to be certain that the designer is not creating a "dead-end corridor." In case of a fire or moving shooter emergency, the automated walls are triggered. This compartmentalizes the facility into horizontal segments, preventing the spread of the fire or the movement of the shooter. In the case of the moving shooter attack scenario, this confines the shooter into a controllable space that can bring the event to a much swifter end with far less potential loss of life.

Automated walls are operated automatically by the fire-alarm system or manually from a security command center. They are typically configured to require two keys to operate to ensure that they are not accidentally or haphazardly deployed.

VEHICLE PORTALS
Standard barrier gates

Vehicles need to be controlled on a property to ensure proper traffic flow (employees here, visitors there, etc.) and to control entry to restricted areas, such as to an employee parking structure. The least expensive and

■ **FIGURE 5.11** Semaphore arm barrier gate. *Image courtesy of Delta Scientific.*

most familiar way to do this is by using a standard lift-arm barrier gate. Lift-arm (or semaphore arm, named after the semaphore flags that were used to guide airplanes onto aircraft carriers) barrier gates come in a variety of configurations depending on the width of the lane and the frequency of opening. The simplest semaphore arm gate comprises a metal stand housing the motor and operating electronics and a gate arm made of wood, aluminum, or PVC plastic (Fig. 5.11).

Semaphore arm gates are typically used in conjunction with a card reader or parking ticket dispenser to open the gate and also to incorporate safety features, including:

- A safety edge on the arm to reverse its direction in case it strikes a vehicle, person, or object while descending
- Buried vehicle loops that detect the presence and position of one or more vehicles including:
- A presence-sensing entry loop at the card reader or ticket dispenser to ensure that the gate is not deployed up when there is no vehicle to enter
- A safety loop under the arm to prevent it from beginning its descent when a vehicle is present
- An after-presence-sensing exit loop to trigger the arm to descend after a vehicle has passed through the gate

The buried loops work in conjunction with a vehicle loop detector usually housed in the enclosure that houses the motor. The loop detector senses the loops and operates by a logic function to operate the gate in response to signals from the loops, a card reader or ticket dispenser, or a remote trigger from the security command center.

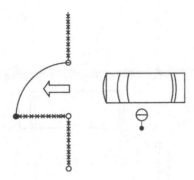

■ **FIGURE 5.12** Vehicle swing gate.

It is common to associate a field intercom station with a card reader at a parking gate so that if a visitor requires assistance the gate can be opened remotely from a security command center where the intercom call is answered. It is also common to associate a video camera viewing the vicinity of the gate and often also the license plate of the car and the face of the driver (three cameras), which are displayed when the driver presses the intercom call button.

Automated vehicle swing gates

Vehicle swing gates are often found at residences and commercial or industrial facilities and are used to control the entry and exit of vehicles to/from the property (Fig. 5.12). A swing gate operator is mounted to a concrete pad and operates an articulating arm or hydraulic ram that opens and closes the gate. These are also normally operated by a card reader or remotely from a facility in conjunction with an intercom. It is common to utilize vehicle loops for safety. Card readers for automated vehicle swing gates are often set back an additional distance to allow for the swing of the gate, so the entire gate and queuing area can become somewhat large. The open/close time of a standard lane-width gate is about 20 seconds. Where the door is used as a security element, it may be advisable to add a magnetic lock to ensure that the door cannot be easily forced open against the operator by potential intruders. This also helps to prevent damage to the operator, which can occur if the door is forced open.

Automated sliding vehicle gates

Another common type of vehicle portal is the sliding vehicle gate (Fig. 5.13). Automated sliding gates use a sliding gate operator and a long chain near the lower part of the sliding gate. The chain is held at

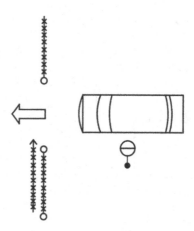

■ **FIGURE 5.13** Sliding vehicle gate.

both ends of the sliding gate and feeds through a gear on the gate operator. As the gate operator rotates its motor, the chain is moved along and it moves the gate, which is attached to the chain.

Sliding gates, like sliding doors, have the advantage of taking very little driveway space, so they are especially useful on sloping driveways where a swing of the gate would conflict with the oncoming vehicle or where it would conflict with the slope of the driveway. All other aspects of automated sliding vehicle gates are similar to automated vehicle swing gates. Sliding gates are also usually equipped with safety edges to prevent them from closing on a vehicle, person, or object. Sliding gates are also ideal where it is desirable to locate the card reader closer to the gate because there is no setback for the swing of the gate, so they often work better in confined areas. The operating time of automated sliding gates is similar to swing gates—about 20 seconds.

Automated roll-up vehicle gates

Roll-up vehicle gates are most often seen at a vehicle entry to a building, although they are also used at fence lines in high-security facilities, most notably at detention facilities (Fig. 5.14). Roll-up doors are more difficult to break into than either swing or sliding gates.

High-security barrier gates

High-security barrier gates are used to prevent crash-through entries into secure facilities such as ports and nuclear facilities (Fig. 5.15).

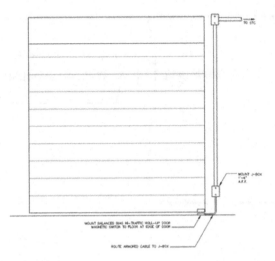

■ **FIGURE 5.14** Roll-up vehicle gate.

■ **FIGURE 5.15** High-security barrier gate collage. *Image courtesy of Delta Scientific.*

Crash-rated gates are available in a variety of designs including:

- Lift-arm barrier gates
- Sliding gates
- Web fabric gates
- Phalanx gates

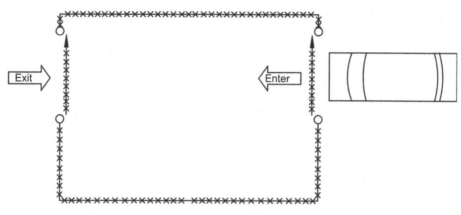

■ **FIGURE 5.16** Sally port.

- Rising bollards
- Uncommon miscellaneous types

Sally ports

Sally ports are a secure, controlled entryway for vehicles having two gates, similar to a man-trap in operation (Fig. 5.16). Sally ports are the vehicular equivalent of a man-trap. The entrance is typically part of an outer perimeter such as a secure fence line or prison perimeter wall and the interior gate is also fortified against easy entry. Sally ports are operated remotely by a security officer from a separate secure space where they can supervise the entry and exit. Only one sally port door is ever opened at a time to ensure that no one can sneak in (to a military compound) or out (of a prison).

CHAPTER SUMMARY

1. Basic portal passage concepts include:
   a. Card entry/free exit
   b. Card entry/card exit
   c. Tailgate detection
   d. Positive access control
   e. 2-Man rule
   f. Schedules
   g. Antipassback
2. Common pedestrian portal types include:
   a. Standard doors
   b. Automatic doors:

       i. Automatic swing doors

       ii. Automatic sliding doors

       iii. Automatic fourfold doors

    c. Revolving doors

    d. Turnstiles

    e. Man-traps

    f. Automated walls

3. Common vehicle portal types include:

    a. Standard barrier gates:

       i. Semaphore arm barrier gates

       ii. Automated vehicle swing gates

       iii. Automated vehicle sliding gates

       iv. Automated roll-up vehicle gates

    b. High-security barrier gates

    c. Sally ports

Q&A

1. *What is an access portal?*

    a. System network access

    b. Organization's security program

    c. Authorization given to access higher security zones

    d. Passageway through which a person must pass to go from one access zone to another

2. *Card Entry/Card Exit*

    a. Is designed to allow entry only to authorized users but allows anyone inside the access zone to exit freely

    b. Is designed to allow entry only to authorized users and to allow exit with authorization inside the access zone

    c. Is the practice of an unauthorized user following an authorized user through an access controlled door by catching the door before it closes

    d. Ensures no single person ever has access to the secured area alone, thus providing a higher level of security by ensuring that one person will always be watching another

3. *Tailgate Detection*

    a. Is designed to allow entry only to authorized users but allows anyone inside the access zone to exit freely

    b. Is designed to allow entry only to authorized users and to allow exit with authorization inside the access zone

    c. Is the practice of an unauthorized user following an authorized user through an access controlled door by catching the door before it closes

      d. Ensures no single person ever has access to the secured area alone, thus providing a higher level of security by ensuring that one person will always be watching another

4. *2-Man Rule*
   a. Is designed to allow entry only to authorized users but allows anyone inside the access zone to exit freely
   b. Is designed to allow entry only to authorized users and to allow exit with authorization inside the access zone
   c. Is the practice of an unauthorized user following an authorized user through an access controlled door by catching the door before it closes
   d. Ensures no single person ever has access to the secured area alone, thus providing a higher level of security by ensuring that one person will always be watching another

5. *Antipassback schemes are designed to prevent one authorized user from allowing his card to be used by a second unauthorized person*
   a. True
   b. False

6. *Standard Doors*
   a. Are used in high-traffic areas, or where materials must be moved
   b. Is the basic unit of an electronic access control system
   c. Are configured as a pair of paddles or glass wings enclosed in a confined pathway
   d. Are a series of two doors on each end of a vestibule where only one door can be opened at any time

7. *Automatic Doors*
   a. Are used in high-traffic areas, or where materials must be moved
   b. Is the basic unit of an electronic access control system
   c. Are configured as a pair of paddles or glass wings enclosed in a confined pathway
   d. Are a series of two doors on each end of a vestibule where only one door can be opened at any time

8. *Turnstiles*
   a. Are used in high-traffic areas, or where materials must be moved
   b. Is the basic unit of an electronic access control system
   c. Are configured as a pair of paddles or glass wings enclosed in a confined pathway
   d. Are a series of two doors on each end of a vestibule where only one door can be opened at any time

9. *Man-Traps*
   a. Are used in high-traffic areas, or where materials must be moved
   b. Is the basic unit of an electronic access control system
   c. Are configured as a pair of paddles or glass wings enclosed in a confined pathway
   d. Are a series of two doors on each end of a vestibule where only one door can be opened at any time

10. *Automated Vehicle Swing Gates*
   a. Are often found at residences and commercial or industrial facilities and are used to control the entry and exit of vehicles to/from the property
   b. Are the least expensive and most familiar way to control entry to restricted areas
   c. Are used to prevent crash-through entries into secure facilities such as ports and nuclear facilities
   d. Are commonly operated remotely by a Security Officer from a separate secure space where they can supervise the entry and exit

11. *Sally Ports*
   a. Are often found at residences and commercial or industrial facilities and are used to control the entry and exit of vehicles to/from the property
   b. Are the least expensive and most familiar way to control entry to restricted areas
   c. Are used to prevent crash-through entries into secure facilities such as ports and nuclear facilities
   d. Are commonly operated remotely by a Security Officer from a separate secure space where they can supervise the entry and exit

12. *Standard Barrier Gates*
   a. Are often found at residences and commercial or industrial facilities and are used to control the entry and exit of vehicles to/from the property
   b. Are the least expensive and most familiar way to control entry to restricted areas
   c. Are used to prevent crash-through entries into secure facilities such as ports and nuclear facilities
   d. Are commonly operated remotely by a Security Officer from a separate secure space where they can supervise the entry and exit

13. *High-Security Barrier Gates*
   a. Are often found at residences and commercial or industrial facilities and are used to control the entry and exit of vehicles to/from the property
   b. Are the least expensive and most familiar way to control entry to restricted areas
   c. Are used to prevent crash-through entries into secure facilities such as ports and nuclear facilities
   d. Are commonly operated remotely by a Security Officer from a separate secure space where they can supervise the entry and exit

Answers: (1) d; (2) b; (3) c; (4) d; (5) a; (6) b; (7) a; (8) c; (9) d; (10) a; (11) d; (12) b; (13) c.

Chapter 6

Life Safety and Exit Devices

CHAPTER OBJECTIVES

1. Learn the basics in the chapter overview
2. Understand the top priority in access control: life safety first
3. Security versus life safety
4. Understand national and local access control codes
5. Understand how locks can affect life safety
6. Understand how exit devices can affect life safety
7. Understand how access control systems depend on fire alarm systems
8. Answer challenge questions on life safety and exit devices

CHAPTER OVERVIEW

This may be the most important chapter in this book. In fact, if someone learns everything else in the book, but does not learn the material in this chapter, that person would be a failure in the security industry. Life Safety is the most important element of security. The most basic core element of all security is to protect life first, whether in regards to access control or preventing workplace violence or terrorism. This chapter discusses the apparent conflict between security and life safety principles and how they should be resolved. We will discuss national and local life safety codes and regulations and how they apply to electronic access control systems. We will talk about how locks and exit devices affect life safety.

LIFE SAFETY FIRST

Let me introduce you to your mantra: "It's all about life safety." Once when I was 17 years old, in the mid-1960s, I went to an office building late in the afternoon where I was hoping to go to a business that had a product I was interested in. I had called earlier and they had said to

Electronic Access Control. DOI: http://dx.doi.org/10.1016/B978-0-12-805465-9.00006-3

"come anytime," which unfortunately I took literally (I was 17—what can I say?). Their offices were on the eighth floor of a 12-story building in downtown Columbus, Ohio. I arrived there just at 6:15 p.m. on a Friday evening and caught the front door of the ground floor lobby as a man was walking out. I took the elevator to the eighth floor (most ceiling lights were off). I wandered down the dark hallway and located the company's office, only to find that they were closed. Oh well, not to worry, I'll come back another time. I'll go back downstairs, get some dinner at a nearby diner, and go home. So I take the elevator back down to the ground floor lobby (lights are off) and I head to the front lobby door, leaning into it. Bang! Door won't open. It is locked. Uh-oh. Building is closed. Somewhere between the time I had arrived and got back downstairs, the last person had left and locked the building. Well, there has got to be a janitor somewhere with keys. I go back to the elevator and start at the top floor working my way down to the basement, checking each floor for a janitor or anyone else with keys. Nobody, I mean nobody, is in the building. I am stuck in a high-rise building in downtown Columbus, Ohio, at seven in the evening and there is nobody here to open the building. Okay Tom, think. There has got to be another exit. I wander around the building. All offices are locked. All perimeter doors are locked from both sides. I am locked in this building. It is Friday. The offices are closed on Saturday.

I tried to get the attention of anyone outside the front door. Nobody (hey, it's Columbus, Ohio, not New York). Two hours pass. It is 9:30 p.m. Finally, a cop passes by and I bang on the door. He hears me. We talk through the door. He says he will get help. A very unhappy building engineer arrives to let me out just after 11:00 p.m. I have spent 5 hours locked in a building. It is a lesson I will never forget. Like the old Arab proverb says: "You don't learn anything the second time you get kicked by a mule."

That was inconvenient (for me and for the poor building engineer). Had there been a fire in that building at that time, it would have redefined the meaning of "inconvenient."

Life safety is about preserving life above every other principle. That building locking arrangement is illegal today. Code officials have learned a few things. It is against every building code in America, Europe, and most developed countries to lock anyone inside a building. (Variances can be acquired for extraordinary circumstances, but more on that later.) When designing a building, one must ensure that all occupants have free, unobstructed egress from every part of the building to the outside.

SECURITY VERSUS LIFE SAFETY

If there is a decision to be made between security and life safety, it is simple. Life safety wins, each time, every time. No matter how the argument is constructed, no matter how extenuating the circumstances, there has to be a provision for occupants to escape quickly in an emergency—freely and with no special knowledge of how to exit.

UNDERSTANDING NATIONAL AND LOCAL ACCESS CONTROL CODES AND STANDARDS
NFPA 101

In the United States, the primary international Life Safety Standard is the Life Safety Code, published by the National Fire Protection Association (NFPA) and is known as NFPA 101. NFPA 101 is revised every 3 years. Despite its title, the document is a standard and not legal code. Its statutory authority is derived from local Authorities Having Jurisdiction (AHJs), typically the local Fire Department. The language of the NFPA 101 is crafted in a form that makes it suitable for mandatory application and is adopted into law by local authorities.

International building code

The second applicable document is the International Building Code (IBC; formerly known as the Universal Building Code, UBC). In other countries a set of model building codes are used, which are often very similar to the IBC. Again, like the NFPA 101 Life Safety Code these are standards, not law. In all countries, local authorities are recommended to reference the relevant standards as a mandatory reference to follow when designing buildings.

Building Codes such as the IBC include instructions beyond issues of life safety, including information about structural safety and resistance to natural disasters such as hurricanes, tornadoes, and floods; however, life safety is at the heart of all building codes. Key life-safety references in the IBC include allowable sizes and placements for exit doors; lengths and widths of hallways; allowable construction materials; use of fire stop compounds in wall penetrations to prevent the spread of fire from one space to another through those penetrations; minimum and maximum room sizes for various types of occupancies; storage requirements for flammables, hazardous materials and the like; and requirements regarding exiting requirements.

NFPA 72

The third relevant standard is NFPA 72, which was also developed by the National Fire Protection Association. NFPA 72 applies to "the application, installation, location, performance, inspection, testing, and maintenance of Fire Alarm Systems, fire warning equipment, and emergency warning equipment, and their components." Like NFPA 101, local AHJs all across the country refer to NFPA 72 as mandatory law.

More on these codes

NFPA 101, NFPA 72, and the IBC are all referred to by Fire Authorities and litigating attorneys and insurance companies, so the code is often referenced for legal action and to defend against insurance claims. NFPA 101, NFPA 72, and the IBC can all be amended by local authorities, so it is important to know not only these standards, but also the local deviations, additions, and modifications to them. Failure to know can be expensive.

I was once called by a building manager of a 30-story building downtown in a large Northern California city. He was in a panic. I had designed the security system for the original building and had recently designed a renovation of the access control system to include locks on all stairwell doors, using a lock type and locking scheme that was approved by the city. We had discussed the matter with the San Francisco Fire Department, submitted plans, and received approval. The building manager told me that the local fire department inspector was instructing him to shut down his building immediately. I asked him to put the fire inspector on the phone. I asked the inspector tell me what the problem was. He explained to me that the security contractor, who the building manager had selected against my advice (a relatively new contractor who put in an extraordinarily low bid), had substituted electric strikes instead of approved fire stairwell locks and was halfway up the building installing these locks. The contractor had begun work without submitting the hardware for the consultant's approval. The net result was that the new security contractor, who was very happy about winning his first big job, had cut holes into all of the fire stairwell door frames to make way for the door strikes and in so doing had violated the fire rating of all of those frames (16 floors, 32 doors). This caused the building to be out of compliance with its certificate of occupancy, so the certificate of occupancy was no longer valid. The building had to be closed until the door frames were returned to code compliant condition. I asked him if he would accept a "fire watch" until the repairs were made. He said he would waive the ruling if a fire watch was put into place. "Thanks" I said. "Put the building manager back on please."

I explained the situation to the building manager and explained that a "fire watch" is a 24/7 guard on every floor whose job it is to watch for fires and evacuate the building if one occurs. Sixteen floors, 16 guards— it will not be cheap. But it will certainly be better than informing all of his tenants that they have to leave the building immediately! He sighed. I gave him the number of a major guard company in his city. He called, they came, the fire inspector left.

The security contractor who was so happy about winning his first big job and who was at first elated and proud that he had "found a way around the requirement for that damned security consultant's very expensive locks" also left. The building manager hired another major security contractor to replace all of the destroyed door frames (at a very significant cost) and the first security contractor was billed for the work and the cost of the 24/7 "fire watch" guards. He filed bankruptcy. It pays to understand Fire Code.

I have talked to many security contractors and consultants whose knowledge of NFPA 101, NFPA 72, and the IBC amount to what they learned from "some seminar I took," or "some article or book summary I read." From the pinnacles of their modest knowledge, these people often say, "I know all I need to know." They don't. Let me say this clearly. If you are going to make a career in the security industry, you need to buy a copy of NFPA 101, NFPA 72, and the IBC. I have talked to security designers and contractors who say: "I don't need to know what is in the IBC because it is the architect's job to design the building, not mine." But sometime in your career, you will be brought into a small project that involves a minor change in the floor plan that seems like a good idea to the owner who will have hired some unlicensed contractor to make the change and has then asked you to design or install access control equipment on the new area. That is a bad idea. You can get sued for designing or installing equipment on noncode compliant structures when there is a fire and the people inside cannot escape. You may have skipped the electrified panic hardware because the owner wanted "cheaper locks." You may find out too late that your work is not in compliance with fire life safety codes. You can pay dearly for what you do not know, and many security designers and contractors have. Buy a copy of each. You will be glad you did.

UL 294

Underwriters Laboratories (UL) is an organization that has been in operation for more than a century. It is an independent, nonprofit organization that writes standards and tests products against those standards to provide an assurance to the marketplace that a given product meets stringent manufacturing and safety standards (UL Standards). UL 294 is the standard for access control systems. It applies to "... the construction, performance and operation of the systems intended to regulate or control entry into a protected area or a restricted area or access to or the use of a device(s) by electrical, electronic or mechanical means" (From Part 1— Scope of UL 294).

Manufacturers submit their access control systems to UL for certification under UL 294 to comply with building codes that require systems to be UL certified. Additionally, most specifiers require UL certification on all of the electronic and electrical equipment submitted for installation under their specifications.

UL 204 covers product assembly, electrical protection, and protection of service personnel, how the enclosure is manufactured, electrical shock protection, corrosion protection, field wiring and grounding connections, internal wiring, electrical components, and performance on many parameters. Additionally, UL 294 requires manufacturing and production line tests, designated markings, outdoor use performance standards, and standards on accessory equipment.

In addition to UL 294, similar standards exist for Europe (the CE Mark) and Canada. The Canadian Standards Association (CSA) is the equivalent of UL in Canada. The CE Mark is the equivalent of UL in the European market. Its mark is required for products to pass freely between European Union member states. Another mark is the ETL Mark. ETL Testing Laboratories is another nationally recognized testing laboratory that is required in certain jurisdictions, notably New York City. ETL is also recognized by the Occupational Safety and Health Act (OSHA).

It is not necessary for designers and contractors to fully understand UL 294 (or its CE/CSA/ETL equivalents) to succeed in the industry. But I think it is foolish of anyone serving the security industry to specify or allow installation of any noncertified equipment when certified equipment is available.

LIFE SAFETY AND LOCKS

We will be looking at electrified locks in great detail in Chapters 9 through 13, but for a moment, let's talk about how electrified locks relate to life safety and life safety codes.

The key factor is that for most installations, a person *must* be able to exit a space with no special knowledge. This means that anyone should be able to walk up to the door and exit with "No Special Knowledge." This includes people of other languages, nationalities, and cultures. It is unacceptable to have a special lock that requires manipulation of some switch that is marked in a language everyone does not understand.

There is a common violation to this rule. It is common to see magnetic locks operated by a nearby red push button. Strictly speaking, this does not comply with NFPA 101.

Ideally, and everywhere possible, locks should be electrified mechanical locks instead of magnetic or electrified dead-bolt locks. Turn a handle or push a panic bar—go out. Stick to that rule and you will *never* have a problem with fire inspectors. To be sure, there will be times when you cannot use a mechanical lock; e.g., on the existing frameless glass front lobby doors of a high-rise building or on a door that is already equipped with an unsuitable mechanical lock. However, there are ways to adhere to the "No Special Knowledge" rule by selecting exit sense devices that do not require any special knowledge to operate.

There is a seemingly infinite array of electrified lock types, and electrified locks are the bane of security contractors. After you finish Chapter 13, Specialty electrified locks, you will be a bona fide expert on electrified locks. You will know about locks that most people (even most architects I have met) do not even know exist. Electrified locks can be categorized by:

- Type of locking mechanism
- Physical interface to the door
- Level of security
- Special/no special knowledge to operate
- Dependability in an emergency
- Silent or audible operation
- Suitable or not suitable for fire exit doors
- Robust or not robust against physical abuse
- Suitability for different types of occupancies
- Suitability for use with various types of exit sensors

But for the purposes of life safety the essential attributes you should be interested in include:

- No special knowledge to operate
- A lock that is suitable for the occupancy
- A dependable means of exit sensing

The two best kinds of electrified locks for life safety are electrified panic hardware (Fig. 6.1) and the electrified mortise lock.

Electrified panic hardware is simply conventional panic hardware (a panic bar with an associated locking mechanism) to which a solenoid has been added to toggle the electrified lock in the panic hardware. There is also an exit sense switch inside the panic bar to tell the electronic access control system that the person opening the door is legally exiting and not breaking into the door. The exit sense function bypasses the door alarm. Electrified panic hardware is a purely mechanical lock that can be remotely electrically unlocked to facilitate access control by means of a card reader.

A mortise lock (Fig. 6.2) is a very strong mechanical lock that fits into a pocket (mortise) in a door. Mortise locks are available in a wide variety of functions including passage, privacy, entry, storeroom, classroom, classroom security, and office. These will all be discussed in detail in

■ **FIGURE 6.1** Electrified panic hardware. IRCO-VonDuprin.

■ **FIGURE 6.2** Electrified mortise lock. *Image courtesy of Security Door Controls.*

Chapter 10, Free egress electrified locks. Additionally, in Chapter 10, Free egress electrified locks, we will discuss electrified panic hardware, which is also available in several versions, the most common of which are the rim, mortise, and concealed/surface vertical rod versions. In Chapter 10, Free egress electrified locks, we will also briefly discuss a couple of specialty locks that provide free mechanical egress. Those will be fully explored in Chapter 13, Specialty electrified locks.

LIFE SAFETY AND EXIT DEVICES

Remember, in most cases the door with an electrified lock in an access control system is also monitored for intrusion. Any time the door opens it is because of a legal entry, a legal exit, or intrusion. Accordingly access controlled doors are almost always equipped with a door position switch, which monitors whether the door is opened or closed.

The switch does not know if the door is opened for a legal entry, legal exit, or for an intrusion. It just reports that the door is open. So the Access Control Panel must make a decision as to whether the door opening is appropriate or a concern to be reported as an alarm. It does this when a valid credential is read (card, keycode, biometric) and reported to the access control panel. After this, the panel executes several actions, including:

- Releasing the door lock
- Recording the access event and the user by association with the credential
- Bypassing the door position switch alarm

When the access panel sees that the door is open from the door position switch, it ignores the opening for some programmable period of time (using an alarm bypass timer). If the door remains open after that time, the panel sees that the programmed time period has elapsed and the door position switch is still reporting the door as open. Thus it triggers another event, a propped door alarm.

When a person exits the door legally another device, a request-to-exit (REX) sensor, senses the person leaving and triggers a similar set of events:

- Releases the door lock
- Records the door opening as a legal egress event
- Bypasses the door position switch alarm
- Sets the alarm bypass timer

- Triggers a propped door alarm if the door does not reclose within the time period

But if someone forces the door open without a valid credential or a valid REX sensor event, the change of state of the door position switch is instantly recognized as an intrusion and an immediate alarm event is triggered.

So REX sensors play an important role in preventing false alarms. There are three primary types:

- Mechanical switch (part of the lock)
- Electronic Presence Detector (usually above the door)
- Mechanical switch (nearby push button)

The best of these are mechanical switches that are part of the lock. These can be ordered as an accessory in most electrified panic hardware locks and many electrified mortise locks. Essentially, as the inside handle is turned a switch is also triggered before the door reaches the fully unlocked mechanical position. For the most part, these operate flawlessly. The switch is wired to the REX sense input on the access control panel, which bypasses the intrusion alarm.

Another common type of REX sensor is in the form of a motion detector usually placed above the door. Typically these are in the form of a small infrared detector that is mounted centerline above the door. The motion detector has a limited field of view, mostly directly down on the door handle, and senses the presence of a person immediately at the door. These are commonly referred to as REX sensors, although technically any switch that responds to an exit maneuver by a user fits that description. REX sensors are not flawless. Because the sensors view the whole area at the door, they can unlock the lock in response to a couple of people standing in the area at the door, thus allowing an unauthorized person outside to open the door and enter without having presented any authorized credential. Additionally, on double glass doors, it is possible to slip an object between the doors from the outside and trigger the sensor, thus opening the door and creating an undetected intrusion event. REX sensors are most often used on doors equipped with magnetic locks. The REX sensor is wired to the REX sense input on the access control panel.

Another common type of REX sensor is the so-called exit push button. These are typically configured as a red, mushroom-shaped button near the door. Many municipalities require labeling these with "push to exit" in 1-inch high contrasting letters. They also trigger the REX sense input on the access control panel, which unlocks the door and bypasses the door alarm.

Both the REX sensors and the Exit Push Buttons have intrinsic problems from a life safety standpoint. In both cases, the common form of wiring requires the access control panel to unlock the door, especially where the lock is a magnetic lock. Let me say this: I do not trust electronics. Electronics is the high priest of false security. Electronics fail. Sooner or later, you may get a call from an irate building owner who has been locked into their lobby. (It seems to always only happen to the owner!) In the early days, I got such a call. I have talked to dozens of security specifiers and contractors who have gotten the call. Do not trust electronics. They will fail. Better designers and contractors always install REX sensors with some kind of a back-up system. Where I must use a REX sensor, I also use a marked exit push button that is wired not only to the access control panel, but also directly to the lock itself so that when the push button is pushed, the lock is unlocked regardless of any potential wiring or electronic failure. Additionally, it is best to use pneumatically operated exit push buttons, which include a small piston that resists returning the button to the "out" position for a defined (adjustable) time period once the button has been pushed. This ensures that the lock will remain unlocked for that time period, and does so purely mechanically, through no electronic means. Thus, the person exiting does not need to hold the lock while he pushes open the door, which on many door installations is a physically impossible task. The user can simply hit the button and use the same hand to push open the door, ensuring that they can exit.

Exit Push Buttons are disapproved in some municipalities because they require special knowledge; i.e., not everyone has seen or been confronted with a locked door that requires any action to open other than by pushing on the door. I have seen elderly people, rural people, and people from developing countries all stop for a couple of minutes until someone nearby explained to them how to exit. This is not a good thing in an emergency.

LIFE SAFETY AND FIRE ALARM SYSTEM INTERFACES

Best practices in security system design call for interfacing all electrified locks in the access control system to the fire alarm system such that all locks of a common area (building, floor, etc.) will unlock automatically in response to a fire alarm in that area. I strongly recommend that security designers and security contractors use this practice as a uniform standard. Even in buildings that do not have a fire alarm system, it is advisable to design and install the access control system with an unused fire alarm interface so that if and when a fire alarm system is installed the system is intrinsically ready to interface with it with no modifications. It can be difficult and expensive to modify an existing access control system to interface with a fire alarm system, but it is inexpensive and easy to design and install a new access control system with a fire alarm interface.

There are several ways to interface an access control system to a fire alarm system. The best way, in my opinion, is to interface the fire alarm system to the main (A/C) power of the lock power supplies ensuring that, when the alarm triggers, all power is dropped to all electric locks. This requires ensuring that lock power is fed from a separate line than power to the access control panels and any other security system equipment. This is inexpensive and easy to do in new construction and even in existing construction on newly installed systems. The interface comprises a relay from the fire alarm system (summary alarm relay) that triggers an interposing high-current, high-voltage relay (120VAC/20 amp, US) or (220VAC/10 amp, European) on the A/C line that provides power to the lock power supplies (Fig. 6.3). For large systems, this may require multiple relays to handle multiple power runs zoned according to the layout of the fire alarm system.

On systems that have previously been installed without a fire alarm interface, adding one can be more complicated. There are a few ways to do this:

- Run a new A/C power line to all lock power supplies of a building, zoned as previously discussed.
- Install an interposing relay from the summary alarm relay of the fire alarm system.

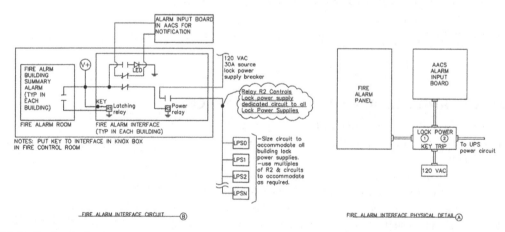

■ **FIGURE 6.3** Fire alarm interface to new system.

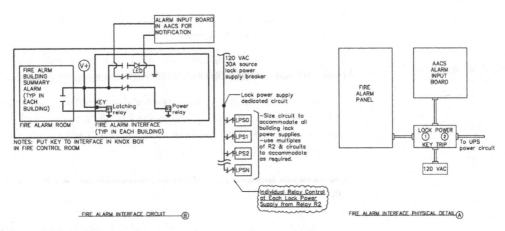

■ **FIGURE 6.4** Fire alarm interface to existing system.

Interface that relay to additional interposing relays (one for each lock power supply) with those relays located at the Lock Power Supplies. Those relays will interrupt A/C power to each lock power supply. This scheme is illustrated in Fig. 6.4.

CHAPTER SUMMARY

1. Life safety concerns precede all other concerns on access control systems.
2. Relevant national and local access control codes and standards include:
   a. NFPA 101
   b. International Building Code
   c. NFPA 72
   d. UL 294
3. The selection of the correct electrified lock is closely tied to the relevant national and local access control codes and standards.
4. The three key attributes you should be interested in for life safety include:
   a. No special knowledge to operate
   b. A lock that is suitable for the occupancy
   c. A dependable means of exit sensing
5. The two best kinds of electrified locks for life safety are electrified panic hardware and the electrified mortise lock.
6. Exit sensors signal to unlock the door and bypass the intrusion alarm on the door.

7. For life safety purposes, the electrified lock should open freely without the use of electronics.

Both REX sensors and common exit push buttons have intrinsic problems from a life-safety standpoint.

Best practices in security system design call for interfacing all electrified locks in the access control system to the fire alarm system such that all locks of a common area (building, floor, etc.) will unlock automatically in response to a fire alarm in that area.

Q&A

1. *IBC includes:*
   a. Information about structural safety and resistance to natural disasters such as hurricanes, tornadoes, and floods
   b. NFPA 101, NFPA 72
   c. Maintenance of fire warning equipment and emergency warning equipment and their components
   d. Product assembly, electrical protection, and protection of service personnel
2. *NFPA 72 includes:*
   a. Information about structural safety and resistance to natural disasters such as hurricanes, tornadoes, and floods
   b. UL 204, IBC
   c. Maintenance of fire warning equipment and emergency warning equipment and their components
   d. Product assembly, electrical protection, and protection of service personnel
3. *UL 204 includes:*
   a. Information about structural safety and resistance to natural disasters such as hurricanes, tornadoes, and floods
   b. NFPA 101, IBC
   c. Maintenance of fire warning equipment and emergency warning equipment and their components
   d. Product assembly, electrical protection, and protection of service personnel
4. *What is an electrified panic hardware?*
   a. Equipped with a door position switch that monitors whether the door is opened or closed
   b. Triggers a propped door alarm if the door does not reclose within the time period
   c. Purely mechanical lock that can be remotely electrically unlocked to facilitate access control by means of a card reader
   d. All of the above

5. *What is an electrified mortise lock?*
   a. Equipped with a door position switch that monitors whether the door is opened or closed
   b. Triggers a propped door alarm if the door does not reclose within the time period
   c. Purely mechanical lock that can be remotely electrically unlocked to facilitate access control by means of a card reader
   d. Very strong mechanical lock that fits into a pocket in a door

Answers: (1) a; (2) c; (3) d; (4) c; (5) d.

Chapter 7

Door Types and Door Frames

CHAPTER OBJECTIVES

1. Discover the basics about doors and security
2. Understand the basic single-leaf door
   a. Hollow metal
   b. Wood
   c. Framed glass
   d. Unframed glass
3. Understand double-leaf doors
4. Learn all about door frames
   a. Hollow metal
   b. Aluminum
   c. Wood
5. Get a line on roll-up doors
6. Understand all about revolving doors
7. Pass a quiz on door types and door frames

CHAPTER OVERVIEW

This chapter expands on the information contained in Chapter 5, Types of access controlled portals. Access control is all about controlling access through portals. For pedestrians, most portals are at a door and door frame.

This chapter explains the basics you need to know about how doors and frames are constructed and how they affect access control portal decisions.

BASICS ABOUT DOORS AND SECURITY

My first job as a security consultant was in Southern California working for one of the top security consultants in the nation. I am very grateful for all he taught me. He was a brilliant man who could win an argument

Electronic Access Control. DOI: http://dx.doi.org/10.1016/B978-0-12-805465-9.00007-5

on any subject. People knew this and so out of great respect often they did not invite him to parties (Variation on a quote from Dave Barry). Once, however, he stated that electronic security was far more important than physical security. Ah ha! I jumped on that immediately.

I asked him to imagine an empty field of grass. Now, I said, imagine a steel structure made of columns and beams in the field in the shape of a single story building. Now please place an entire security system including card readers, door position switches, locks, exit sensors, motion detectors, and video cameras all mounted to the columns and beams. Now, please imagine wiring all of these system elements to a desk in the middle of this structure with a computer and monitors to monitor the conditions in the security system. Got it? OK? What do you have? You have a working security system and no security whatsoever. Security begins with physical security. There is no security without physical security. He conceded the argument. (The only time that ever happened, so the moment is hard to forget.)

Electronic security supplements physical security. To secure physical assets, one must have a container to which one can control access to authorized people and restrict access from unauthorized people. The original and continuing reasons people chose to live in caves and later in structures are to provide shelter from weather and to protect themselves from wild beasts and plundering tribes. But if buildings secure assets, one must provide a way to get into and out of these protective structures. And so was created the door.

Doors have a tough job. A door can be evaluated on a number of important criteria:

- They must block passage
- They must open and close easily
- They should be robust against intrusion
- They should fit the visual aesthetics of the environment in which they are mounted.

For access control applications, physical security is usually a given. Except for the rare total glass lobby doors, doors used in access control applications should be robust and able to withstand at least a moderate physical attack.

But a door is just a panel in a wall hung from a frame. So the frame is also an important part of the security package. A robust door hung in a light wood frame will be of little use in securing what is inside. We will evaluate doors and frames in this chapter to review their capabilities and limitations.

STANDARD SINGLE-LEAF AND DOUBLE-LEAF SWINGING DOORS

The basic door is the standard single- and double-leaf swinging door. This door includes a frame, hinges, door(s), a lock, and sometimes other hardware including a door closer, door coordinator (closes one door, and then the other), and kick plates. Swinging doors are available in a wide variety of types of construction including:

- Hollow metal doors
- Solid core wood doors
- Framed glass doors
- Unframed glass doors
- Total doors

This list excludes doors not common to commercial environments. Let's look at each type of door:

Hollow metal doors

Hollow metal (Fig. 7.1) doors are the workhorse of security. The basic hollow metal door includes an internal metal frame and metal skin. There are usually cutouts on the frame for hinges and for a lock. The three most common types of access controlled locks on hollow metal doors are mortise, electrified panic hardware, and magnetic locks. Hollow metal doors are fitted from the manufacturer for the type of lock they are to receive. Doors that receive mortise locks are prefitted with a mortise

■ **FIGURE 7.1** Hollow metal door. *www.idighardware.com.*

pocket, whereas a door that is to be fitted with electrified panic hardware may receive either concealed or surface-mounted vertical rods.

Assets: Hollow metal doors are robust—some more than others—and are available in a variety of security (penetration test) ratings and even in blast- and bullet-resistant versions. They also score very high on fire resistance and are available in UL-listed fire ratings up to 2 hours (including the frame).

Liabilities: Hollow metal doors in their basic form are not pretty; however, they can be laminated with appearance coverings that emulate fine woods or other finishes.

Caveat: Hollow metal doors can be mounted to hollow metal, aluminum, or wood frames; they are only as secure as the frame they are mounted to.

Suitable Hardware: Hollow metal doors can be equipped with electrified mortise locks, electrified panic hardware (with surface-mounted vertical rods), and magnetic locks. These doors can be cored to run the lock power to a mortise lock from the hinge using an electrified hinge that has wires running through it from the frame to the door. This allows for invisible wiring to the mortise lock. Power can be run to electrified panic hardware using either a flexible or coiled cable, flexible conduit, or through an electrified hinge sleeve.

Solid core wood doors

Solid core doors (Fig. 7.2) are primarily used in finished office spaces because of their superior appearance. They are manufactured using a wooden frame. Unlike hollow core wood doors, which use a honeycomb of cardboard as an insert into the frame, Solid Core doors use an engineered wood filler, fully filling the wood frame inside. The door is then provided with a basic wood veneer skin and then with a fine wood or plastic veneer. A router is used to create hinge insert locations and a mortise pocket if appropriate. Then a hole is drilled for the door handle. Solid core doors are available with UL-listed fire ratings up to 90 minutes.

Assets: Solid core doors are beautiful and can be somewhat less expensive than a hollow metal door.

Liabilities: Solid core doors are robust, but not as robust as a similar hollow metal door. They are more susceptible to flexing if torqued (say, by a fool who holds the door open by placing an object between the door and frame at the hinge area) and can be susceptible to misfittings due to building settling, although this is usually because

■ **FIGURE 7.2** Solid core wood door. *Image courtesy of Security Door Controls.*

many solid core doors are fitted into aluminum frames, which are more susceptible to this.

Suitable Hardware: Solid core doors can be equipped with electrified mortise locks, electrified panic hardware (with surface-mounted vertical rods), and magnetic locks. These doors can be cored to run the lock power to a mortise lock from the hinge using an electrified hinge that has wires running through it from the frame to the door. This allows for invisible wiring to the mortise lock. Power can be run to electrified panic hardware using either a flexible or coiled cable. These doors are not suitable for an electrified hinge sleeve.

Framed glass doors

Framed glass doors (Fig. 7.3), also called "Storefront Doors," are found on the front doors of merchant stores and small office buildings. Framed glass doors are manufactured using an aluminum frame with a tempered glass insert. Typically the glass is a single pane, completely filling the frame, but you will occasionally see frames that have upper and lower glass inserts, allowing for a space to mount a mortise lock or electrified panic hardware.

Framed glass doors should not be considered security doors as it is possible to easily break the glass and enter the area with simple tools (or even a well-placed kick).

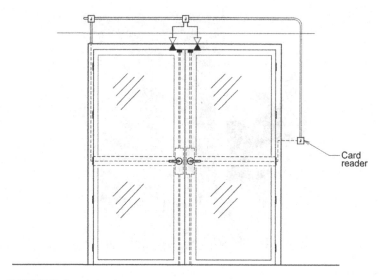

Card
reader

■ **FIGURE 7.3** Framed glass door.

Assets: Framed glass doors provide a view into and out of the secured space. They are the least expensive type of glass door.

Liabilities: A well-placed kick can get an intruder inside the space. Framed Glass Doors carry no fire rating. They should be considered a psychological, rather than a physical, barrier.

Suitable Hardware: Framed glass doors typically use either electrified panic hardware with surface-mounted vertical rods or magnetic locks. A few framed glass doors that have a middle aluminum section can also be fitted with a mortise lock, but these are rare.

Unframed glass doors

Class A Office Buildings often use unframed glass doors (Fig. 7.4) as the centerpiece of their main lobby entry. Frameless glass doors are made by creating a thick glass plate, typically between 1/2″ and 3/4″ thick, depending on the size, which is usually hung into a frameless glass frame (a series of glass panels to the sides and above the doors). It is not unusual to see a large beam above the doors supporting the top pivots.

Assets: The key asset of Unframed glass doors is their appearance. The look of an uninterrupted sheet of glass on the front of an iconic building is architecturally spectacular.

Liabilities: Unframed glass doors are strong due to their thickness, but there have been cases when they have been tweaked by an

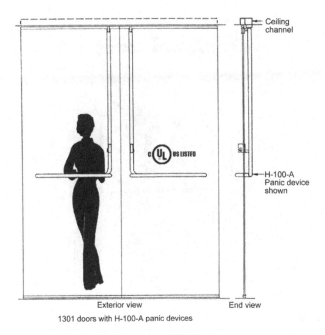

Ceiling channel

c(UL) US LISTED

H-100-A Panic device shown

Exterior view   End view

1301 doors with H-100-A panic devices

■ **FIGURE 7.4** Unframed glass door. *Photo courtesy of CRL Blumcraft.*

aggressive person pulling hard against a locked door and the door has broken; shattered actually into thousands of small glass pebbles because they are made of tempered glass. It is spectacular when it happens. Due to their ability to break, unframed glass doors should not be considered security doors as they present a psychological, rather than physical, barrier.

Most Suitable Hardware: Except for the smallest examples of unframed glass doors, which you will occasionally see on the front of a small office suite, unframed glass doors are mounted using Pivots instead of Hinges. Because they are completely glass with no structure to mount a lock to, Unframed Glass Doors cannot mount a mortise lock or conventional electrified panic hardware (although on one occasion I saw an awkward example of such). The most common electrified lock on unframed glass doors is the magnetic lock.

Total doors

Total doors are a real "find." These doors solve an amazing array of architectural and security problems, so it is surprising that many architects do not know about them (Fig. 7.5). Total Door is the name of a firm

■ **FIGURE 7.5** Total door. *Total doors.*

that makes a complete integrated door including doors, frame, continuous hinge, and integral locking hardware.

Assets: This door is amazing. It is suspended by a long wire hook from an area near the top of the frame (hinge side) to an area near the bottom of the door, similar to a long wire coat-hanger type of hook. That wire supports the entire weight of the door. The door swings on a full-height, semiconcealed hinge that fits flat and can rotate up to 180 degrees, folding the door flat into a pocket against the wall. The hinge takes up less space than a conventional door so there is greater clear width. The lock is equally unique. It comprises a full-height locking channel that cocks at an angle when open and snaps shut over the full length latch stop when closed. It provides exceptional security on single doors and eliminates vertical rods, floor hardware, coordinators, astragals, and flush bolts on pairs and double-egress doors (where each door of a pair swings in an opposite direction— very common in hospitals). The lock can be equipped with a flush mounted panic fire exit device that does not extend into the door opening when open, fitting flush into the door. The doors are also very robust and forgiving of frame alignment issues. Total doors are available in a variety of fire/smoke ratings depending on the materials used in A Label (up to 3 hours), B Label (1−1/2 hours), and C Label (90 minutes). Total doors are also available in a hurricane-rated door system in both in- and out-swinging versions. OK, I admit it; I like these doors.

Liabilities: Total doors are more expensive than conventional doors.

Suitable Hardware: All hardware is part of the Total door assembly, not added on like conventional doors.

Pivoting doors

Pivoting doors (Fig. 7.6) are highly unusual in the United States; they are more common in Europe, Asia, and the Middle East. A pivoting door is a double width door that swings from the middle, not from the side. When a pivot door opens, both sides open—one side in and the other side out. Pivot doors are typically of framed glass door construction, though you may also see them made of elegant wood construction.

Assets: They are unique architecturally, so they provide an interesting feature to the space they enclose.

Liabilities: Pivoting doors can surprise people nearby as the door swings inward as a person walks out. Additionally, pivoting doors have been known to trap small children in the pivot space (between the door and the frame, (opposite the side of the door equipped with handles)).

Suitable Hardware: These doors are typically not locked with electrified hardware, but when they are, it is usually with magnetic locks.

■ **FIGURE 7.6** Pivoting door. *IRCO.*

Balanced doors

Balanced doors (Fig. 7.7) were invented by Ellison Bronze in 1927. They are architecturally interesting doors that are hinged using an articulated hinge such that the door requires much less force to open than a normal door of its weight. The weight of the door is carried on a pair of balancing beams (one at the top and one at the bottom) such that when the door opens, about two-thirds of the door swings outward while one-third of the door near the frame swings into the space as a person opens the door outward. The door travels in an elliptical arc rather than the circular path of most doors. This allows external wind loads to assist rather than resist the opening of the door.

> *Assets:* Using the balancing mechanism, very heavy doors can be used that a normal person would have great difficulty opening. It is possible to open these doors literally with fingertip operation, which can eliminate the need for power-assisted openers. So these doors are sometimes balancing monumental doors. Balanced doors are used almost entirely on the front of Class A office buildings.
>
> *Liabilities:* Because a portion of the door near the hinge swings inward when the door swings out, it is possible for someone to catch their hand in the space between the door and the frame so that as the door

■ **FIGURE 7.7** Balanced door. *Image courtesy of Ellison Bronze Inc.*

closes again it closes on the person's hand, possibly breaking bones. Sensing hardware is available to stop the door from closing on fingers. *Suitable Hardware:* Balanced doors usually use magnetic locks.

DOOR FRAMES AND MOUNTINGS

Once we understand the types of doors available, it is important to understand the different types of door frames and mountings. Door frames are rated by frequency of use and ability to withstand impact.

Hollow metal—high-use and high-impact

Hollow metal frames are the most common type of door frame in commercial buildings. Hollow metal frames are robust and suitable for frequent use and environments where doors receive abuse by impact from carts and so forth. Rooms such as those at building perimeters, delivery and service doors, toilets, classrooms, and office suite doors are high-use rooms and are suitable for hollow metal frames. These frames are structurally sound and are not prone to flexing with minor building settling as much as aluminum frames, thus, the doors stay aligned longer than lesser frames.

Hollow metal frames are commonly built with ANSI mounting areas for hinges, frames, and door strikes. The doors can also be fitted with integral electrical enclosures for electric strikes, locks, door position switches, and electrified hinges.

These frames can be configured for single or double openings, either in the same or opposite directions (double swing door—one in/one out) with features for sound control, ballistic resistance, and fire and smoke penetration resistance. They can be configured to accept from two to many hinges or a continuous hinge.

Hollow metal frames/doors can carry fire penetration ratings from 90 minutes up to 3 hours. They are suitable for hollow metal doors, solid core wood doors, or framed glass doors.

Aluminum—medium-use and medium-impact

Aluminum frames are suitable for moderate use and impact. These frames are less expensive and lighter weight. They are typically used in medium-use and medium-impact applications, such as on an office suite.

Aluminum frames can carry fire penetration ratings for 20, 45, or 60 minutes. Aluminum frames are suitable for solid core wood doors or framed glass doors.

Wood—light-use and light-impact

Wood frames are not normally used in modern commercial applications, but they are sometimes seen on older and historic buildings. They are strictly for light duty and light impact. They are not normally fire rated.

Door mounting methods

Doors are normally mounted to frames using individual or continuous hinges, but they can also be mounted using pivots or a balanced door mechanism. Hinges are inexpensive and easy to apply (many installers know how to mount doors using hinges). Continuous hinges are rarer but are extremely robust since the door cannot easily be torqued in the frame. Pivots and balanced door mechanisms are typically used to mount framed glass or monumental doors.

OVERHEAD DOORS
Roll-up doors

Roll-up doors are common in vehicle and freight entries and are rarely used in industrial door openings for pedestrian doors. Roll-up doors are mounted into a pair of "C" channels to the left and right of the door opening and a spring-loaded roller mechanism receives the roll-up elements at the top of the door.

Roll-up doors are suitable for moderate security, but can be cut open with a high-speed rotary saw.

Paneled overhead doors

Paneled overhead doors (overhead garage doors) are comprised of a single or double panel that folds up into the space above the door opening. Double-leaf overhead doors simply fold the two leaves flat against each other at the top of the opening. These doors are relatively easy to break into using simple tools.

REVOLVING DOORS

Revolving doors consist of three or four leaves (or wings) centered on a central pivoting shaft that has a vertical axis. These are enclosed in a round enclosure with openings on the inside and outside to allow passage as the door rotates within the enclosure. Revolving doors may be manually or automatically operated. Large diameter revolving doors can accommodate rolling luggage.

Revolving doors provide a draft block for high-rise buildings that prevents the chimney effect that would otherwise allow the air to circulate from bottom to top of building, creating energy costs for heating and air conditioning.

Revolving doors can be equipped for special functions such as one-way passage to prevent a person from bypassing airport security checkpoints. Those doors use a braking mechanism to prevent opposite flow, then the door automatically revolves backwards to allow the offender to exit from the entry side.

A version of revolving doors is used in factory and subway environments where the leaves of the door are made from horizontal pipes that interleaf with a barrier to allow entry in only one direction. These doors ratchet to prevent reverse direction traffic.

In 1942 at a popular nightclub called the Cocoanut Grove in Boston, Massachusetts, a fire broke out killing nearly 500 people. This happened because there was only one revolving door at the front to allow passage and people became jammed into the opening with many dying of smoke inhalation. Since then a change in the fire code has required flanking outward swinging doors or a mechanism that makes the revolving door leaves collapsible, allowing people to pass on either side. All American revolving doors are now collapsible and most jurisdictions also require flanking panel doors.

Revolving doors can be equipped with magnetic locks for access control.

SLIDING PANEL DOORS

Sliding doors simply slide rather than swing to create an opening. Sliding panel doors comprise a single- or double-leaf setup that slides along a track to open and close. Sliding panel doors are used in industrial and office environments (usually where there is a high passage count). Sliding doors may be manually operated (usually in industrial environments) or automated (usually in office environments). Automated sliding doors may be single- or double-leaf (think grocery storefront door).

Automated sliding doors can be easily interfaced to access control systems. The interface comprises tying the door lock relay of the access control panel to the trigger input of the automatic door operator.

BIFOLD AND FOURFOLD DOORS

Bi- and fourfold doors are also used both for industrial and office environments. Bifold doors comprise two leaves that fold together and fourfold doors comprise two sets of bifold doors. These are almost always automated in both applications. Both can be interfaced to access control systems in the manner described in the section sliding panel doors.

CHAPTER SUMMARY

1. Electronic security supplements physical security.
2. Standard single- and double-leaf swinging doors include:
   a. Hollow metal doors
   b. Solid core wood doors
   c. Framed glass doors
   d. Unframed glass doors
   e. Total doors
   f. Pivoting doors
   g. Balanced doors
3. Standard door frames and mountings include:
   a. Hollow metal—high-use and high-impact
   b. Aluminum—medium-use and medium-impact
   c. Wood—light-use and light-impact
4. Doors may be mounted using:
   a. Standard individual hinges
   b. Continuous hinges
   c. Pivots
   d. Balanced door mechanisms
5. Overhead door types include:
   a. Roll-up doors
   b. Paneled overhead doors
6. Revolving doors consist of three or four leaves (or wings) centered on a central pivoting shaft that has a vertical axis. These are enclosed in a round enclosure with openings on the inside and outside to allow passage as the door rotates within the enclosure. Revolving doors may be manually or automatically operated. Large diameter revolving doors can accommodate rolling luggage.
7. Sliding panel doors comprise a single- or double-leaf setup that slides along a track to open and close.
8. Bifold and fourfold doors comprise two leaves that fold together, and fourfold Doors comprise two sets of bifold doors. These are almost always automated in both applications.

Q&A

1. *What is the basis of security?*
   a. Physical security
   b. Electronic security

    c. Both physical and electronic security
    d. Neither physical nor electronic security
2. *What are the two kinds of pedestrian doors?*
    a. Single-leaf and double-leaf doors
    b. Hollow metal doors and solid core doors
    c. Framed glass and unframed glass doors
    d. Hollow metal and total doors
3. *What common kind of door is the most secure?*
    a. Solid core wood doors
    b. Hollow metal doors
    c. Framed glass doors
    d. Unframed glass doors
4. *Pivot doors are typically of what kind of construction?*
    a. Hollow metal
    b. Solid core
    c. Framed glass
    d. Unframed glass
5. *Balanced doors are mounted using*
    a. Common hinges
    b. Continuous hinges
    c. Articulated hinges
    d. Pivots
6. *Which kind of frames are the most secure?*
    a. Aluminum
    b. Wood
    c. Hollow metal
    d. None of the above
7. *Hollow metal door frames are for*
    a. High use and medium impact
    b. Medium use and medium impact
    c. Low use and medium impact
    d. High use and high impact
8. *Aluminum door frames are for*
    a. Low use and medium impact
    b. Medium use and medium impact
    c. Low use and low impact
    d. High use and high impact
9. *Wood door frames are for*
    a. Low use and medium impact
    b. Medium use and medium impact
    c. Low use and low impact
    d. High use and high impact
10. *Overhead doors include:*
    a. Hollow metal and solid core
    b. Hollow metal and roll-up
    c. Paneled and solid core
    d. Roll-up and paneled

11. *Revolving doors*
    a. Comprise three or four leaves around a central horizontal axis
    b. Comprise three or four leaves around a central vertical axis
    c. Can be equipped with magnetic locks for access control
    d. Both b and c
12. *Sliding panel doors*
    a. Are made of one or two panels
    b. Can be automated and operated by an access control system
    c. Both a and b
    d. Neither a nor b
13. *Fourfold doors*
    a. Comprise four sets of individually operating panels
    b. Comprise two sets of bifold doors
    c. Neither a nor b
    d. Both a and b

Answers: (1) a; (2) a; (3) b; (4) c; (5) c; (6) c; (7) d; (8) b; (9) c; (10) d; (11) d; (12) c; (13) b.

8

Doors and Fire Ratings

CHAPTER OBJECTIVES

1. Chapter overview
2. What are fire ratings?
3. Door assembly ratings
4. Fire penetration ratings
5. "Path of egress" doors
6. Electrified locks and fire ratings
7. Chapter summary

CHAPTER OVERVIEW

If you are going to have a career designing, installing, or maintaining access control systems, you will need to thoroughly understand doors and fire ratings. Fire ratings are the metrics used by licensed underwriting agencies, government entities, door/frame manufacturers, installers, and fire code officials to judge the worthiness of a given door and frame assembly to withstand a fire for a certain time period.

Doors and frames are fire rated together as an assembly, not individually as pieces. The integrity of the door assembly is critical to the success of the assembly to achieve and maintain its fire rating. Any penetrations into the door or frame will affect the ability of the assembly to hold its rated resistance to fire and smoke penetration.

Certain doors in a facility are considered "path of egress" doors. These are the doors that one must go through to get from the fire emergency to the safety outside. Such doors are dictated by building code and are designated as "path of egress" doors by the architect.

Electrified locks can have an effect on door assembly fire ratings. It is essential to understand how security door/frame modifications can affect door assembly fire ratings to ensure that no changes are made to the door assembly that could void the door assembly's fire rating.

Electronic Access Control. DOI: http://dx.doi.org/10.1016/B978-0-12-805465-9.00008-7

All information in this chapter should be taken as informational only and must be reviewed for approval with the local Fire Authority Having Jurisdiction (AHJ). Information from The Steel Doors Institute was used as a resource for much of this chapter.

WHAT ARE FIRE RATINGS?

Basic fire egress concept

According to Section 7.1.10.1 of NFPA 101, Life Safety Code, 2006 edition: "Means of egress shall be continuously maintained free of all obstructions or impediments to full instant use in the case of fire or other emergency." This is clear and unambiguous language indicating that the security system must take second place to life safety during a fire emergency.

How should this be done?

NFPA 72, Section 6.16.7.1, The National Fire Alarm Code (NFAC), 2007 edition, states: "Any device or system intended to actuate the locking or unlocking of exits shall be connected to the fire alarm system serving the protected premises." Additionally, NFPA 72, 2007 edition, Section 6.16.7.2, states: "All exits connected in accordance with 6.16.7.1 shall unlock upon receipt of any fire alarm signal by means of the fire alarm serving the protected premises." NFPA 101, 2006, Section 7.1.9, Life Safety Code, clarifies this further: "Any device or alarm installed to restrict the improper use of a means of egress shall be designed and installed so that it cannot, even in case of failure, impede or prevent emergency use of such means of egress, unless otherwise provided in 7.2.1.6 and Chapters 18, 19, 22, and 23" (A good article on this subject appeared in *Electrical Contractor Power & Integrated Building Systems* online magazine, April 2008. This article, titled "Low-Voltage Opportunities in Healthcare" by Allan B. Colombo, a 32-year veteran of the security and life safety markets, explains these principles in excellent detail.).

Exceptions

There are two broad exceptions to these standards. The first of these includes "free mechanical egress" electrified locks and the second is called "delayed egress locks."

Free mechanical egress locks do not rely on any electrical signal to allow egress; i.e., one can simply turn the handle or push a bar and exit, no matter which state the electric lock is in (locked or unlocked). Three types of locks fall under this classification:

- Electrified mortise or electrified cylinder locks
- Door strikes
- Electrified panic hardware

The second class involves a special lock called a "delayed egress lock." This class of lock inserts a short delay, accompanied by an alarm before exit is allowed. Use of these locks requires a special variance granted by the local Fire AHJ. We will cover all these locks in Chapter 13, Specialty electrified locks.

FIRE PENETRATION RATINGS

Door assembly ratings are related to wall fire ratings. There is no need to place a fire-rated door into an unrated wall. Walls are rated by the "time" (in minutes or hours) they can withstand fire or smoke penetration (Much of this chapter contains information from The Steel Doors Institute.).

Wall ratings are:

- 4-hour:
  - Walls that separate buildings or divide a single building into individual fire areas
- 2-hour:
  - Walls that enclose vertical spaces such as stairwells and elevator hoist-ways
  - Exterior walls where there is a potential for severe fire exposure from the exterior of the building
- 1-hour:
  - Walls that separate occupancies within a building
  - Corridors or room partitions
  - Exterior wall that could be exposed to a light or moderate fire from the exterior of the building
  - Walls where smoke and draft control is required

Similarly, door assemblies (frames, doors, hinges, locks, etc.) are also rated for their ability to withstand fire and smoke penetration. The fire rating of the wall into which the door will be fitted determines the required fire rating of the door assembly.

Door assembly ratings are:

- 3-hour (180-minute) doors:
  - Used on 4-hour Walls

- 1−1/2-hour (90-minute) doors:
  - ❑ Used on 2-hour walls that enclose vertical spaces such as stairwells
  - ❑ Used on 2-hour walls where there is chance of severe exposure to a fire from the outside of the building
- 1-hour (60-minute) doors:
  - ❑ Used on 1-hour walls that divide occupancies within a building
- 3/4-hour (45-minute) doors:
  - ❑ Used on 1-hour walls where there are openings or room partitions
  - ❑ Used on 1-hour walls where there is a chance of light to moderate exposure to a fire from the outside of the building
- 1/3-hour (20-minute) doors:
  - ❑ Used on 1-hour walls where smoke and draft control is required
  - ❑ Not Fire Rated—doors rated "20 Minute Tested without Hose Stream."

Hose stream test

The additional rating category of "20 Minute Tested without Hose Stream" should not be confused with a "20 Minute Fire-rated Door." The hose stream test is required on all doors rated 45 minutes or higher and is required by ASTM E 119, ASTM E 814/UL 1479, and ASTM E 1966/UL 2079, all of which affect fire stopping measures. The hose stream test was developed to protect firemen from collapsing structures resulting from metal doors, frames, and stairwells becoming brittle when exposed to fire. The hose stream test is conducted by applying a stream of water from a fire hose to a burning door, stair assembly, or fire stop compound (at a wall penetration).

The hose stream test provides an indication of impact, erosion, and cooling exposure of a stream of water on rated wall penetrations and structures. When fire ratings were first implemented it was discovered that some assemblies could withstand the fire exposure test but would fail when sprayed with a stream of water. In particular, fire stop compound including mortars that are applied to seal wall penetrations may withstand the fire test, but may become so structurally weak from the fire that when sprayed with water they crumble and allow fire to spread through the rated wall. Doors that are rated "20 Minute Tested without Hose Stream" are not considered fire-rated doors. Such doors may include doors with glass openings.

DOOR ASSEMBLY RATINGS

The three-fourths rule

Doors are rated for three-fourths of the rating of the surrounding wall. For a 4-hour wall, a 3-hour door assembly is used; for a 2-hour wall, a 1−1/2-hour door assembly is used, and so forth. However, it is possible to use a door assembly rating that exceeds the rating of the wall, but never lower than three-fourths of the rating of the wall.

Doors with glass

Doors with glass may be rated if they include a vision light kit including 1/4″ wire mesh glass (something like chicken wire between two layers of glass) or ceramic glass. The industry uses a chart to define the amount of exposed glass allowable in a door opening for a given fire rating. The vision light kit is generally approved for door assemblies up to 90 minutes with 1/4″ wire glass and up to 180 minutes for doors using ceramic glass. Three-hour doors are not generally allowable with any type of glass (rare exceptions are made by local AHJ).

Temperature rise doors

Certain doors, such as those in fire stairwells, need to protect the occupants of the stairwell as they transit down the building to safety. In so doing, they may pass from a floor above a fire to floors below the floor where the fire originated. If heat is transmitted through a fire stairwell door, it can prevent the occupants from going past the door to safety, trapping them above the floor that has the fire.

In such cases, these doors need a special fire rating, called a "Temperature Rise Door." These doors are constructed using a compound inside the door that resists transmission of heat through the door. Temperature rise ratings are made in addition to the normal fire rating of the door assembly and include 250, 450, and 650°F. These indicate the maximum rise in temperature above the ambient temperature as measured on the side of the door that is not exposed to the fire during the first 30 minutes of the standard fire test. The 250°F test is the most stringent, because it allows the least rise in temperature.

Louvers

Certain doors have louvers in them to allow for the free flow of air between spaces. It is common to see louvered doors in industrial spaces to provide ventilation, especially for spaces that could process volatile or hazardous fumes. The idea of louvers and fire ratings do not go hand in hand, however it is possible to order fire doors with fusible-link type louvers to a maximum size of 24 × 24″ in a 90, 60 or 30-minute rated door. Doors that may not be equipped with louvers include 20-minute rated doors or any door that could be part of a smoke and draft assembly and doors with glass lights or equipped with fire exit devices such as panic bars.

FIRE DOOR FRAMES AND HARDWARE

Fire door frames are rated according to the type of wall they will be used in and the type of door they can support. Frames intended for installation into masonry walls are rated for use with 3-hour doors, whereas those intended for installation into drywall walls are intended for use with a maximum 1−1/2-hour fire door. Fire doors intended for 3-hour doors may be fitted with doors that are rated for less.

Door hardware on fire-rated doors must be consistent with the ratings of the frame and door. Underwriters Laboratories (UL) publishes a *Building Materials Directory* that lists door hardware, hinges, and latching devices and their suitability for use on rated doors. Similar lists are also published by other ratings agencies.

Latching devices

All fire-rated doors must be equipped with a labeled latching device that latches the door when it closes. Deadlocks can be provided also, but must be in addition to the latching device. For fire egress doors, if a deadlock is used, it must be interfaced to the latchbolt so that both retract with the latchbolt. Dead-bolts may not be used instead of latchbolts.

Latches must be of a given size, depending on the fire rating of the door assembly. For double doors, the latchbolt must be longer than for single doors to allow for the fit of the doors. The minimum length of the latchbolt is stipulated on the fire door label.

Fire exit hardware

Fire exit hardware (approved panic bars) must be used on all fire egress doors and all rooms with an occupancy of 50 or more (local codes may vary). Fire-rated doors intended for use with panic hardware (fire exit hardware) must carry a label stating that they are suitable for that use, because they must be reinforced to carry the extra force of loading that occurs when being opened repeatedly by panic hardware. Fire exit hardware also carries size and hourly ratings restrictions, and they must be labeled as fire exit hardware.

It is a common mistake to find unrated electrified panic hardware fitted sometime after the building is built to a fire-rated door, thus voiding the fire assembly rating.

PAIRS OF DOORS

Pairs of doors are available in either active leaf/passive leaf or double active leaf configurations. Typically door closers are required on both leaves, except for mechanical rooms, where the closer can be omitted from the passive leaf if it is acceptable to the AHJ.

Latching hardware

For paired doors, the active leaf can use either labeled fire exit hardware, or any labeled latch that can be opened with a single operation from the egress side.

Inactive leaf on pair of doors

For paired doors on equipment rooms, maintenance rooms, and so forth, the inactive leaf may be secured shut using manual flush bolts or surface-mounted bolts. The NFPA recommends that inactive leaves do not receive a knob or other visible hardware that could lead a person to think that they can turn a handle and exit.

Labeled fire exit devices are required for exits unless the local AHJ gives specific approval for the use of labeled self-unlatching and -latching devices. Examples of these are automatic flush bolts on the inactive leaf. The self-unlatching feature must work only when the active leaf is opened.

Double egress pairs

Double egress doors are doors in which one door swings out and the other door swings inward. Typically the right-hand door swings away from the direction of travel of a person approaching the pair and the left-hand door swings toward the person approaching the door. These doors allow convenient travel in either direction, especially with carts and hospital gurneys, and they also ensure that a door that separates two occupancies or attached buildings will allow fire exit in either direction.

Surface-mounted or concealed vertical-rod-type fire exit device hardware is required for double egress fire doors.

Astragals

An astragal is a lip at the edge of one door of a double leaf set against which the other door will come to rest. The astragal will seal the door so that nothing can be inserted between the doors when closed.

Fire egress doors, if equipped with an astragal, must be equipped with one that inhibits the free use of either leaf. These are typically an automatically operated astragal that hinges away to allow the doors to open and closes automatically when both doors are closed. Pairs of 3-hour doors must always use an astragal per NFPA 80.

Check with your local Fire AHJ regarding the specifics for your community on use of astragals.

Smoke and draft control

Doors that open onto a rated egress corridor may be required to have a smoke and draft control rating. Such doors are tested for both fire resistance and air leakage.

"PATH OF EGRESS" DOORS

Remember, fire-rated doors come in two types: Nonegress Rated and Fire Egress Rated. Doors that are in the path of egress must be fire rated and must carry fire-rated hardware rated for fire egress, in addition to the nominal hourly ratings from 20 minutes to 3 hours.

Security designers, contractors, installers, and maintenance technicians must be able to recognize which doors in a facility are fire egress doors and, if they are modified with electrified locks, hardware appropriate to the application must be used. Also, remember that modifying a fire-rated door assembly can cause that assembly to lose its fire rating. Fire door assemblies are tested *as an entire assembly*, and changing the lock on that assembly can void the fire rating of the entire assembly. If this is the case, there are several ratings agencies that send inspectors to a facility to rerate the door based upon the approvals of the individual components and the way they were installed. Be certain to check with a ratings agency before modifying any fire-rated door, especially any fire-rated door that is also a fire egress door.

If a fire-rated door, especially a fire-rated path of egress door, needs to be modified with an electrified lock, it is best to:

■ Look at the label on the door and frame.
■ Note its fire rating and if listed as an egress door.
■ Evaluate the application and determine the best type of electrified door hardware used to comply with the ratings.
■ Take photos of the door, labels, and data sheets on proposed and alternate hardware to the AHJ and submit them and ask if the door can be rerated by a ratings agency with the suggested hardware, getting approval in writing if possible.
■ Install the approved hardware in an approved fashion (see the section Electrified Locks and Fire Ratings).
■ Get the new assembly rerated by a ratings agency and submit the rerating to the local Fire AHJ for approval.
■ This process is only one of many variations that may be required by the local Fire AHJ in the local community. Be certain to verify the exact process before beginning the approval process.

ELECTRIFIED LOCKS AND FIRE RATINGS

The mere act of installing an electrified lock on a fire-rated door can void the door's fire rating. As previously mentioned, it is important to use locks that are suitable to the rating of the door assembly. It is also imperative that the lock be installed in such a way to allow for the door assembly to be rerated after the lock is installed.

Remember: any drilling, cutting, or nonoriginal penetration to the door or frame whether to mount a lock or door position switch can void the rating, as can using improperly rated locks. Typically on fire doors having ratings of 45 minutes or higher, any electrified lock must be mounted so that it does not allow fire or smoke to penetrate. This is usually done in originally rated assemblies by ensuring that wiring goes into ANSI electrical pockets that are built into the door or frame. That's right, I am talking about a little electrical box configured right into the frame itself. And yes, there is also a little box in the top of hollow metal doors to receive the magnet for a door position switch.

If you are getting the idea that fire code officials are serious about not having fire and smoke penetrate their rated doors, you would be right. They are dead serious about it and any designer, contractor, or installer who does not understand how to get a retrofit to meet the same strict standards as the originally rated door assembly can be out of a lot of money when it comes to replacing the entire rated assembly, door, and frame—at his own expense—times the number of doors on which he got it wrong. That is a lesson you will never forget if it ever happens to you. So don't let it happen. Do it right the first time and every time.

ADDITIONAL REFERENCES

- NFPA 80–1999: Standard for Fire Doors and Fire Windows
- NFPA 252–1995: Standard Methods of Fire Tests of Door Assemblies
- Underwriters Laboratories Standard for Safety UL 10B: Fire Tests of Door Assemblies
- Underwriters Laboratories Standard for Safety UL 10 C: Positive Pressure Fire Tests of Door Assemblies
- Uniform Building Code Standard 7–2: Fire Tests of Door Assemblies
- International Building Code 2000

CHAPTER SUMMARY

1. The basic Fire Egress Concept is contained in NFPA 101, Life Safety Code, 2006 edition: "Means of egress shall be continuously maintained free of all obstructions or impediments to full instant use in the case of fire or other emergency."
2. NFPA 72, Section 6.16.7.1, 2007 edition describes how this should be done: "Any device or system intended to actuate the locking or unlocking of exits shall be connected to the fire alarm system serving the protected premises."
3. Further to that, NFPA 72, 2007 edition, Section 6.16.7.2 states: "All exits connected in accordance with 6.16.7.1 shall unlock upon receipt of any fire alarm signal by means of the fire alarm serving the protected premises."
4. And this is also clarified by NFPA 101, 2006 Edition, Section 7.1.9: "Any device or alarm installed to restrict the improper use of a means of egress shall be designed and installed so that it cannot, even in case of failure, impede or prevent emergency use of such means of egress, unless otherwise provided in 7.2.1.6 and Chapters 18, 19, 22, and 23."
5. The two broad exceptions to the rules above include free mechanical egress locks, which do not rely on any electrical signal to allow egress, and delayed egress locks.
6. Door assembly ratings are related to wall fire openings.

7. The Hose Stream Test provides an indication of impact, erosion, and cooling exposure of a stream of water on rated wall penetrations and structures.
8. Door assembly ratings:
   a. Doors are rated for three-fourths of the rating of the surrounding wall.
   b. Doors with glass may be rated if they include a vision light kit including 1/4″ wire mesh glass (something like chicken wire between two layers of glass) or ceramic glass.
   c. Certain doors, such as those in fire stairwells, need to protect the occupants of the stairwell as they transit down the building to safety. In such cases, these doors need a special fire rating, called a "Temperature Rise Door."
   d. It is possible to order fire doors with fusible-link type louvers to a maximum size of 24 × 24″ in a 90, 60 or 30 minute door.
   e. Fire door frames are rated according to the type of wall they will be used in and the type of door they can support.
   f. All fire-rated doors must be equipped with a labeled latching device that latches the door when it closes.
   g. Fire Exit Hardware (approved Panic Bars) must be used on all Fire Egress Doors and all rooms with an occupancy of 50 or more (local codes may vary).
   h. Pairs of doors are available in either active leaf/passive leaf or double active leaf configurations.
   i. For paired doors, the active leaf can use either labeled fire exit hardware or any labeled latch that can be opened with a single operation from the egress side.
   j. For paired doors on equipment rooms, maintenance rooms, and so forth, the inactive leaf may be secured shut using manual flush bolts or surface-mounted bolts.
   k. Double egress doors are doors in which one door swings outward and the other door swings inward.
   l. Surface-mounted or concealed vertical-rod-type fire exit device hardware is required for double egress fire doors.
   m. An astragal is a lip at the edge of one door of a double leaf set against which the other door will come to rest. The astragal will seal the door so that nothing can be inserted between the doors when closed.
   n. Doors that open onto a rated egress corridor may be required to have a smoke and draft control rating. Such doors are tested for both fire resistance and air leakage.
   o. There are two types of fire-rated doors: Nonegress Rated and Fire Egress Rated.

9. Security designers, contractors, installers, and maintenance technicians must be able to recognize which doors in a facility are fire egress doors; and if modified with electrified locks, hardware appropriate to the application must be used.

10. The mere act of installing an electrified lock on a fire-rated door can void the fire rating on the door, so it is important to use locks that are suitable to the rating of the door assembly.

Q&A

1. *What is the basic Fire Egress Concept?*
   a. "Means of egress shall be continuously guarded for full instant use in the case of fire or other emergency."
   b. "Means of egress shall be continuously maintained free of all obstructions or impediments to full instant use in the case of fire or other emergency."
   c. "Means of egress shall be well lighted by egress signage at every egress door."
   d. "Means of egress shall include automated doors at the exits on all commercial buildings."

2. *This is clear and unambiguous language that:*
   a. The security system must manage life safety during a fire emergency
   b. The security system must work closely with life safety during a fire emergency
   c. The security system must take second place to life safety during a fire emergency
   d. Security must be considered before life safety during a fire emergency

3. *Common fire penetration ratings for Wall Assemblies include:*
   a. 30 minutes, 1 hour, 2 hours, and 4 hours
   b. 45 minutes, 1 hour, 2 hours, and 4 hours
   c. 1 hour, 2 hours, and 4 hours
   d. 45 minutes, 2 hours, and 4 hours

4. *Common fire penetration ratings for Door Assemblies include:*
   a. 1 hour, 2 hours, and 4 hours
   b. 30 minutes, 1 hour, 90 minutes, and 3 hours
   c. 45 minutes, 1 hour, 2 hours, and 4 hours
   d. 20 minutes, 45 minutes, 1 hour, 90 minutes, and 3 hours

5. *The additional rating category of "20 Minute Tested without Hose Stream"*
   a. Is the same as the 20-minute fire rating on the door
   b. Applies to all doors fire rated lower than 45 minutes
   c. Should not be confused with a 20-minute fire-rated door
   d. Is used to test doors that will never be in the area of a hose stream

6. *Doors are rated*
   a. The same as the surrounding wall
   b. For half the rating of the surrounding wall

    c. For three-fourths of the rating of the surrounding wall

    d. Whenever glass doors are used

7. *Temperature Rise Doors are commonly used*

    a. At building perimeters

    b. On office suites

    c. Where temperatures may rise and fall during the day

    d. In stairwells

8. *Fire Door Frames and Hardware*

    a. Are rated according to the type of wall they will be used in and the type of door they can support

    b. Are frames intended for installation into masonry walls rated for use with 3-hour doors while those intended for installation into drywall walls are intended for use with a maximum 1−1/2-hour fire door

    c. Are fire doors intended for 3-hour doors may be fitted with doors that are rated for less

    d. Are all of the above

9. *Latching Devices for Fire-Rated Doors*

    a. All fire-rated doors must be equipped with a labeled latching device that latches the door when it closes

    b. Deadlocks can be provided also, but must be in addition to the latching device; for Fire Egress doors, if a deadlock is used, it must be interfaced to the latchbolt so that both retract with the latchbolt

    c. Dead-bolts may not be used instead of latchbolts

    d. All of the above

10. *Fire Exit Hardware (Panic Bars)*

    a. Must be used on all fire egress doors and all rooms with an occupancy of 50 or more (local codes may vary)

    b. Must be used on all fire egress doors and all rooms with an occupancy of 100 or more (local codes may vary)

    c. Must be used on all Fire Egress doors and all rooms with an occupancy of 200 or more (local codes may vary)

    d. None of the above

11. *Double Egress Doors*

    a. Are doors in which two or more people may exit simultaneously

    b. Are doors in which one door swings outward and the other door swings inward

    c. Both of the above

    d. Neither of the above

12. *Fire-Rated Doors are of two types:*

    a. Hollow Metal or Solid Core

    b. Nonegress Rated and Fire Egress Rated

    c. Always Fire Egress Rated

    d. All of the above

13. *The mere act of installing an electrified lock on a fire-rated door can void the fire rating on the door.*

    a. True

    b. False

Answers: (1) b; (2) c; (3) c; (4) d; (5) c; (6) c; (7) d; (8) d; (9) d; (10) a; (11) b; (12) b; (13) a.

Chapter **9**

Electrified Locks—Overview

CHAPTER OBJECTIVES

1. Discover why we use electrified locks
2. Learn all about the different types of electrified locks
3. Understand how electrified locks work
4. Uncover insights into lock power supplies
5. Understand electrified lock wiring considerations
6. Learn how Ohm's law makes lock circuits work—or not
7. Understand how wire gauge controls lock circuits
8. Learn about the different types of electrified lock controls
9. Get to know types of locks that are not recommended
10. Answer questions about electrified locks

CHAPTER OVERVIEW

The selection of the correct electrified locks is critical to the success of every project. Where improper locks are used, the project suffers from any of the following problems:

- Failure to comply with code
- Requirement by code officials to remove locks and reinstall proper locks
- Improper function, causing traffic queues
- Unsafe exiting conditions
- Unreliable lock operation resulting in frequent maintenance
- Persons trapped within a space during an emergency

This chapter explains all about electrified locks and sets the stage for you to be able to select locks that will be code compliant, reliable, and easy to use. It will also cover information about how to select the right lock power supply and electrified lock controls as well as help you to understand how choosing the correct wiring for an electrified lock can make the difference between a reliable lock and an unreliable one that is in

Electronic Access Control. DOI: http://dx.doi.org/10.1016/B978-0-12-805465-9.00009-9

constant need of repair. Additionally, this chapter covers the basic rules for selecting the right lock and which ones to avoid.

WHY ELECTRIC LOCKS?

From the early days walls provided physical barriers, doors allowed passage through those barriers, and locks ensured that unauthorized people did not use those doors to pass into the protected zone. This all worked fine as long as only a limited number of people needed to pass through the locked door. But as a larger number of people were authorized to pass through, it was necessary to provide someone at the door to unlock the door, or else give everyone authorized a key to the door. As late as the 1980s, it was a combination of these methods that secured most doors in the world, but both approaches were problematic.

Guards at doors are expensive in any economy. At 2015, a simple door guard costs about $9.00–$12.00 per hour to the client. Assuming 40 hours/week and 52 weeks/year, the cost is about $16,640–$24,960 per year just to guard one door. That had better be a pretty important door!

Distributing keys to all authorized users for a highly used door is also a problem. Employee turnover makes rekeying costly. The original cost of a key is about $1.00 each. If 5000 people need to use a door each day the cost of the original key distribution is $5000 plus the cost of actually distributing the keys. If the time to distribute each key is only five minutes, that will take 416 hours (52 days) to distribute 5000 keys. At a simple cost of $8.00 per hour for labor plus 12 keys distributed per hour, the cost per hour rises to $20 per hour. Multiply that by 416 hours and the total cost to distribute the keys is $8320.

Oh, and that is assuming that nobody gets fired or leaves employment. On average, a business employing 5000 people with a turnover of only 15% per year (an exceptionally low employee turnover) will lose 750 people per year and hire 750 people per year. That is another 750 keys to be distributed (another 62 hours and $1250 in employee and key costs). So the first year cost for keys for that door is $9570.

And there are two other problems. This is not the only door to which employees may need a key in the facility. Remember, keys must be distributed to only the employees who need them and no others, and different doors need keys distributed to different people. So if the average employee needs keys to perhaps 12 doors, now the first year cost of keys is $114,840. It adds up quickly, don't you agree?

And we have not even talked yet about the cost of lost keys. It is hard to calculate how many keys might be lost in a year among a population of 5000, but every key that is lost requires not only the issuance of another key to that person, but also a new key to all 5000 employees. This is because that lost key is a key to a door that protects assets that the company is already willing to pay $114,840 to protect, so they are not going to let a key be lost somewhere out there, probably on a key ring with a key-tag given free at last year's 4th of July barbeque with the company's name or logo on it. And yes, that does happen. Companies *are* that stupid (search Google images for "key-tag company logo"). The company cannot just replace the key to the lost employee, they have to replace all 5000 keys just to be sure that the guy out there who picks up the key ring with the company's name on it does not just decide to try out the "found" keys someday (and yes, that also happens). There have been cases where employees have reported keys lost just to cause the company the havoc of having to reissue keys to all employees. For the record, I do not advocate anyone doing anything that is in any way illegal (unless it is hilarious).

So just for fun, let's assume that of the 5000 employees, a mere 5% are too incompetent or stupid to keep track of their keys (and I am probably being gracious about the number of stupid people here). That is another 250 keys per year *or virtually one each and every workday*. So essentially the company has to give out another 5000 keys every day they are open just to keep up with the number of lost keys. This would cost $8320 (cost to distribute 5000 keys) times 250 (keys lost per year), or $2,080,000 a year. And that is just for one door. Security management is complicated isn't it? It kind of makes the guard look cheap at only $24,960 per door per year, doesn't it?

Of all the dumb ideas spawned in the 1960s and 1970s (and there were a lot of them), the idea of companies issuing keys to large numbers of employees was at the top of the list. It was not long until the gigantic line item in the security budget for keys was overwhelming the entire budget for executive martinis and golf. This got the attention of accounting departments who thought to themselves: "I think this is a bad idea. This is costing me martinis." There had to be a better way to secure the facility.

Then they heard about electronic access control systems—enter the 1970s and electronic access control systems. It is pretty clear to see how an early access control system operating from a file-cabinet-sized computer serving less than 16 doors at a cost of a hundred thousand dollars or so (1970s costs) is a tremendous bargain compared to the cost of issuing and maintaining keys for a large population of employees.

So, while electrified locks were primarily used in prisons, jails, and in very limited commercial use to control access to a single or double door, access control systems presented a ready market for electrified locks. As the access control market grew, so did the need for electrified locks. Soon there were numerous types of electrified locks with each having its own purpose (or not). The industry developed some really loony ideas about electrifying locks for a while until fire code officials put a stop to it.

TYPES OF ELECTRIFIED LOCKS

I have seen some pretty basic descriptions of the types of electrified locks over the years. But if you want to be able to sort through lock types and find the correct lock for a given application, it is essential that you understand all the ways that locks can be categorized. Here is the complete list.

All electrified locks fall into only a few common type categories:

- Electrified mortise locks
- Electrified panic hardware
- Magnetic locks
- Electric strikes
- Electrified cylinder locks
- Electromechanical dead-bolts
- Combinations of the above

These can be further categorized into those locks that provide "free mechanical egress" (just push the bar or turn the handle and go), and those that require some other method to exit:

Free mechanical egress locks

- Electrified mortise locks
- Electrified panic hardware
- Electric strikes
- Electrified cylinder locks
  Other means of egress
- Magnetic locks
- Electromechanical dead-bolts
- Combinations of the above may fall into either category

Electrified locks can be further categorized into two other groups:

- Locks that use mechanical devices to achieve the lock/unlock function
- Locks that use purely electromagnetic means to do so

Other categories:

- Those that can be used with existing door locks and those that must be used alone
- Locks that are visually aesthetic and those that are not
- Locks that sit on the frame and locks that sit in the door
- Locks that work well in emergencies and locks that might be slower to use
- High-security, medium-security, and low-security locks
- Locks that make noise when they operate and those that are nearly silent
- Locks that indicate their lock/unlock status and those that do not
- Fail-safe locks and fail-secure locks

As we progress through this book, you will see how understanding all of these different categories of locks will help you make the correct decision on which lock to select for a specific function.

HOW ELECTRIFIED LOCKS WORK

Electrified locks are mechanical or electrical devices that hold a door closed. All electrified locks use electromagnetism to achieve this function. Electrified locks fit into two broad categories:

- Locks that use electromagnetism to activate a mechanical lock mechanism (electromechanical locks)
- Locks that use electromagnetism as the direct means of locking (magnetic locks)

Let's look at electromechanical locks first. This category represents by far the largest assortment of electrified locks. First, let's analyze how mechanical locks operate.

All electromechanical locks are based on a mechanical locking principle; i.e., almost all electromechanical locks are based on a lock that was first a mechanical lock with no electrical components. The exception to this are electric strikes, which have no operating mechanical equivalent. Their equivalent is a manual strike plate. Manual strike plates are simply a bent brass plate that has no operational elements. The elements of a manual strike plate are the faceplate and the strike opening into which the door lock latch engages (Fig. 9.1).

■ **FIGURE 9.1** Electric strike. *Image courtesy of Security Door Controls.*

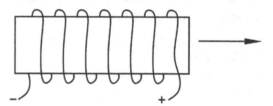

■ **FIGURE 9.2** Solenoid function.

Electric strikes

Electric strikes function by creating an operating element where none existed before. In the manual strike plate, the door latch engages into the strike opening (or pocket). In the electromechanical version there is a strike opening that is engaged and released by a solenoid (Fig. 9.2). A solenoid is a small electromagnet that pushes or pulls a plunger that can operate a function. In this case, it holds the strike opening closed or allows the strike opening to swing open, thus allowing the door's lock latch to open without the lock latch being retracted. When a door strike is released, it is possible to pull the door open from the outside while the mechanical door lock latch is still engaged (locked). The strike opening just swings away and there is no longer anything holding the latch in the latch pocket of the frame.

Electrified mortise locks

Electrified Mortise Locks are simply standard mortise locks equipped with a solenoid that keeps the latchbolt from retracting (usually only from the unsecure side). That is, the latchbolt can be retracted from the secure side simply by turning the handle, but on the outside of the door, the handle will not turn unless the solenoid is disengaged. This function ensures that anyone inside can get out simply by turning the handle, but outsiders must use a key or electric control of the lock solenoid. Electrified mortise locks typically have a mechanical override function so that they can be opened by using a key or by turning the handle.

Electrified mortise locks (Fig. 9.3) are available in either fail-safe or fail-secure modes. In fail-safe mode, power must be applied to lock the lock, whereas in the fail-secure mode power must be applied to unlock the lock. Thus when power to the lock is disengaged or "fails", the lock will go to its fail-safe (unlocked) or fail-secure (locked) mode. Most electrified mortise locks are ordered as fail-safe.

■ **FIGURE 9.3** Electrified mortise lock. *Image courtesy of Security Door Controls.*

Electrified panic hardware

Electrified panic hardware provides an instantaneous means of egress in every emergency along with the ability to unlock the door remotely or by an access control system. Electrified panic hardware is available in three locking types (Figs. 9.4–9.6):

■ **FIGURE 9.4** Panic hardware with mortise lock. *IRCO-VonDuprin.*

■ **FIGURE 9.5** Panic hardware with rim latch. *IRCO-VonDuprin.*

■ **FIGURE 9.6** Panic hardware with vertical rods. *IRCO-VonDuprin.*

- Combines panic hardware with an electrified mortise lock
- Panic hardware with an integral latchbolt to lock against a rim mounted latch plate
- Panic hardware that operates vertical rods that latch into strike pockets at the top (and often also at the bottom) of the door

Electrified cylinder locks

Electrified cylinder locks are simple knob sets (no mortise lock) that have a small solenoid inside to control the lock function. They are also available in fail-safe and fail-secure versions. Electrified cylinder locks are among the least secure locks available. They cannot withstand even a mild attack of force even without tools. Additionally, electrified cylinder locks cannot stand constant duty cycles due to their lightweight construction. They have been known to fail often. I do not recommend using electrified cylinder locks.

Magnetic locks

Magnetic locks are among the most common electrified locks used by the security system industry. This is because they can be fitted onto any door (not always legally) and they require no modifications to the door or frame (although they are a legal modification to the door and frame, which makes them illegal to use on certain types of fire doors). There are two common types of magnetic locks: surface-mounted (Fig. 9.7) and shear (usually concealed).

All magnetic locks work by applying an electrical current to an electromagnet (on or in the door frame). There is also an armature (a metal plate) attached to the door. When the door closes, the armature comes into contact with the electromagnet that, when energized, holds the armature (thus the door) tight against the electromagnet.

Surface-mounted magnetic locks position the operating surface of both the electromagnet and the armature in a vertical configuration so the armature comes to rest against the electromagnet when the door closes.

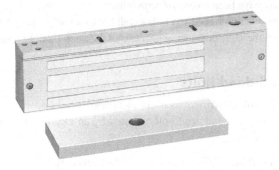

■ **FIGURE 9.7** Surface-mounted magnetic lock. *Image courtesy of Security Door Controls.*

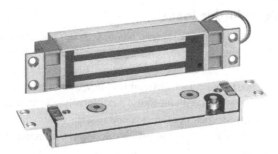

■ **FIGURE 9.8** Magnetic shear lock. *Image courtesy of Security Door Controls.*

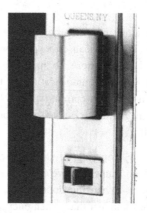

■ **FIGURE 9.9** Electrified dead-bolt. *Securitech.*

Magnetic shear locks (Fig. 9.8) place the electromagnet inside the top of the door frame with the electromagnet facing down, not against the swing of the door. There is an armature inside a pocket in the top of the door that floats on coil springs and is lifted up by the electromagnet when the door is closed and locked. There have been some problems with these locks over the years when the armature gets misaligned and fails to fall back into its pocket when power is released on the electromagnet, which causes the door to fail to open.

Electrified dead-bolts

Another class of electrified locks is electrified dead-bolts. These locks comprise a simple round dead bolt operated by an electric solenoid in the top of the door frame.

Electrified dead-bolts (Fig. 9.9) are installed similarly to magnetic shear locks; i.e., the lock is in the top of the door frame and there is a pocket in the top of the door into which the dead-bolt is received. When the solenoid is engaged the bolt drops into a bolt pocket in the top of the door. When the solenoid is disengaged the solenoid retracts the bolt up out of the bolt pocket in the top of the door. The bolt is magnetized, and the pocket is equipped with a permanent magnet that repels the dead-bolt upward out of the pocket in case the solenoid fails to retract it fully.

These see little use now for safety reasons, but you may run into some that were installed many years ago. The problem with electrified dead-bolts is that if the door becomes misaligned due to building settling or poor door maintenance, the dead-bolt can become jammed inside the bolt pocket in the top of the door, locking all occupants inside the space.

Another version of this lock is a simple, surface-mounted electrified dead-bolt, usually mounted on the door, with a receiver mounted on the door frame. At 2015, electrified dead-bolts are only approved for use on storage rooms and other nonoccupied spaces.

Paddle-operated electromechanical dead-bolts

The last class of common electrified locks is paddle-operated electromechanical dead-bolts (Fig. 9.10). These are rare but can be quite useful in the right circumstances.

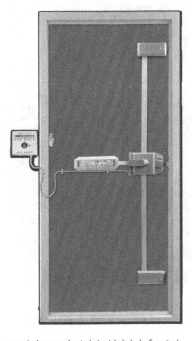

■ **FIGURE 9.10** Paddle-operated electromechanical dead-bolt lock. *Securitech.*

They are typically configured with a paddle to unlock the dead-bolts from the inside, allowing for free mechanical egress and work in conjunction with a door position switch so that when the door is closed, the access control panel sends a signal to the lock to reengage the dead-bolt. Paddle-operated electromechanical dead-bolt locks operate by using a solenoid to throw a lever/cam that rotates the dead-bolts into their locked position. They can be ordered with one, two, three, four, or five dead-bolts, all operated by a single solenoid lever/cam. Thus it is possible to see up to five points of dead-bolts locking a single door. Now that is security! These locks combine very high-security with good egress safety.

LOCK POWER SUPPLIES

Lock power supplies (Fig. 9.11) supply the power to electrified locks. With these all being electrified locks, somehow it figures that they require an electrical power source, right? Lock power supplies range from the simple to the sublime. First, the simple.

■ **FIGURE 9.11** Good quality lock power supply. *Image courtesy of Altronix Corp.*

Perhaps the simplest lock power supply can be found on AC-type electric strikes. You can spot these quickly because they "buzz" when they unlock. Where is the buzz coming from, you ask? You can hear it because the solenoid on those strikes (and on AC locks) is operated from alternating current (AC) rather than from direct current (DC). AC reverses its polarity 60 times per second (in the United States—and 50 times per second in most other parts of the world). The sound you hear is the solenoid rattling each time the current is reversed. The power supply for AC-type electric strikes is nothing more than a simple low-voltage transformer.

Most electrified door locks are operated by DC. These are nearly silent in most cases. The lock power supply for these locks can range from a simple "plug-pack" type arrangement (which includes a low-voltage transformer and a diode to convert AC power to DC), or a more sophisticated arrangement that powers multiple locks. The best of these are enclosed in an electrical cabinet and contain a fuse for the whole assembly, a transformer to convert mains power to low voltage, a rectifying circuit to convert AC power to DC power, individual fuses for the number of locks that the power supply will support, and a terminal strip for connecting the lock power cables to the locks and lock control device(s). Additionally, better power supplies are all rated by a licensed ratings agency such as Underwriters Laboratories (UL), or the European Standards Association (CE) or rated by the Canadian Standards Association (CSA). They will also have a cabinet lock, and the best are also equipped with a power-on lamp and tamper switch to alert the security system in case anyone opens the cabinet. These should be wired through a metallic conduit to prevent unauthorized tampering with the lock power wiring.

ELECTRIFIED LOCK WIRING CONSIDERATIONS

Electrical devices operate on the principle of Ohm's Law. Ohm's Law is the basic law of electricity. Those who know it well rarely ever have problems with their installations. Those who do not understand it suffer just like those who do not understand gravity. Ohm's Law works on three variables: voltage (E), current (I), and resistance (R). The basic statement of Ohm's Law is $E = IR$ (voltage is equal to current times resistance). Ohm's Law can also be stated as $R = E/I$ (resistance is equal to voltage divided by current) or $I = E/R$ (current is equal to voltage divided by resistance).

The power supply provides voltage and current to a load (electrified lock) through a conductor (the wire). This is controlled by a switch (access control panel lock relay or request-to-exit sensor). Together, these comprise the electrical circuit. The amount of current that flows through the circuit is equal to the voltage supplied by the power supply divided by the resistance of the electrified lock and the wire. Don't overlook that last part—the wire is part of the circuit and it affects the performance of the lock. Both the lock and the wire present an electrical load (resistance) to the power supply (the source of voltage). The amount of current flowing through the circuit is equal to the power supply's voltage divided by the total resistance of the lock and the wire together. Each (the lock and the wire) contributes its own resistance to the voltage.

The lock contributes a certain resistance. So does the wire. The total resistance of the circuit is the resistance of the lock and the wire together. Remember, the lock is rated to operate at a given voltage (usually 12 or 24 volts) and the power supply supplies one of those voltages (12 or 24 volts). We are depending on the wire to present very little resistance or else there will not be enough voltage to operate the lock.

Ohm's Law is important to understanding locking circuits because of a function called Voltage Drop. The greater the resistance of an electrical load the less current will flow. Remember the version of Ohm's Law that said $E = I/R$ (voltage = current divided by resistance)? Here is the critical part. Each part of the circuit "drops" a certain amount of voltage across the entire power supply voltage depending on the amount of resistance of that component. Add more resistance to the circuit and you get less current across the same voltage.

Each wire adds resistance to the circuit, and in wire, resistance is determined by the diameter (cross-section) of the wire (the larger the wire, the lower its resistance; the smaller its wire gauge, the higher its resistance). There is also another factor—the longer the wire, the higher its resistance and the shorter the wire, the lower its resistance.

Remember, both the lock and the wire have their own resistance. Lock power supplies are "constant voltage" devices; i.e., they provide a constant voltage to whatever is connected to them. Their current will vary according to the resistance of the total load (lock and wire combined). The total voltage of the power supply is "dropped" across the total resistance of both the wire and the lock. If the lock requires 12 volts to operate and we are dropping 3 volts across the wire from a 12-volt power supply, that will exactly 9 volts for the 12-volt lock to operate on. This is not enough, so the lock will not operate. We need to drop as close to zero volts across the wire as possible.

Table 9.1 Resistance of Copper Wire at 77°F

Size (AWG)	Diameter (in.)	Area (cir. Mils)	Weight (lb/1000 ft)	Resistance (Ohm/1000 ft)
12	0.081	6530.0	19.77	1.62
14	0.064	4107.0	12.43	2.58
16	0.051	2583.0	7.82	4.09
18	0.040	1624.0	4.92	6.51
20	0.032	1022.0	3.09	10.40
22	0.025	642.4	1.95	16.50

Circular 31, U.S. Bureau of Standards. Ideally, we would connect the lock to the power supply with very short, very big wires that would add very little resistance. But we often find wire runs that are too long or using wire diameter that is too small to properly power the lock. In practice, the longer the "wire run," the higher the probability that the lock may not work. Refer to Table 9.1 to calculate the voltage drop across the wire. To do that, add the resistance of the lock and the resistance of the cable (determined by its length) to find the total resistance. Remember, total wire resistance includes both strands of wire in a pair. Add the resistance of each strand together. Divide the power supply voltage by the total resistance to get the total current. Divide the current by the wire resistance to get the voltage drop across the wire. The amount calculated is the voltage left to operate the lock. It is not enough to use a larger wire diameter or place the power supply closer to the lock. Either method will reduce the effect of wire resistance on the overall circuit.

Voltage drop example

It is essential to understand how to calculate voltage drops to ensure that electrified locks will have sufficient voltage to operate. Let's look at a sample case. Suppose we have an electrified lock that draws 0.3 amps connected to 1500 ft of 18-gauge wire. From Table 9.1, we see that 18-gauge wire has a resistance of 6.51 ohm at 1000 ft (at 77°F). Then 1500 ft of 18-gauge wire will have a resistance of 9.765 ohm (6.51 ohm $\times$ 1.5). This cable will drop 2.91 volts (6.51 $\times$ 1.5 $\times$ 0.3 amp = 2.91 volts). If we power a 12-volt lock from a 12-volt power supply we will lose 2.91 volts to cable voltage drop, leaving only 9.09 volts available to power the electrified lock, which is not enough. Most electrified locks have a 10% engineering factor (they will operate at $\pm$ 10% of their rated voltage). Thus a 12-volt lock may not operate at a voltage below 10.8 volts (12 volts minus 10%).

ELECTRIFIED LOCK CONTROLS

Few contractors or installers spend much time thinking about electrified lock controls, but maintenance technicians know all about them. They get to figure out how to make the thing work after the system stops working due to an improperly configured lock control.

In theory, lock controls should be simple in practice, but in practice, they are not so simple.

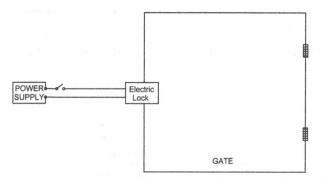

■ **FIGURE 9.12** Simple electrified lock control.

It is simple to control a lock—you have an electrified lock, a power supply, and a switch of some sort. Close the switch and the lock energizes as the electrical circuit is completed. Open the switch and there is no power to the lock as the electrical circuit is opened. That is the theory (Fig. 9.12).

In practice here is how they actually work:

First Case (lock control from the access control panel): This *is* actually rather simple. The circuit includes the power supply, the lock, the cable, and the relay (electronically controlled switch) that allows power to the lock. All that it takes for this circuit to work properly is to be certain that the power supply is appropriate to the lock, the wire gauge is appropriate to prevent too much voltage drop, and that the access control panel relay is rated to carry the current of the circuit. If the lock circuit presents more current than the relay can support, you can add an "Interposing Relay" to the circuit. An interposing relay is a large relay that is controlled by a small relay. Its purpose is to allow a small relay to control a circuit carrying more current than the small relay can handle. Essentially, the small relay (the access control panel lock control relay) will switch the larger (interposing) relay, which will carry the actual circuit. This may be true, e.g., if one relay must control magnetic locks to six doors.

Second Case (lock control from the request-to-exit sensor): In this case the circuit comprises the power supply, the electrified lock, the wiring, an access control panel lock relay, and a request-to-exit sensor at the door. Now, in theory, we have two devices that can control the lock: the access control panel lock relay and the request-to-exit sensor at the door. However, in practice in most installations we find that the request-to-exit sensor is not wired to control the lock directly, instead it is wired to an exit sensor input on the access control panel where the sensor switch state is sensed and then the access control panel makes a decision to unlock the door (using the previously mentioned lock relay). It also ignores the door-open alarm signal sent by the door position switch when the door is opened. This all works fine—until it doesn't. If the lock is a magnetic lock with no mechanical way to open the lock, and if the access control panel stops working (and yes, that does sometimes happen), there will be somebody locked inside the space because the lock power never got interrupted by the access control panel lock relay.

Solution: I recommend using an interposing relay on the request-to-exit sensor. In this case, the interposing relay is not controlling a larger relay; instead it is providing two contacts (one controls power to the lock and the other signals the request-to-exit sensor input on the access control panel). In this way, it is ensured that the lock will unlock when the request-to-exit sensor triggers and the access control panel is relegated to bypassing the alarm. It will also interrupt power to the lock, but by the time it figures out to do that, the lock is already unlocked by the interposing relay. Everything previously discussed about calculating the correct wire gauge still applies.

Third Case (lock control from a remote location, like a security console or desk): In this case, we have a power supply, an electrified lock, and a remotely operated control panel. This may be near to the lock or some distance away, such as in another building. This arrangement is common; e.g., at a remote entry door where a visitor may approach the door and use an intercom to call the security console and the lock is opened after interviewing the visitor via the intercom.

- Where the control point is close to the lock, the control can be directly interfaced with the lock power circuit as described with the request-to-exit sensor.
- Where the control point is not practical to wire directly, it can be controlled via an access control panel lock relay. In this case, the lock control panel switches are wired to the request-to-exit sensor of the access control panel and when the release button is pressed at the console, the access control panel will release the lock.
- In all cases, comply with codes regarding electrified locks.

TYPES OF LOCKS NOT RECOMMENDED

There are some types of electrified locks that I do not recommend using for either security or safety reasons. Let's be clear, this is just one guy's opinion, so you can use these locks at your own risk.

I do not recommend these locks for security reasons:

Electrified Cylinder Locks: Electrified cylinder locks are cute. They pretend to be real locks, but of course they are not in the sense that an insurance gecko or mouse can kick the door open while using these. These locks are so weak that even children have been known to kick through doors locked with cylinder locks. And since electrification adds no strength whatsoever to a cylinder lock, it is still a cylinder lock and is mostly used for humorous rather than security reasons.
Electrified Strikes: See Electrified Cylinder Locks. These might require two mice; otherwise I do not recommend them for the same reason. There are a few high-security electrified strikes, but there are better ways to lock a door in most cases.

I do not recommend (or recommend caution using) the following locks for safety reasons:

Electrified Dead-Bolts: Simple electrified dead-bolts are an almost sure way to get somebody trapped behind a door. If there is a person and a door and the door is locked with an electrified dead bolt, that person will eventually be very unhappy. Oh, by the way, he will also be the owner of the building or somebody else who can make your life unhappy. If multiple people use the door, the lock will wait until the single most important and angry man in the organization is there before it will fail (permanently... and in the secure mode).
Watch out for Magnetic Locks: Magnetic locks are wonderful. They are simple, reliable, and easy to install. Almost too easy. So they get installed on all sorts of doors that they should not be used on; e.g., fire egress doors that already have mechanical panic hardware on them. The installer will ensure that the magnetic locks will always unlock by placing a motion-operated request-to-exit sensor above the door, so that when a person comes to the door and pushes on the panic bar, the magnetic locks will already be unlocked. Except that they will not be unlocked. The motion-operated request-to-exit sensor will fail, leaving the magnetic locks locked and the person with a stunned look on his face while he pushes on the panic bar. This will happen to the same important angry man who just got locked inside a room last year by an electrified dead bolt. He will be even more important in the organization now because he intimidates everyone in the organization into getting his own way, and of course he just happens to generate millions of dollars in revenue so the organization does whatever he wants. And he is still angry... at you! Or, if he is out of town, this will happen during a fire. Don't use magnetic locks

on fire exit doors. And where you do use them always provide at least two methods for unlocking them.

- ❑ Magnetic locks can be used with a request-to-exit sensor within the panic bar. That would be better, but unlike electromechanical locks, there is no mechanical override for a magnetic lock, so if the request-to-exit sensor fails (and everything fails sooner or later), somebody is going to be locked inside. And it will happen to that important angry guy or when there is a fire.
- ❑ There is a version of magnetic locks called a "delayed egress lock" that actually meets a specific provision of NFPA 101 for use on fire exit doors. This application combines a magnetic lock with an exit bar sensor, a countdown timer, and a local alarm and signage. The door is usually locked and not used as a normal passage door. It is a special fire exit door, only used in case of an emergency. In case of a fire, power to the lock is integrated into the fire alarm system so that it unlocks automatically, allowing everyone to use the door to exit. In any other emergency, a person can push on the panic bar, holding it pushed for the time delay duration (usually 15 seconds or so; Fig. 9.13) as a counter on the magnetic lock counts down to zero. During this entire time, a local alarm is sounding at the door. When the counter gets to zero the lock unlocks and the person can exit the door. Delayed egress locks are used to ensure security while still allowing people to exit in an emergency, unless someone is firing a gun at you. Then you're dead.
- ❑ Remember, magnetic locks *are not* free egress locks. Oh, sure somebody will argue that they are used on delayed egress locking systems, and those are on fire exit doors so... Forget it. That is a special case and it requires a Fire Code Variance, and as I said if somebody is trying to get through the door to escape a workplace violence event, well it is not going to be so good for them.

■ **FIGURE 9.13** Delayed egress hardware. *Image courtesy of Security Door Controls.*

Generally, magnetic locks are dangerous because their Request-to-Exit sensors can fail. It is important to provide at least two means of exit sensing with magnetic locks.

❏ Lastly, magnetic shear locks have been known to jam up in the door as the building settles even when all electronics worked perfectly. I suggest you do not use them at all.

❏ For the record, I am not saying do not use magnetic locks at all. I am saying that when you use magnetic locks, you must be certain that they fully comply with code and are installed so they will fully accommodate Murphy's Law. Always use at least two means of request-to-exit sensing.

❏ I often couple magnetic locks up with a motion-sensing, request-to-exit switch and a latch position sensor in the latch pocket. Neither of these requires special knowledge and both are wired in series so that the activation of either one will unlock the lock. Oh, I also use an interposing relay on these so that they both actually interrupt power to the lock as well as sending an unlock signal to the exit sensor input of the access control panel. Belts and suspenders. Keep your pants up.

CHAPTER SUMMARY

1. All electrified locks fall into several categories:
   a. Electrified mortise locks
   b. Electrified panic hardware
   c. Magnetic locks
   d. Electric strikes
   e. Electrified cylinder locks
   f. Electromechanical dead-bolts
   g. Combinations of the above
2. These can be further categorized into those locks that provide "free mechanical egress" (just push the bar or turn the handle and go), and those that require some other method to exit:
   a. Free mechanical egress locks
      i. Electrified mortise locks
      ii. Electrified panic hardware
      iii. Electric strikes
      iv. Electrified cylinder locks
   b. Other means of egress
      i. Magnetic locks
      ii. Electromechanical dead-bolts
   c. Combinations of the above may fall into either category

3. Electrified locks can be further categorized into two other groups:
   a. Locks that use mechanical devices to achieve the lock/unlock function
   b. Locks that use purely electromagnetic means to do so
4. Other categories include:
   a. Those that can be used with existing door locks and those that must be used alone
   b. Locks that are visually aesthetic and those that are not
   c. Locks that sit on the frame and locks that sit in the door
   d. Locks that work well in emergencies and locks that might be slower to use
   e. High-security, medium-security, and low-security locks
   f. Locks that make noise when they operate and those that are nearly silent
   g. Locks that indicate their lock/unlock status and those that do not
   h. Fail-safe locks and fail-secure locks
5. Electrified locks fit into two broad categories:
   a. Locks that use electromagnetism to activate a mechanical lock mechanism (electromechanical locks)
   b. Locks that use electromagnetism as the direct means of locking (magnetic locks)
6. Electrified locks operate on the principle of Ohm's Law.
7. It is essential to understand how to calculate voltage drops to ensure that electrified locks will have sufficient voltage to operate.
8. Basic types of electrified lock controls include:
   a. First Case (Lock Control from the Access Control Panel)
   b. Second Case (Lock Control from the Request-to-Exit Sensor)
   c. Third Case (Lock Control from a remote location, like a Security Console or Desk)
9. Types of locks not recommended include:
   a. Electrified cylinder locks
   b. Electrified strikes
   c. Electrified dead-bolts
   d. Use care when using magnetic locks
10. Delayed egress hardware can allow the use of magnetic locks in conditions where security is high and egress is essential.

Q&A

1. *Electrified Locks are used because:*
   a. Guards at doors are expensive
   b. Distributing keys to all authorized users for a highly used door is also a problem
   c. Employee turnover makes rekeying costly
   d. All of the above

2. *Three common types of Electrified Locks include:*
   a. Electrified Paneled Locks, Electrified Nonmagnetic Locks, Electrified Pin-Grinding Locks
   b. Electrified Mortise Locks, Electrified Panic Hardware, Magnetic Locks
   c. Electrified Paneled Locks, Electrified Panic Hardware, Magnetic Locks
   d. None of the above

3. *Locks can be categorized as:*
   a. Free Mechanical Egress Locks
   b. Other Means of Egress
   c. Both Types
   d. Neither Type

4. *Electrified Locks can be further categorized into two other groups:*
   a. Locks that use nonmetallic materials and locks that require remote release
   b. Locks that allow handicapped people to enter easily and locks that do not
   c. Locks that comply with international codes and locks that only comply with local codes
   d. Locks that use mechanical devices to achieve the lock/unlock function and locks that use purely electromagnetic means to do so

5. *Electrified Locks are:*
   a. Mechanical or electrical devices that hold a door closed
   b. Purely electrical devices that hold a door closed
   c. Purely electrical devices that unlock a door
   d. None of the above

6. *Electrified Mortise Locks:*
   a. Are available only in fail-safe mode
   b. Are available in either fail-safe or fail-secure modes
   c. Are available only in fail-secure mode
   d. None of the above

7. *Electrified Panic Hardware:*
   a. Provides an instantaneous means of egress in every emergency along with the ability to unlock the door remotely or by an access control system
   b. Causes people to panic
   c. Keeps people from panicking
   d. Automatically opens when people panic

8. *Magnetic Locks*
   a. Are not recommended in occupied areas
   b. Are the first choice of every designer
   c. Are among the most common electrified locks used by the security system industry
   d. Are dangerous and should never be used
9. *Electrified Dead-Bolt Locks:*
   a. Are not recommended in occupied areas
   b. Are the first choice of every designer
   c. Are dangerous and should never be used
   d. See little use now for safety reasons
10. *Lock Power Supplies always:*
    a. Always make electrified locks buzz when they open
    b. Supply the power to electrified locks
    c. Should be placed in an area where there is a fire suppression system because they sometimes catch on fire
    d. Require constant monthly maintenance
11. *Wiring length and gauge must be considered because:*
    a. The total voltage of the power supply is "dropped" across the total resistance of the wire
    b. The total voltage of the power supply is "dropped" across the total resistance of the lock
    c. The total voltage of the power supply is "dropped" across the total resistance of both the wire and the lock
    d. The total voltage of the power supply is "dropped" across the resistance of the wire minus the resistance of the lock
12. *Certain Electrified Locks are not recommended for either safety or security reasons. These include:*
    a. Electrified Cylinder Locks
    b. Electrified Strikes
    c. Electrified Dead-Bolts
    d. All of the above

Answers: (1) d; (2) b; (3) c; (4) d; (5) a; (6) b; (7) a; (8) c; (9) d; (10) b; (11) c; (12) d.

Free Egress Electrified Locks

CHAPTER OBJECTIVES

1. Learn the basics in the chapter overview
2. Learn about electrified mortise locks
3. Understand the ultimate free egress lock—"panic" hardware
4. Learn about electric strikes
5. Learn about electrified cylinder locks
6. Answer questions about free egress locks

CHAPTER OVERVIEW

There is a reason that free egress locks are the first type to learn about—they should be the first type of lock you reach for in any circumstance. The best design rule is to use another kind of lock only if there is a good reason that you cannot use a free egress lock.

Free egress locks are called that because they allow any person familiar with the operation of a normal door to walk up to the door and turn the handle or push on the panic bar and go out of the door, no matter what else is happening—no matter if the door is locked from the outside, no matter if it there is a fire emergency or not, no matter what. With this lock a person can always exit the door, and there is "no special knowledge" required to do this.

The person exiting does not have to read a sign and does not have to know something that everyone else does not know (such as how to use a push button to unlock the door), or the ability to read any particular language.

Free egress locks are the first and best solution for most electrified locking requirements.

Electronic Access Control. DOI: http://dx.doi.org/10.1016/B978-0-12-805465-9.00010-5

TYPES OF FREE EGRESS LOCKS

There are three common types of free egress locks. These include electrified mortise locks, electrified panic hardware, and electric strikes. (Okay, technically a strike is not a lock; forgive me for categorizing them together.)

For each of these I will first discuss the lock itself, how it works, variations available, and how it is fitted into doors and frames. Then the types of doors they work best in and what types they should not be used in will be discussed. Finally, their overall benefits and limitations will be reviewed.

ELECTRIFIED MORTISE LOCKS

An electrified mortise lock is a mortise lock to which has been added a solenoid mechanism that can prevent the handle from turning, thus disallowing normal passage through the door to those who are not authorized to enter without a key.

A mortise lock is a very strong type of lockset. It is set deeply into a pocket in the door (a mortise). These locks are well known for their strength and security. Mortise locks (and thus electrified mortise locks) are available in many different configurations and can be handed for left- or right-handed doors that swing either in or out.

For most commercial doors in new construction, and many doors in retrofits, electrified mortise locks are the best and safest locks to use. First, let's discuss how electrified mortise locks work. To do that, let's look at a nonelectrified mortise lock (a basic mechanical mortise lock; Fig. 10.1).

You can see three views in Fig. 10.1 (outside, inside, and lock front) show the working elements of the mortise lock. From the outside view, we can see that the lock with a knob set, mortise cylinder (key-way), and the escutcheon plate that we see as we approach the door.

Behind the escutcheon plate you can see the outline of the mortise lock, and extending from that you can see the latch (bottom) and the auxiliary dead-latch, and at the top, you can see the dead-bolt.

On the inside view, you can see the lever handle with its rose (small circular metal disk around the lever set), turn knob, and again the outline of the mortise lock body.

From the Lock Front view, you can see the mortise lock body, knob, lever, key, turn knob, dead-bolt, latch, and dead-latch.

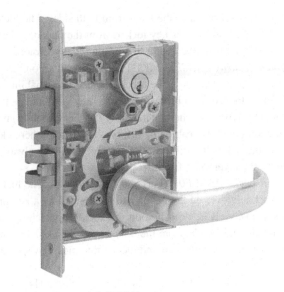

■ **FIGURE 10.1** Mechanical mortise lock. *IRCO-Schlage.*

The mortise lock has many useful functions not found in a cylinder lock (discussed later in this chapter) Mortise Lock Types. Some common examples are included in the list below.

Mortise latch only—no lock

Passage Set—ANSI F01: This is a nonkeyed function. The passage set is equipped with levers or knobs on both sides and a latch. There is no dead-bolt on the passage set. Turning the knob or lever retracts the latch from either side.

Mortise locks with no dead-bolt

Office Lock—ANSI F04: This is a keyed lockset. The lock is equipped with knobs/levers on both sides and a key lock on the outside. It is also equipped with stop works buttons below the latch on the strike face. When the stop works is set to unlock, the lock can be opened by turning the knob/lever from either side. When the stop works is set to lock, egress is free by turning the inside knob/lever but the lock can only be operated from the outside by inserting and turning a key, which retracts the latch.

Classroom Lock—ANSI F05: This is a keyed lockset. The lock includes a knob/lever on both sides and is keyed outside. The knob/lever always operates the latch from both sides. The lock is always unlocked unless it is key locked from the outside. In all cases, egress is enabled from the inside by turning the lever/knob.

Classroom Security—ANSI (None): This is a variation on the classroom lockset that adds a key lock to the inside, enabling the lockset to be locked from both sides with a key. The latchbolt is automatically retracted when operated by the lever/knob from either side unless the lock is enabled by use of a key from either side. In that case the lock is locked from the outside, but it can always be opened by the lever/knob from the inside.

Storeroom Lock—ANSI F07: This is a keyed lockset. The lock includes a knob/lever on the inside and is operated by a key only on the outside. There is no dead-bolt on this lock. The latch operates as a dead-latch; i.e., it locks from the outside as it latches. The inside knob/lever is always unlocked and the outside must always be opened by a key.

Institution Lock—ANSI F30: This is a keyed lockset. The lock is equipped with a knob/lever on both sides and is keyed on both sides. The knob/levers are always fixed and do not operate the lock. An auxiliary latch deadlocks the latchbolt when the door is closed and it can be locked by a key from either side.

Mortise locks with dead-bolts

Storeroom Lock with Dead-Bolt—ANSI (None): This is a keyed lockset. The lock is equipped with a knob/lever on both sides, a key on the outside, and a thumb-turn on the inside. It has both a latchbolt and dead-bolt. The latchbolt is retracted by the key outside or by the lever or knob inside. The outside knob/lever is unmovable. The dead-bolt is thrown or retracted by a key outside or thumb-turn inside. Both the dead-bolt and latchbolt can be retracted automatically by turning the inside lever/knob. An auxiliary latch deadlocks the latchbolt when the door is closed. The inside lever/knob is always free for egress.

Corridor Lock—ANSI F13: This is a keyed lockset. The lock includes a knob/lever on the inside and outside and a dead-bolt that can be retracted by a key from the outside or a thumb-turn on the inside. Throwing the dead-bolt locks the outside knob/lever. Turning the inside knob/lever will retract the latchbolt and the dead-bolt and automatically unlocks the outside knob/lever. The inside knob/lever is always free for immediate egress.

Store/Utility Room Lock—ANSI F14: This is a keyed lockset. The lock includes a knob/lever on both sides, is keyed from both sides, and includes a latchbolt and dead-bolt. The latchbolt can be retracted by the knob/lever from either side. The dead-bolt can be thrown or retracted using a key from either side.

Privacy with Dead-Bolt—ANSI F19: This is also a nonkeyed function. The privacy set is equipped with levers or knobs on both sides. It also has a dead-bolt and a turn knob to operate the dead-bolt on the inside and a special emergency coin-turn to operate the dead-bolt from the outside. For the Privacy set, the latch is retracted by the knob or lever from either side. Extending the dead-bolt locks the outside lever. Turning the knob/lever from the inside retracts the latch and dead-bolt and unlocks the outside lever. The Privacy set can be opened in an emergency by inserting a coin into a slot on the outside and using it as a thumb-turn to retract the dead-bolt, which also enables the outside knob or lever.

Entry—ANSI F20: This is a keyed lockset. The entry set includes knobs or levers inside and outside and also a key lock outside and dead-bolt operated by the key outside or a thumb-turn inside. The lock also has stop works buttons below the latch. The latch is retracted by the knob/lever from either side unless the outside knob/lever has been locked by the stop works buttons or the thumb-turn. The dead-bolt can be extended or retracted by key (outside) or thumb-turn (inside). The outside knob/lever is locked when the dead-bolt is extended. The auxiliary dead-latch deadlocks the latch when the door is closed. The inside knob/lever retracts the latch and dead-bolt simultaneously, allowing for quick free egress under any circumstances.

Dormitory Lock—ANSI F21: This is a keyed lockset. The lock includes a knob/lever on both sides, a key on the outside, and a thumb-turn on the inside. The lock also includes both a dead-bolt and a latchbolt. The latchbolt is retracted by the knob/key from either side. The dead-bolt can be thrown or retracted by a key outside or the thumb-turn inside.

Prison Function Lock—ANSI (None): This is a keyed lockset. The lock includes a knob inside and outside, a key cylinder outside, and no thumb-turn inside. It also includes both a latchbolt and dead-bolt. This type of lock is designed to work with two types of keys (Guard and Prisoner). The outside knob is free-spinning. The dead-bolt is only thrown by the guard's key. When the dead-bolt is thrown the inside knob becomes fixed. The prisoner's key only operates the latchbolt and not the dead-bolt. Thus, the prisoner can be provided with a key to his/her cell that allows daytime entry but no exit when the guard's key locks the dead-bolt.

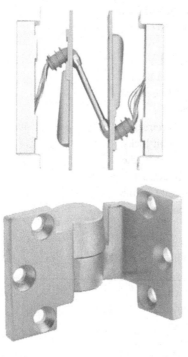

■ **FIGURE 10.2** Hinge/pivot/EPT/flexible spring montage. *IRCO-VonDuprin.*

All of these can be electrified, except for the passage set, which has no lock function. Electrification involves adding a solenoid inside the lock that prevents the outside lever from turning. When the solenoid is retracted, it ensures that the outside lever can be turned to open the lock, operating additional lock features if necessary to do so.

Electrified mortise locks are wired through the back of the lock (opposite the strike face) and require that the door be cored (wood doors) or that a tube be threaded through the door (hollow metal doors) so that wiring can extend from the hinge side of the door to the lock completely through the depth of the door from hinge to strike. Wiring is run from the frame to the door using an electrified hinge, electrified pivot, an Electric Power Transfer (EPT), or a frame/door mounted flexible spring (Fig. 10.2).

Door frame considerations

Electrified mortise locks may require different wiring provisions depending on the type of frame into which the door is set. Doors are commonly fitted into frames so that the door swings 90 degrees to full open. Hinges perform this function and, for electrification of the mortise lock, the installer can use either electric hinges or the EPT unit. However, doors can also be fitted into frames so that they swing to full open at 180 degrees, flat against the wall. In such cases, a special type of hinge is used or the door may open on pivots (both of which place the hinge point outside the frame to allow the door to swing 180 degrees). In either case, electrified hinges and EPTs will not work. In these instances, one can use either electrified pivots (the top pivot is wired) or the pocketed coil (through which wiring is threaded). The pocketed coil will wrap around the frame to accommodate the fully open door.

Additional lock switch fittings

Electrified mortise locks can also be fitted with additional features including:

- Door position switch
- Latch position switch
- Dead-bolt position switch
- Interior lever handle release switch

The door position switch includes a magnetic switch that is installed in the lock and matches up with a magnet in the strike plate. This functions to tell the system that the door is closed. These are unusual and rare. The latch position switch signals that the latch is engaged into the strike pocket. This functions to tell the security system that the lock is latched.

The dead-bolt position switch signals that the dead-bolt is extended. This functions to tell the security system that the door is deadlocked. The interior lever handle release switch signals when the interior handle has been turned, releasing the latch to open the door. It functions as a request-to-exit sensor. Whenever possible, use an electrified mortise lock that is properly equipped with appropriate internal switches, especially the internal request-to-exit sensor.

Door handing

A note about door handing is in order. Doors open inward or outward and the lock may be on the right or the left. When you order a lock, it is important to order it correctly so that it will work in the correct door configuration (Fig. 10.3).

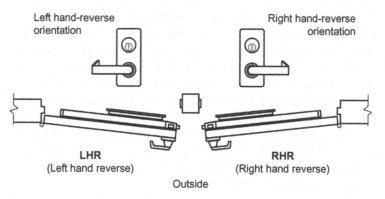

■ **FIGURE 10.3** Door handing diagram. *IRCO-VonDuprin.*

ELECTRIFIED "PANIC" HARDWARE

The second major category of free egress locking devices is electrified panic hardware. This is also one of the most familiar locks in the world, seen everywhere on public buildings, but mostly on exit doors.

Again, before we talk about electrified panic hardware, let's look first at mechanical panic hardware. The purpose of panic hardware is to get a lot of people out of a building quickly in an emergency. Were it not for the need to quickly exit large numbers of people in emergencies, there might be little need for panic hardware.

Panic hardware are known as Exit Devices in code documents. So we will call them that from now on in this chapter.

Rim exit devices

Rim exit devices comprise one or a pair of panic bars with a latch built into the panic bar. In the case of a pair of doors, rim devices require a center mullion so that the panic bar latch in each leaf can have a strike edge to close against. Rim exit devices are not common on double doors because it is undesirable to have the mullion separating the two halves of the double entryway.

The electrified rim device (Fig. 10.4) is operated by a solenoid within the panic bar that retracts the latch in response to an electric signal. When the latch is retracted by the solenoid, the door can be opened from the outside, and exit is always possible by pushing on the panic bar. A pneumatically controlled version is also available. Additionally, a variety of

■ **FIGURE 10.4** Rim exit device. *IRCO-VonDuprin.*

■ **FIGURE 10.5** Mortise lock exit device. *IRCO-VonDuprin.*

"electrical options" are available. See the section "Electrical Options" for the complete list.

Mortise lock exit devices

The Mortise Lock Exit Device (Fig. 10.5) comprises a panic bar connected to a modified conventional mortise lock. There is a horizontal cutout in the mortise lock into which a lever from the panic bar extends, engaging a lever in the mortise lock.

When electrified, the same solenoid that operates in the rim exit device is used to operate the panic bar lever that extends into the mortise lock. When the solenoid is activated, the latchbolt on the mortise lock retracts, allowing the door to be opened from the outside, but one can always exit freely by pushing on the panic bar. As in the rim device, a pneumatically controlled version is also available (see the section "Electrical Options").

Surface-mounted vertical rod exit devices

Surface-mounted Vertical Rod Exit Devices (Fig. 10.6) are most often used on double doors where it is undesirable to have a center mullion that would otherwise divide the double entryway into two. To latch the doors a pair of vertical rods extends from the center case of the Panic Bar where they terminate into latches at the top and bottom of the door. The latches latch into a top strike at the top of the door, which is surface mounted, and into a bottom strike at the bottom of the door, which is fitted into the threshold.

■ **FIGURE 10.6** Surface-mounted vertical rod exit device. *IRCO-VonDuprin.*

■ **FIGURE 10.7** concealed vertical rod exit device. *Irco-VonDuprin.*

When electrified, the same solenoid discussed earlier is used to operate a cam in the center case of the panic bar, which causes the vertical rods to retract from their home (extended) position. When extended, they are engaged with the top and bottom strike plates. The vertical rods can be retracted by pushing on the panic bar on the inside of the door, depressing a lever or latch on the outside or by activating the solenoid in the panic bar when it is activated by an access control panel's lock relay.

One can always exit freely by pushing on the panic bar. As in the rim device, a pneumatically controlled version is also available (see the section "Electrical Options").

Concealed vertical rod exit devices

Concealed Vertical Rod Exit Devices (Fig. 10.7) operate on exactly the same principle as the surface-mounted vertical rod exit devices; however, on these units, the working mechanism and top/bottom latches are built into the door. This approach provides a much more visually esthetic presentation than with surface-mounted vertical rod exit devices. These are typically built into hollow metal doors; however, a special version of these is also made for solid core wood doors.

In all other respects, concealed vertical rod exit devices operate exactly the same as their surface-mounted cousins.

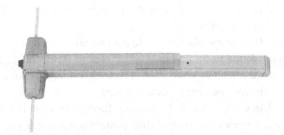

■ **FIGURE 10.8** Three-point latching exit device. *IRCO-VonDuprin.*

Three-point latching exit device

The Three-Point Latching Exit Device (Fig. 10.8) is a more secure version of the surface-mounted vertical rod exit device. In the three-point latching exit device a rim latch is added to surface-mounted vertical rods to create three points of latching instead of only one (rim) or two (surface-mounted vertical rods). The three-point latching exit device must be used either as a single door or with a center mullion between a pair of doors.

In the electrified version, the same familiar panic bar solenoid operates all three latches.

Exit device functions

Normal Function: In the normal function of the exit device egress is always free by pushing on the panic bar and entrance is controlled by the latch. After the door opens, it will latch again when it closes.
Latch Dogging: The term "dogging" refers to holding the latch(es) of a panic device retracted to create a push/pull function (Lori Green—Ingersoll Rand Company.). When the latch is dogged, the latch remains retracted so that the door can close, but not latch. Dogging of latches is not legal on fire egress doors because the doors are required by code to latch when they close; however, it is legal to dog the dead-bolt on specialized locks, if one exists. Fire exit hardware can be dogged electrically when interfaced with the fire alarm system.

Electrical options

Request-to-Exit (RX): The Request-to-Exit switch is an internal switch operated by the Panic Bar. This signals the access control panel that the door is being opened for a legal exit and is not from the outside as an intrusion.

Latchbolt Monitoring (LX): The latchbolt monitoring kit is a switch located inside the Panic Bar and is interfaced to the latchbolt, alerting whenever the latchbolt is retracted. The latchbolt monitor kit is a signaling device and cannot operate a load.

Electric Latch Retraction (EL): Electric Latch Retraction is the basis for remotely unlocking an electrified panic hardware equipped door. An electrical signal operates a solenoid within the panic bar, which retracts the latchbolt, allowing the door to be opened from the outside. Electric latch retraction can also be used to coordinate the electrified panic hardware with automatic door openers to ensure that the door is unlatched before the automatic opener attempts to open the door. In such cases a short delay (250 ms) should be inserted between the command to unlatch the panic bar and energize the automatic door opener.

Signal Switch (SS): A Signal Switch combines the RX switch with the LX, combining both functions into a single signal. Because both are monitored, the door is tamper monitored, making it more secure. The signal switch is functionally used as a request-to-exit sensor, signaling the access control panel to allow passage.

Electric Mortise Lock Device (E7500): Both this and the Electric Rim Device (see below) are abbreviated with an E. Should they be the same? Combining panic hardware with an electrified mortise lock, the electrified mortise lock device allows the mortise lock to be electrically controlled to remotely unlock the door. When unlocked, the door remains latched, preserving the fire rating. This combination is very useful where codes permit locking, but require unlocking during an emergency (such as for approved fire stairwells). The electric mortise lock device is typically also combined with the RX switch to signal an authorized exit.

Electric Rim Device (E99): The Electric Rim operates the mechanical rim type panic hardware by use of a solenoid within the panic bar. This function also combines a switch to monitor the outside trim condition (locked or unlocked) and a second switch to monitor the latchbolt.

Alarm Kit (ALK): The Alarm Kit is a battery-operated alarm designed to deter unauthorized use of the door. Although a person can exit using the panic bar, an alarm will sound at the door unless it has been bypassed by a key switch on the panic bar. The alarm horn provides an audible alert that the door is being opened. This deters the use of the door, although it is still available for emergencies.

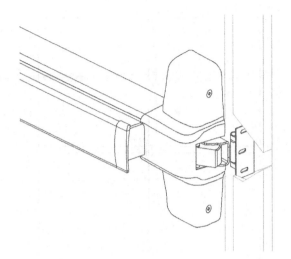

■ **FIGURE 10.9** Rim devices—same direction. *IRCO-VonDuprin.*

Pneumatic Controlled Devices (PN): The pneumatic option operates the same as the solenoid function, but is safe for hazardous areas and areas where a spark could cause an explosion (intrinsically safe operation).

Popular double door applications

Several popular double door applications are shown in the illustrations below. These include:

These illustrate how the various lock types are applied to common exiting applications (Figs. 10.9–10.13).

ELECTRIC STRIKES

Electric Strikes are access control devices that can be used with conventional mechanical locks to allow the door to be opened regardless of the lock/unlock status of the mechanical lock. The Electric strike replaces the fixed strike faceplate normally found on door frames, into which the mechanical lock latchbolt extends when it closes, thus latching the door.

Just like a fixed strike faceplate, the Electric strike normally presents a ramped surface for the mechanical latch to close against. As the door closes, the latchbolt retracts momentarily to accommodate the ramped surface, and then springs back to extend again into the latch pocket when

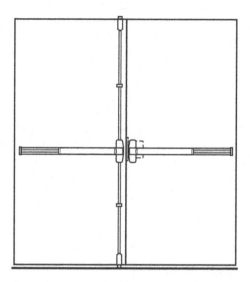

■ **FIGURE 10.10** Mortise lock and vertical rods—same direction. *IRCO-VonDuprin.*

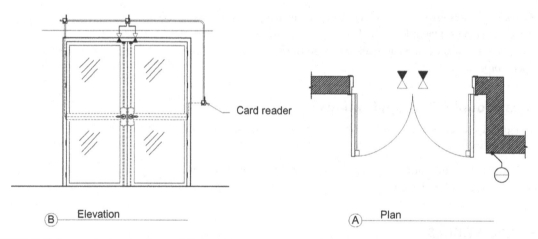

Card reader

(B) Elevation

(A) Plan

■ **FIGURE 10.11** Two vertical rods—same direction. *T. Norman.*

the door is fully closed. The latch is kept in place by the back of the ramped surface, called a latch keeper.

However, unlike a conventional fixed strike faceplate, the electric strike has a solenoid that controls the position of the latch keeper. In its quiescent state, the latch keeper is fixed, like the keeper of a fixed strike faceplate. But when the solenoid is activated, the latch keeper can swing

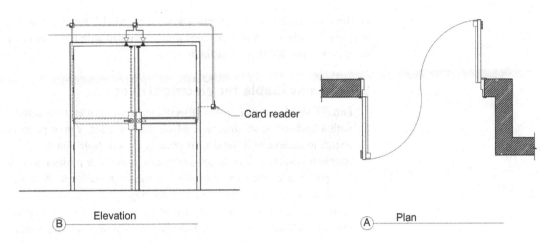

Card reader

ⒷElevation   ⒶPlan

■ **FIGURE 10.12** Double egress with vertical rods. *T. Norman.*

■ **FIGURE 10.13** Upper-only vertical rods. *IRCO-VonDuprin.*

aside, allowing a space where the extended latch is free to open with the door. Thus the door can be opened while the lock is still locked.

Electric strikes are available in either fail-safe or fail-secure functions. Fail-safe, also called fail-open, causes the electric strike to lock when power is applied. Fail-safe unlocks the electric strike when power is applied.

Electric strikes are available in either alternating current (AC) or direct current (DC) versions. AC versions "buzz" when operated, while DC versions are nearly silent, except for a quiet "click" when the lock is released. DC electric strikes are sometimes equipped with a buzzer to signal a person outside that the strike is open.

Switches available for electric strikes

Latch-Bolt Monitoring Switch: Electric strikes can often be ordered with a latchbolt position sensor, which many use like a door position switch to determine if the door is open or closed. Note that the latchbolt position sensor is not equivalent to the door position switch. It is possible to place a wad of paper in the latch pocket holding the latchbolt position sensor closed, thus fooling the access control system into thinking that the door is closed when it is actually open.

Dead-Bolt Monitoring Switch: Electric strikes built to accommodate mechanical locks with dead-bolts may also be equipped with a dead-bolt position monitoring switch.

Lock Status Monitoring Switch: This switch identifies the condition of the electric strike's locking mechanism, telling the access control system if the strike is locked or unlocked.

You should exercise caution when modifying an existing door frame to accept an electric strike to be certain that you are not doing so on a fire-rated door frame. Such a modification would void the fire rating of the frame.

ELECTRIFIED CYLINDER LOCKS

Electrified Cylinder Locks (Fig. 10.14) are mechanically similar to a standard mechanical cylinder lock, but with the addition of a small internal solenoid that remotely releases the lock. Electrified cylinder locks are

■ **FIGURE 10.14** Electrified cylinder lock. *Image courtesy of Security Door Controls.*

inexpensive and are common on light-duty doors under direct supervision. Like all cylinder locks, electrified cylinder locks lack physical strength, and it is inadvisable to use one as a security device.

Electrified cylinder locks are generally available in either 12 or 24 volt versions and also as AC or DC activated. AC versions create a buzzing sound that alerts users outside that the lock is unlocked.

SELF-CONTAINED ACCESS CONTROL LOCKS

The final category of free egress locks is a specialty category commonly found on hotel room doors. Increasingly, however, these are also on stand-alone specialty doors where connection to an electronic access control system is not practical for economic reasons.

The lock comprises an electrified cylinder lock controlled by a card reader, keypad, or fingerprint reader that is part of the lock itself. These are usually battery operated. When a valid card, keycode, or enrolled fingerprint is presented, a small processor within the lock is activated and processes the request. If the credential presented is valid in the small lock database, a solenoid is activated, releasing the lock. An indicator light on the lock indicates that the lock can be opened. These locks typically have a data connection port allowing upload and/or download of stored credentials and transaction records (Fig. 10.15).

CHAPTER SUMMARY

1. There are three common types of free egress locks: electrified mortise locks, electrified panic hardware, and electric strikes.
2. A mortise lock is a very strong type of lockset.
3. An electrified mortise lock is a mortise lock to which has been added a solenoid mechanism that can prevent the handle from turning, thus disallowing normal passage through the door to those who are not authorized to enter without a key.
4. Electrified mortise locks may require different wiring provisions depending on the type of frame into which the door is set.
5. Electrified mortise locks can also be fitted with additional features including:
   a. Door position switch
   b. Latch position switch
   c. Dead-bolt position switch
   d. Interior lever handle release switch

■ **FIGURE 10.15** Self-contained access control locks.

6. Door Handing: When you order a lock, it is important to order it correctly so that it will work in the correct door configuration.
7. The second major category of Free Egress locking devices is Electrified Panic Hardware.
8. The third category of Free Egress locking devices is Electrified Strikes.
9. Electric strikes are access control devices that can be used with conventional mechanical locks to allow the door to be opened regardless of the lock/unlock status of the mechanical lock.
10. Electrified Cylinder Locks are mechanically similar to a standard mechanical cylinder lock, but with the addition of a small internal solenoid that is used to remotely release the lock.
11. The final category of Free Egress locks is a specialty category, commonly found on hotel room doors.

Q&A

1. *Three common types of Free Egress Locks include:*
   a. Electrified Mortise Locks, Magnetic Locks, and Electrified Dead-Bolt Locks
   b. Electrified Mortise Locks, Magnetic Locks, and Electrified Panic Hardware

    c. Electrified Mortise Locks, Magnetic Locks, and Electric Strikes
    d. Electrified Mortise Locks, Electrified Panic Hardware, and Electric Strikes

2. *An Electrified Mortise Lock*
    a. Is a Mortise Lock to which has been added a solenoid mechanism that can prevent the handle from turning, thus disallowing normal passage through the door to those who are not authorized to enter without a key
    b. Is a very strong type of lockset
    c. Is the best and safest lock to use for most commercial doors in new construction
    d. All of the above

3. *Electrified Mortise Locks*
    a. May require different wiring provisions depending on the type of frame into which the door is set
    b. Are commonly fitted into frames such that the door swings 90 degrees to full open
    c. Can use electrified hinges or Electric Power Transfer (EPT) units to provide power to the lock
    d. All of the above

4. *Electrified Panic Hardware*
    a. Causes people to panic
    b. Keeps people from panicking
    c. Allows large quantities of people to exit quickly in an emergency
    d. None of the above

5. *Electric Strikes*
    a. Are access control devices that can be used with conventional mechanical locks to allow the door to be opened regardless of the lock/unlock status of the mechanical lock
    b. Are access control devices that can be used with conventional mechanical locks to allow the door to be opened only when the electric strike is mechanically bonded to the top of the door
    c. Both b and c
    d. Neither a nor b

6. *Electrified Cylinder Locks*
    a. Are mechanically similar to a standard Mortise Lock, but with the addition of a small internal solenoid that is used to remotely release the lock
    b. Are mechanically similar to a standard panic hardware, but with the addition of a small internal solenoid that is used to remotely release the lock
    c. Both a and b
    d. Neither a nor b

7. *Self-Contained Access Control Locks*
    a. Are used in large quantities in commercial buildings
    b. Are most common in frameless glass doors
    c. Both a and b
    d. Neither of the above

Answers: (1) d; (2) d; (3) d; (4) c; (5) a; (6) d; (7) d.

Magnetic Locks

CHAPTER OBJECTIVES

1. Learn the basics in the chapter overview
2. Learn about standard magnetic locks
3. Understand magnetic shear locks and when to use them
4. Answer questions about magnetic locks

CHAPTER OVERVIEW

Magnetic locks are the workhorse of electrified locking mechanisms for most security contractors. They are easy to install, versatile, provide good security, and are generally dependable. Magnetic locks can be fitted to any door with a metal or wood surface near the top of the door, so they can be fitted to more than 90% of doors installed in commercial installations.

Magnetic locks work by using an electromagnet on the door frame and an armature mounted to the door. When the door closes, the armature comes into contact with the electromagnet. Magnetic locks can also function as a door position switch and lock engaged switch, when so configured. This is done by sensing the magnetic bond between the electromagnet and the armature. When the bond is made, the lock is locked, and thus the door is closed. All surface-mounted magnetic locks should be mounted on the secure side of the door.

STANDARD MAGNETIC LOCKS

The Standard Magnetic Lock (Fig. 11.1), also called a magnetic plate lock, is the primary version of the magnetic lock. The Magnetic lock is a blissfully simple device. It comprises only an electromagnet and an armature. Connect the electromagnet to the door frame and the armature to the door, add power, and you have yourself a locked door.

Electronic Access Control. DOI: http://dx.doi.org/10.1016/B978-0-12-805465-9.00011-7

■ **FIGURE 11.1** Standard magnetic lock. *Image courtesy of Security Door Controls.*

The standard magnetic lock places the electromagnet and armature both in vertical orientation so that the armature closes directly against the electromagnetic. This provides a large contact area and ensures freedom of the door when the magnetic lock is released. In an attack on a standard magnetic lock, the attacker must overcome the tensile force of the lock to break the magnetic bond.

There is a seemingly endless combination of door configurations that quite honestly drives many sane people crazy. No, I mean it. If you ever want a 3-month holiday from work, just read up on door lock configurations and you will finish the article in a state that can get you institutionalized by any psychiatrist. After more than 30 years in the security industry, I still find new door configurations. Magnetic locks solve most of those problems by using an assortment of adapters to work in a large variety of mounting and use configurations.

Magnetic locks work with direct current only. Most magnetic locks can be ordered with an optional door/lock position/engagement sensor that alerts when the door is open. They are also useful for doors with misalignment problems or buildings with foundation problems resulting in frequent misaligned doors. Due to the nature of the design, standard magnetic locks can suffer some misalignment and still function well. (This is not true for magnetic shear locks.)

Because magnetic locks have no mechanical locking elements, they require power to work, meaning that they will unlock if power is lost. Accordingly, a back-up power source such as an uninterruptable power supply is required if the lock is used in a security-critical application.

Standard magnetic locks come in a variety of holding forces, the most common is 650 pounds (light security), then 1200 pounds (medium security), and 1500 pound and higher (high security). A magnetic lock with a holding force of only 650 pounds should be used for traffic control only. These locks can easily be defeated. Locks with holding force of 1200 pounds are sufficient for aluminum framed glass doors, because when 1200 pounds of force is exerted on such a door the glass will break. These are useful where the doors are not likely to engage aggressive attacks. For hollow metal doors and doors where aggressive use of force is possible, such as a door securing a psychiatric ward of a hospital, high security magnetic locks are called for.

Standard magnetic locks are not the perfect one-size-fits-all solution that some contractors think they are (see the section "Cautions About Magnetic Locks").

Magnetic locks have several basic forms:

- Basic surface-mounted magnetic lock
- Magnetic lock with escutcheon filling the top of the door frame for a more aesthetic look
- Magnetic gate locks
- Magnetic locks built into the door frame

Standard magnetic lock applications

The following illustrations show the three most common applications for magnetic locks. These include (Figs. 11.2–11.4):

These illustrations indicate how surface-mounted magnetic locks are applied to the most common door swing applications.

MAGNETIC SHEAR LOCKS

The second major variation of magnetic locks is the Magnetic Shear Lock (Fig. 11.5). Unlike the vertically oriented standard magnetic locks in which the armature pulls directly away from the electromagnet, the magnetic shear lock is vertically oriented so that if someone is attacking

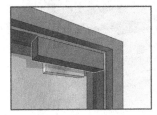

■ FIGURE 11.2 Outswinging door configuration. *Image courtesy of Security Door Controls.*

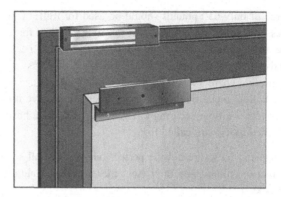

■ **FIGURE 11.3** Inswinging door configuration. *Image courtesy of Security Door Controls.*

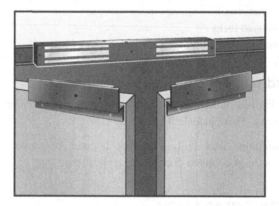

■ **FIGURE 11.4** Double door configuration. *Image courtesy of Security Door Controls.*

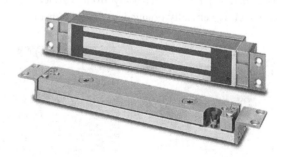

■ **FIGURE 11.5** Magnetic shear lock. *Image courtesy of Security Door Controls.*

the door, they must overcome the shear force of the magnetic lock, which is much greater than the tensile force of a standard magnetic lock. Magnetic shear locks are available in holding forces reaching up to 3000 pounds. These are very strong indeed.

In hollow metal doors, magnetic shear locks can be mounted inside the door frame with the armature mounted inside a pocket in the top of the door, making for a completely invisible lock when the door is closed.

Magnetic shear locks can also be used with aluminum frame glass doors, frameless glass doors, and solid core wood doors. In most of these cases, the armature will be mounted to the door and will be visible.

Most magnetic shear locks are concealed; easy concealment is one of their strong points. However, to ensure complete safety and to provide the higher security only available with the shear lock version, surface-mounted versions are available. These can be mounted to accommodate either inswinging or outswinging doors (see the section "Cautions About Magnetic Locks").

Magnetic shear lock applications

See Figs. 11.6 and 11.7.

MAGNETIC GATE LOCKS

Indoor and outdoor gates can also be locked using magnetic locks. Magnetic Gate Locks (Fig. 11.8) provide high security and, if mounted correctly, are generally very reliable. Since gates do not align as precisely

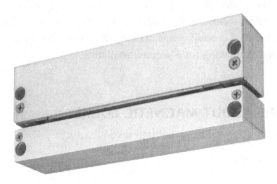

■ **FIGURE 11.6** Surface-mounted magnetic shear lock.

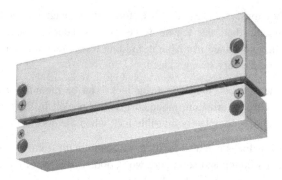

■ **FIGURE 11.7** Top jam mount surface magnetic shear lock. *Image courtesy of Security Door Controls.*

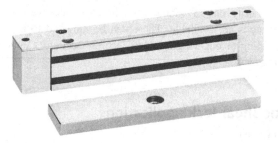

■ **FIGURE 11.8** Magnetic gate locks—swing and sliding. *Image courtesy of Security Door Controls.*

as doors, their natural ability to accommodate some misalignment makes them a good choice.

When mounting magnetic gate locks outdoors, it is important to ensure that wiring is secure from weather and is unaffected by animals and unavailable to passersby. Some magnetic gate locks are available already fitted with threaded conduit connectors to make a seal-tight connection to the rigid conduit, making for a secure installation.

CAUTIONS ABOUT MAGNETIC LOCKS

Problems with magnetic locks fall into several areas:

- Egress assurance
- Operational warnings
- Maintenance warnings

Egress assurance

Magnetic locks are *not* free mechanical egress locks; i.e., power to the lock must be dropped for the lock to unlock. And this is where the problem begins. To exit through a door that is locked with a magnetic lock, the door must be equipped with some kind of request-to-exit sensor, and request-to-exit sensors are not all created equal (Fig. 11.9).

There are two broad types of request-to-exit sensors: those that require no special knowledge to operate and those that require the users to know how to operate the request-to-exit sensor. Examples of those that require no special knowledge include infrared or infrared/microwave motion detectors over the door and a request-to-exit switch inside a panic bar. Examples of request-to-exit sensors that require special knowledge include exit push buttons and emergency pull station exit switches. There are several kinds of request-to-exit sensors, and some are more reliable than others.

Another way to categorize request-to-exit sensors is by reliability. A switch that directly controls power to the magnetic lock has 1 degree of separation, whereas a motion detector that notifies an exit sense input to an access control panel has 7 degrees of separation. The motion sensor must first sense motion with an infrared circuit (1), and it must process that detection (2) and activate a notification relay (3). This will be communicated to an access control panel across wiring (4), where it will be sensed on a request-to-exit input (5). The access control panel must process the request (6) and activate the lock relay (7), which will drop power

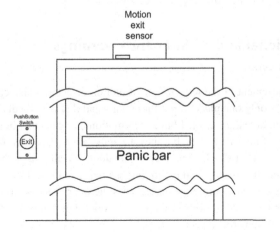

Motion
exit
sensor

PushButton
Switch
(Exit)

Panic bar

■ **FIGURE 11.9** Montage of request-to-exit sensors. *Image courtesy of Security Door Controls.*

to the magnetic lock. (The final cable is not included because if it fails, there would be no power to drop.)

I just described eight chances for failure. The more chances, the more likelihood of failure. Things can go wrong with electronics. And they do. I have been locked behind a door where an access control panel request-to-exit input failed unexpectedly; and with the building manager of course. Luckily, we were on a construction inspection tour and the contractor was there to open the lock and release power.

For this reason, I strongly recommend that you interface every motion detector request-to-exit sensor through an interposing relay having two contacts, such that one contact signals the access control panel and the other contact directly controls power to the lock. An even better option is to use a request-to-exit switch that has two contacts, thus eliminating the relay that could be another point of failure.

Because request-to-exit sensors can cause magnetic locks to fail locked, I also recommend having two means to release the lock from the inside. Typically, this includes one method that requires no special knowledge and a second one that does. Common combinations include:

- Exit motion detector and exit push button
- Panic bar switch and exit push button
- Exit motion detector and panic bar switch

You might be asking why you would have a panic bar on a door with a magnetic lock. This is a common practice on doors that do not have latching locks. In such cases, the panic bar is not mechanical; there is just a panic bar with an integral request-to-exit switch.

Operational and maintenance warnings

Safety Concerns:

- The placement of the request-to-exit motion detector is also critical. There is only one place to put these units and that is immediately centered above the door. I have seen motion exit detectors in a variety of unusual places, but they are most often placed on the ceiling 6–10 ft (2–3 m) in front of the door. Usually this is because someone in authority in the building has requested it so that the lock will be unlocked before anyone reaches the door. This might work fine for someone walking straight to the door, but that is not always the case. Sometimes people stop for a moment and close out a conversation before leaving through the door. In such cases, as neither person is in the area sensed by the motion sensor, they will find the lock engaged and not unlocked as they would imagine.

- When using exit push buttons it is important to remember when wired as described earlier, if the access control panel should fail to unlock the door, it will be up to the exit push button to directly release power to the magnetic lock to unlock the door. In such cases, the exit push button must be able to hold power from the magnetic lock long enough for the user to reach for the door handle and open the door. The best type of exit push button for this function is a pneumatically controlled exit push button. This type of exit push button has a small, air-filled cylinder activated by the push of the button. The cylinder holds the button in by means of a vacuum generated by pushing the button long enough to allow the door to be opened. Although pneumatic exit buttons are pricy, they are the best solution possible when using exit push buttons.

Security Concerns:

- Request-to-exit motion sensors on double frameless glass doors have another concern. Double frameless glass doors do not come together perfectly. There is always a small gap between them when closed. This gap is large enough to allow a person outside to pass a yellow note tablet through the gap where it can be sensed by the motion detector, triggering the access control system to unlock the door. This obviously presents a serious security concern. I once had a client whose Security Command Center manager gleefully demonstrated his "superior knowledge" by telling many employees about this security exploit. The building in question had many elevator lobby doors equipped with double frameless glass doors. While it might have made him momentarily popular with other employees, it was not the case with upper management who may have set his career path on a different course.
- It is important to mount all surface-mounted magnetic locks on the secure side of the door. When the lock is mounted on the unsecure side of the door, access to the lock by unauthorized users is possible, resulting in a compromise of the lock, allowing illegal entry. It is startling to see the number of magnetic locks that are mounted on the unsecure side of the door.

Maintenance Concerns:

- Magnetic locks may use a reverse-mounted diode to dissipate the electromagnetic force (back EMF) that the magnet generates back onto the power supply when it disengages power. I have seen this cause the lock to delay opening by up to half a second when commanded to open. Most modern magnetic locks use a Metal Oxide Varistor (MOV) instead of a reverse-mounted diode to dissipate the back EMF. The MOV does not cause the same delay as the diode.

■ Magnetic locks cannot have anything on the surface of the electromagnet or its armature (i.e., cellophane tape, paper, lacquer, etc.) as this can cause the magnet to fail to lock securely. Magnetic locks should be cleaned annually with a moist, soapy cloth and then wiped with a clean moist cloth.

CHAPTER SUMMARY

1. The Standard Magnetic Lock, also called a magnetic plate lock, is the primary version of the magnetic lock.
2. The standard magnetic lock places the electromagnet and armature both in vertical orientation so that the armature closes directly against the electromagnet.
3. Magnetic locks have several basic forms including:
   a. Basic surface-mounted magnetic lock
   b. Magnetic lock with escutcheon filling the top of the door frame for a more aesthetic look
   c. Magnetic gate locks
   d. Magnetic locks built into the door frame
4. The second major variation of magnetic locks is the Magnetic Shear Lock.
5. Unlike the vertically oriented standard magnetic locks in which the armature pulls directly away from the electromagnet, the magnetic shear lock is vertically oriented so that if someone is attacking the door, they must overcome the shear force of the magnetic lock, which is much greater than the tensile force of a standard magnetic lock.
6. Indoor and outdoor gates can also be locked using magnetic locks.
7. Problems with magnetic locks fall into several areas:
   a. Egress assurance
   b. Operational warnings
   c. Maintenance warnings
8. Magnetic locks are *not* free mechanical egress locks; i.e., power to the lock must be dropped in order for the lock to unlock.
9. There are two broad types of request-to-exit sensors: those that require no special knowledge to operate and those that require the users to know how to operate the request-to-exit sensor.
10. Because request-to-exit sensors can cause magnetic locks to fail locked, I also recommend having two means to release the lock

from the inside. Typically this will include one method that requires no special knowledge and a second that does. Common combinations include:

a. Exit motion detector and exit push button
b. Panic bar switch and exit push button
c. Exit motion detector and panic bar switch

11. Safety Concerns:
    a. The placement of the request-to-exit motion detector is also critical.
    b. When using exit push buttons it is important to remember when wired, as described previously, if the access control panel should fail to unlock the door, it will be up to the exit push button to directly release power to the magnetic lock in order to unlock the door.

12. Security Concerns:
    a. Request-to-exit motion sensors on double frameless glass doors can be easily compromised.
    b. It is important to mount all surface-mounted magnetic locks on the secure side of the door.

13. Maintenance Concerns:
    a. Magnetic locks may use a reverse-mounted diode to dissipate the electromagnetic force that the magnet generates back onto the power supply when it disengages power.
    b. Magnetic locks cannot have anything on the surface of the electromagnet or its armature.

Q&A

1. *The Magnetic Lock comprises:*
    a. An armature and a plate
    b. A plate and the door frame
    c. An electromagnet and a plate
    d. An electromagnet and an armature
2. *The _____ mounts to the door frame*
    a. Armature
    b. Electromagnet
    c. Door
    d. Hinges
3. *The Standard Magnetic Lock is the primary version of the magnetic lock*
    a. True
    b. False
4. *The Standard Magnetic Lock places the electromagnet and armature both in _____ orientation*
    a. Triangular to the door frame
    b. Vertical

    c. Horizontal

    d. Away from the door

5. *Standard Magnetic Locks come in a variety of holding forces including:*

    a. 650 pounds

    b. 1200 pounds

    c. 1500 pounds

    d. All of the above

6. *Magnetic _____ Locks are the second major variation of magnetic locks*

    a. Hollerith

    b. Wiegand

    c. Proximity

    d. Shear

7. *Magnetic Shear locks are oriented _____ to the door*

    a. Horizontally

    b. Vertically

    c. Triangular to the door frame

    d. Away from the door

8. *Most Magnetic Shear locks are concealed*

    a. True

    b. False

9. *_____ and _____ gates can also be locked using magnetic locks*

    a. Open and closed

    b. Swinging and open

    c. Sliding and open

    d. Indoor and outdoor

10. *Problems with magnetic locks fall into several areas:*

    a. Egress assurance

    b. Operational warnings

    c. Maintenance warnings

    d. All of the above

11. *Magnetic locks are free mechanical egress locks*

    a. True

    b. False

12. *There are two broad types of Request-to-Exit sensors:*

    a. Those that require no special knowledge to operate and those that require the user to use a code

    b. Those that require special knowledge to operate and biometric

    c. Those that require no special knowledge to operate and those that require the users to know how the operate the Request-to-Exit sensor

    d. All Request-to-Exit sensors require special knowledge

13. *Request-to-Exit sensors can include:*

    a. Infrared

    b. Switch inside a Panic Bar

    c. Push-to-Exit Button

    d. All of the above

Answers: (1) d; (2) b; (3) a; (4) b; (5) d; (6) d; (7) a; (8) a; (9) d; (10) d; (11) b; (12) c; (13) d.

Electrified Dead-Bolt Locks

CHAPTER OBJECTIVES

1. Learn about surface-mounted electrified dead-bolt locks
2. Learn about concealed direct-throw mortise dead-bolt locks
3. Understand a real problem solver—the hybrid mortise lock with dead bolt
4. Top-latch release bolt—for concealed vertical rod release
5. Gate dead bolt—for sliding and swinging gates
6. Learn when and when not to use electrified dead bolts
7. Answer questions about electronic dead-bolt locks

CHAPTER OVERVIEW

For security, nothing beats a steel door and a dead-bolt lock. Unlike latches that can be manipulated using special tools, a dead bolt is very secure against entry by tools. So the idea of electrifying dead-bolt locks is attractive from a security standpoint. Electrification of a dead-bolt lock enables remote operation, potentially reducing manpower costs.

This chapter discusses a variety of electrified dead-bolt locks including surface-mounted dead bolts, concealed direct-throw dead bolts, mortise locks equipped with a dead bolt, dead bolts used to release concealed vertical rods, and dead bolts for gates.

Electrified dead-bolt locks must be used with caution for safety reasons. There are many applications where electrified dead-bolt locks are not appropriate or safe to use.

SURFACE-MOUNTED ELECTRIFIED DEAD-BOLT LOCKS

There is little doubt that the dead bolt is the most secure kind of lock. Unlike latching locks that latch automatically when a door closes, a dead-bolt lock must be intentionally locked. The dead bolt cannot be

Electronic Access Control. DOI: http://dx.doi.org/10.1016/B978-0-12-805465-9.00012-9

■ **FIGURE 12.1** Surface-mounted electrified dead-bolt lock. *Image courtesy of Security Door Controls.*

easily kicked, sprung, or manipulated with a credit card or hand-tools like latching locks. Electrified dead bolts allow for manual remote release or operation by an electronic access control system (Fig. 12.1).

The simplest kind of dead bolt is the surface-mounted electrified dead-bolt lock. This comprises a solenoid or motor operated dead bolt, a receiver, and enclosures for both. The lock is typically mounted to the door frame and the receiver is surface mounted on the door. This lock should only be used on unoccupied rooms and never on fire exit doors.

CONCEALED DIRECT-THROW MORTISE DEAD-BOLT LOCK

For a more elegant installation, and one where the designer wants to ensure that the door locks when the door is closed, Concealed Direct-Throw Mortise Dead-Bolt Locks (Fig. 12.2) are a good choice. Like the surface-mounted version discussed earlier, the concealed version includes a solenoid or motor-operated dead-bolt lock and a receiver. But that is where the similarity ends. The concealed direct-throw mortise dead-bolt lock is mounted into the door frame and the bolt extends into a receiver pocket mounted into the door.

Typically, the dead bolt is equipped with a nylon sleeve to allow it to move smoothly without binding. Often the dead bolt is also equipped with a magnet at its tip, and there is an opposing magnet in the receiver pocket to repel the dead bolt in case the solenoid or motor needs help. To ensure that the lock relocks automatically when the door closes, some of these locks are also equipped with a ball-switch in the body of the lock that engages a flat face on the lock receiver when the door closes. This action pushes the ball-switch in, energizing the lock to ensure that it does not extend again until the door is closed again. This also enables automatic relocking on door closure.

Like its surface-mounted cousin, this lock should never be used on an occupied room. Its application is appropriate for storage rooms only.

■ **FIGURE 12.2** Concealed direct-throw mortise dead-bolt lock. *Securitech.*

■ **FIGURE 12.3** Dead-bolt equipped electrified mortise lock. *IRCO-VonDuprin.*

DEAD-BOLT EQUIPPED ELECTRIFIED MORTISE LOCK

With all this talk about electrified dead-bolt locks not being useful on occupied rooms, wouldn't it be good to hear about one that *is* legal for occupied rooms? Well, there are a few available and the most basic is the dead-bolt equipped electrified mortise lock (Fig. 12.3).

Imagine an electrified mortise lock to which a dead bolt has been added. Several manufacturers make such a lock. In most cases, the dead bolt is not operated by an electrical mechanism. For most such locks, the operation of the dead bolt is semimechanical.

In this lock, the latch set of the lock operates like a normal electrified mortise lock with its dead bolt retracted by an electrical signal. However, for most such locks the dead bolt cannot be thrown to a locked position electrically.

TOP-LATCH RELEASE BOLT

It is possible to electrify panic hardware with concealed vertical rods to allow for their remote or timer release or release by an electronic access control system. Security Door Controls (SDC) manufactures a lock called the Top-Latch Release Bolt (Fig. 12.4), which is used to control entry from the outside only (remember, there is always free mechanical egress on panic bar equipped doors). On panic hardware doors equipped with concealed vertical rods, the rods at the top rail are designed to engage a mechanical latch built into the frame header. SDC replaces this latch with a solenoid-operated latch release, allowing it to be electrically operated to permit entry from the outside.

The lock is also equipped with a ball-switch to hold the lock open until the door is reclosed, enabling an automatic relock function. The top-latch

■ **FIGURE 12.4** Top-latch release bolt. *Securitech.*

release bolt should generally be used on doors having vertical rods on the top only.

ELECTRIFIED DEAD-BOLT GATE LOCKS

Electrified Dead-Bolt Gate Locks (Fig. 12.5) work exactly like surface-mounted electrified dead-bolt locks. However, they are more robust and resistant to weather, dirt, and water than their indoor-rated cousins.

Electrified dead-bolt gate locks should not be used on pedestrian gates that are in the path of egress.

Electrified dead-bolt lock safety provisions

Remember, life safety is always the primary concern in access control systems. Accordingly most electrified dead-bolt locks should not be used in occupied areas.

Electrified dead-bolt locks are subject to changes in the alignment of the door and the frame. While a lock may work perfectly when installed,

■ **FIGURE 12.5** Electrified dead-bolt gate lock. *Image courtesy of Security Door Controls.*

building settling and/or weather changes can cause the lock to misalign months later. When that happens, the lock may not engage, or worse, may not release. This can and does happen. There was a time that electrified dead bolts were seen as a good solution for locking exterior doors. This is not the case today after many failed lock installations. Let me be clear, this is not the fault of the lock, but rather a bad application for the lock.

CHAPTER SUMMARY

1. The dead bolt cannot be easily kicked, sprung, or manipulated with a credit card or hand-tools like latching locks.
2. The simplest kind of dead bolt is the surface-mounted electrified dead-bolt lock.
3. For a more elegant installation, and one where the designer wants to ensure that the door locks when the door is closed, concealed direct-throw mortise dead-bolt locks are a good choice.
4. Several manufacturers make electrified mortise locks with an integral dead-bolt function.
5. It is possible to electrify panic hardware with concealed vertical rods to allow for their remote or timer release or release by an electronic access control system. This is called the top-latch release bolt. The top-latch release bolt should generally be used on doors having vertical rods on the top only.
6. Electrified dead-bolt gate locks work exactly like surface-mounted electrified dead-bolt locks. However, they are more robust and resistant to weather, dirt, and water than their indoor-rated cousins.
7. Remember, life safety is always the primary concern in access control systems. Accordingly most electrified dead-bolt locks should not be used in occupied areas.

Q&A

1. *Common Electrified Dead-Bolt Locks include:*
   a. Surface-mounted Electrified Dead-Bolt Locks
   b. Concealed Direct-throw Mortise Dead-Bolt Locks
   c. Dead-Bolt Equipped Electrified Mortise Locks
   d. All of the above
2. *Common Electrified Dead-Bolt Locks include:*
   a. Surface-mounted Electrified Dead-Bolt Locks
   b. Top-Latch Release Bolt
   c. Both a and b
   d. Neither a nor b

3. *Electrified Dead-Bolt Gate Locks should not be used on:*
   a. Vehicle gates
   b. Farm animal gates
   c. Aircraft tarmac gates
   d. Pedestrian gates that are in the path of egress
4. *Electrified Dead-Bolt Gates are:*
   a. Appropriate for all pedestrian gates
   b. More robust and resistant to weather, dirt, and water than their indoor-rated cousins
   c. Both a and b
   d. Neither a nor b
5. *Most Electrified Dead-Bolt Gates:*
   a. Should never be used to control large animals
   b. Should not be used in occupied areas
   c. Can be used underwater to control sharks and dolphins
   d. Should not be used near exploding nuclear weapons
6. *When an Electrified Dead-Bolt Lock traps people indoors:*
   a. It is the fault of the lock
   b. It is a bad application for the lock
   c. It is because of terrorists
   d. It is because of marauding gangs

Answers: (1) d; (2) c; (3) d; (4) b; (5) b; (6) b.

13

Specialty Electrified Locks

CHAPTER OBJECTIVES

1. Learn about unique problem-solving locks in the chapter overview
2. Electrified dead-bolt-equipped panic hardware
3. Discover securitech locks
4. Learn about delayed egress locks
5. Learn about a forgotten problem solver—Hi-Tower locks
6. Get to know a rare but useful lock—Blumcraft door locks
7. Pass a quiz on specialized electronic locks

CHAPTER OVERVIEW

Sometimes there just isn't a solution to a locking problem. When you have looked through all of the normal locks that are made by all of the manufacturers and you cannot find a single lock solution to a nagging problem, it would be good to know that there are a few locks out there that just do not fit into the normal boring categories of most locks. These are weird and wonderful locks that can be real problem solvers.

In this chapter we will introduce Electrified Dead-Bolt-Equipped Panic Hardware, which allows for very secure locking of fire exit doors. We will also look at the bizarre and wonderful world of Securitech locks, which are some of the most creative and useful locks in the industry. The third section will be on Delayed Egress Locks, which provide for locking doors that could not be otherwise locked due to fire codes. The last two sections will deal with Hi-Tower locks, which are a good solution for fire-rated stairwells, and finally, we will introduce you to Blumcraft Door Locks, which are an elegant solution for frameless glass doors.

Electronic Access Control. DOI: http://dx.doi.org/10.1016/B978-0-12-805465-9.00013-0

ELECTRIFIED DEAD-BOLT-EQUIPPED PANIC HARDWARE

Remembering that Fire Exit Hardware (Panic Bars) must be used on all fire egress doors and all doors on any room with an occupancy of 50 or more, this would automatically rule out all electrified dead-bolt locks, wouldn't it? Well, perhaps not. If the need involves both security and safe egress to protect a room or corridor with important assets or higher security risks, a solution exists in the form of the Von Duprin Series HS99 Devices. This lock combines a Rim Exit Hardware device with a Sargent and Greenleaf 8470 Automatic Dead Bolt. The result is a high-security lock that provides single-action emergency egress from the inside. The lock can be used in a wide variety of situations requiring both emergency exiting and high security such as school buildings, pharmaceutical drug rooms, money counting areas, or Secured Compartmented Information Facilities. The lock can be dogged down during school hours to reduce maintenance and locked at night for security, while still providing emergency exiting.

When the lock is opened by pressing on the panic bar, both the panic bar rim latch and the automatic dead bolt retract, thus allowing exit. As the door recloses and relatches, a latchbolt sensor detects that the door is latched and power is enabled to relock the automatic dead bolt. These locks can also be operated by locking hardware from the outside, such as a key, combination lock, or keypad. A variance should always be requested from the Authority Having Jurisdiction.

The lock can be used for exit only, key entry, combination lock entry, keypad entry, and electrically controlled entry.

SECURITECH LOCKS

Securitech is a most unusual lock manufacturer. The company began by building extremely robust locks for the New York Housing Authority, where housing residents were driving up the maintenance costs of multiple family housing units by destroying every lock the Housing Authority could find to put on their doors. It was common to find residents destroying lever sets by stressing them beyond their physical limits. The first Securitech locks used very robust mechanisms and clutches to allow the handle to rotate past its apparent limit without breaking it mechanically. When returned to its normal position by only the twist of the hand, it continued to function normally, thus confounding the most determined vandals.

It was not long before Securitech began electrifying their locks upon the request of their clients.

Today, Securitech also makes very high-security electrified locksets that meet fire codes. These include electrified multipoint dead-bolt locks that include panic bars, fitting the need for emergency egress on very high security doors.

DELAYED EGRESS LOCKS

Where high security is required on a seldom-used required exit door, the answer may be to use a Delayed Egress Lock (Fig. 13.1). NFPA 101 makes a provision for this type of lock.

A delayed egress lock is a specially modified magnetic lock that includes the magnetic lock, a countdown timer, exit sense device, a local alarm, Delayed Egress Signage, and a Fire Alarm Interface. The delayed egress lock is designed to delay exit through a fire egress door for a predetermined time period (usually 15–30 seconds). A user approaches the door, pushes on a panic bar equipped with an exit sense switch, and holds the panic bar in while the countdown timer counts down to zero from the preset time. During that time, a local alarm at the door sounds to alert anyone nearby that a person is exiting the door. The alarm signals anyone nearby if a person is exiting while there is no emergency to facilitate compliance with the organization's security policy.

Delayed egress locks must be interfaced with the fire alarm system so that it unlocks automatically in event of a fire alarm in the area. It is usual to also see a fire alarm pull station located next to the delayed egress door and to see a wall-mounted keypad or card reader to bypass the delayed egress door delay and alarm to allow for passage of authorized users.

■ **FIGURE 13.1** Delayed egress lock. *Image courtesy of Security Door Controls.*

■ **FIGURE 13.2** Special school QID lock. *Courtesy of Securitech.*

Specialize school locks to protect against active shooters

Securitech and some other lock manufacturers also make a special lock to protect against active shooters. The lock uses a red pushbutton on the classroom side, which instantly locks the deadbolt, prohibiting access to the classroom from the corridor side. The lock can be quickly unlocked by a key by school officials and first responders in an emergency, assuring that if the shooter is already in the classroom, they cannot hold the class hostage against first responders.

The red pushbutton is instantly recognizable for its locked/unlocked state. The lock functions as a normal classroom lock when the deadbolt is not thrown (Fig. 13.2).

HI-TOWER LOCKS

Security Door Controls manufactures the problem-solving Hi-Tower Lock (Fig. 13.3). This lock was originally intended to accommodate

■ **FIGURE 13.3** Hi-Tower lock. *Image courtesy of Security Door Controls.*

stairwell locking for access control, but its uses far exceed that application. The lock has the advantage of accommodating virtually any wood or hollow metal door application, including those where electric strikes, magnetic locks, and electrical bolt locks are not permitted.

For stairwell applications, electrified locksets and electrified exit devices are the only types of locks that can effectively lock the door and comply with fire and Life-Safety codes. Electric strikes are not acceptable because they do not latch when unlocked. Magnetic locks are not acceptable because they require a separate request-to-exit device to release the lock.

The Hi-Tower mortise lock is also unique because it does not require coring of the wood door or a cross-tube in a hollow metal door to accommodate wiring from an electric hinge to the mortise lock. The lock is a specially adapted standard mortise lock (available using many different manufacturers' lock bodies). The adapted mechanical mortise lock engages a specially designed "Frame Actuator Controlled" strike plate that has a small solenoid operated push button that sits flush with the strike plate when the door is open and retracts when the door is closed and locked. When the lock is commanded to release the latchbolt, the frame actuator pushes the corresponding button on the mechanical lock body, which releases the exterior handle to turn, enabling it to open the door. The inner door handle allows free mechanical egress at all times.

Hi-Tower locks will work in almost any wood or hollow metal door application, requiring minimal modification to existing doors and frames while providing a fail-secure application that can easily pass the review of just about all fire code officials.

CRL-BLUMCRAFT PANIC HARDWARE

Blumcraft of Pittsburgh is a company that makes tempered glass doors and panic exit devices for tempered frameless glass doors. These are beautiful and elegant locks designed to fit on beautiful and elegant doors.

Frameless glass doors are often used as the main entry doors of Class A commercial buildings. The doors are made from thick tempered glass plates usually having only a thin steel bottom and top rail. Front entry doors get a lot of use, so it is important to have hardware that allows for a great deal of activity and strength (consider durability).

A common way to secure frameless glass doors is by using magnetic locks; however, when panic devices are appropriate, rather than using conventional panic hardware on these beautiful doors, which would look out of place, CRL-Blumcraft of Pittsburgh manufactures a variety of panic bars that are stylish, strong, function well, and are mechanically and electrically lockable.

The first set shows variations of the top-latching Model DB-100 Series. The second variation shows the bottom-latching Model DB-110 Series. For the top-latching DB-100 Series the handles appear to be continuous, but the operating portion of this device pivots at the top and middle of the door hinge side, operating the mechanism by slightly moving the handle over the actuator post, which is just above the exterior pull bar. This retracts the latchbolt, allowing the door to open. When the door closes, the bolt relatches automatically. All Blumcraft top-latching panic handles have a built-in dogging device, and are compatible with top-mounted electric strikes to allow for operation by an electronic access control system. Blumcraft's bottom-latching DB-110 Series handles are not compatible with electric strikes.

Blumcraft of Pittsburgh was recently purchased by C.R Laurence Co., Inc., which now markets these products under the name of CRL-Blumcraft.

CHAPTER SUMMARY

1. If the need involves both security and safe egress to protect a room or corridor with important assets or higher security risks, a solution exists in the form of the Von Duprin Series HS99 Devices. This lock

combines a rim exit hardware device with a Sargent and Greenleaf 8470 Automatic Dead Bolt. The result is a high-security lock that provides single-action emergency egress from the inside.

2. The first Securitech locks used very robust mechanisms and clutches to allow the handle to rotate past its apparent limit without breaking it mechanically. When returned to its normal position by only the twist of the hand, it continued to function normally, thus confounding the most determined vandals.

3. Today, Securitech also makes very high security electrified locksets that meet fire codes. These include electrified multipoint dead-bolt locks that include panic bars, fitting the need for emergency egress on very high security doors.

4. Where high security is required on a seldom-used required exit door, the answer may be to use a Delayed Egress Lock. NFPA 101 makes provision for this type of lock.

5. Security Door Controls manufactures the problem-solving Hi-Tower Lock. This lock was originally intended to accommodate stairwell locking for access control, but its uses far exceed that application. The lock has the advantage of accommodating virtually any wood or hollow metal door application, including those where electric strikes, magnetic locks, and electrical bolt locks are not permitted.

6. Blumcraft of Pittsburgh (now CRL-Blumcraft) is a company that makes tempered glass doors and panic exit devices for tempered frameless glass doors. These are beautiful and elegant locks designed to fit on beautiful and elegant doors.

Q&A

1. *Electrified Dead-Bolt-Equipped Panic Hardware*
   a. Can be used on any room with an occupancy of 50 or more with a variance
   b. Can be used on any room with an occupancy of 50 or more without a variance
   c. Cannot be used on any room with an occupancy of 50 or more
   d. Cannot be used in areas occupied by handicapped people

2. *Securitech locks manufacture*
   a. Electrified cylinder locks
   b. Simple magnetic locks
   c. Locks that can be used on frameless glass doors
   d. Electrified multipoint dead bolt locks that include panic bars

3. *Delayed Egress Locks are approved by*
   a. Every fire marshall
   b. Only in California
   c. NFPA 80
   d. NFPA 101

4. *A Delayed Egress Lock comprises*
   a. A magnetic lock, a countdown timer, an exit sense device, and a local alarm
   b. A magnetic lock, a watch-dog timer, an exit push button, and a local alarm
   c. A magnetic lock, a watch-dog timer, a keypad, and a local alarm
   d. A mortise lock, a watch-dog timer, an exit sense device, and a local alarm

5. *Delayed Egress Locks*
   a. Count down exactly 30 seconds
   b. Delay exit through a fire egress door for a predetermined time period
   c. Should not be used on fire egress doors
   d. None of the above

6. *Hi-Tower Locks*
   a. Were originally designed to provide locking access to fire-rated stairwell doors
   b. Were originally designed for frameless glass doors
   c. Were originally designed for use on doggie doors
   d. None of the above

7. *Hi-Tower Locks*
   a. Require the door to be cored for their wiring
   b. Do not require the door to be cored for their wiring
   c. Require a separate magnetic lock
   d. None of the above

8. *Hi-Tower Locks*
   a. Can be used on glass doors
   b. Can be used on hollow metal doors
   c. Can be used on solid core doors
   d. Can be used on either hollow metal or solid core doors

9. *CRL-Blumcraft Panic Hardware*
   a. Was designed to provide locking access to fire-rated stairwell doors
   b. Was designed for frameless glass doors
   c. Was designed for use on doggie doors
   d. None of the above

10. *CRL-Blumcraft Panic Hardware*
   a. Is available for hollow metal doors
   b. Is available for bottom latching
   c. Is available for top latching
   d. Is available in either top or bottom latching

Answers: (1) a; (2) d; (3) d; (4) a; (5) b; (6) a; (7) b; (8) d; (9) b; (10) d.

14

Selecting the Right Lockset for a Door

CHAPTER OBJECTIVES

1. Learn the basics in the chapter overview
2. Get to know standard application rules
3. Look at several sample examples of how to select locks for various kinds of doors
4. Answer questions about standard door/lock combinations

CHAPTER OVERVIEW

I get questions regularly from consultants, integrators, architects, installers, and repair technicians about which hardware is appropriate for different kinds of doors. Accordingly, for this book I have developed a number of lists that identify proper combinations of doors and locks for different applications (subject to local municipality codes, which always prevail. This rule applies throughout this book. Local municipality codes always prevail.).

The lists will include information on Hollow Metal, Solid Core, Framed Glass, Frameless Glass, and Specialty Doors. Variations will include fire-rated doors and normally handed and reverse handed doors for each type of lock, with notes on preference.

STANDARD APPLICATION RULES

Here are the priorities I suggest (subject to local municipality codes, which always prevail):

1. As always, life safety first.
2. For fire doors on corridors and occupancies 50 and over—Electrified Panic Hardware
3. For fire doors off corridors and occupancies less than 50—Electrified Mortise Locks

Electronic Access Control. DOI: http://dx.doi.org/10.1016/B978-0-12-805465-9.00014-2

4. For nonfire doors:
   a. Electrified Mortise Locks
   b. Magnetic Locks with safety provisions

These priorities will be adjusted by door type in the list below.

HOW TO SELECT THE RIGHT LOCK FOR ANY DOOR

The variables to consider when choosing door-locking hardware include:

- Is the door fire rated?
- What kind of door is it?
- What is the level of traffic through the door?
- Is the door standard or reversed swing? (A standard door opens toward you from the outside while a reverse swing door opens toward you from the inside.)

Let's try an exercise to see how this works. First we will describe a door and begin with all of the possible types of locks available, eliminating some each time we describe an attribute of the door. This will leave us with a shortlist of possible locks to use on the door.

Description of door

This is a pair of double doors leading to an information technology server room, off a fire corridor. The room has occupancy of less than 50. Frequently, heavy carts are moved into and out of the room by two or more people. Accordingly, the door is opened by an automatic operator. The room is equipped with an FM-200 Fire Suppression System, so quick egress is required. The doors are recessed into a pocket so that the doors can swing out toward the hallway without extending into the hallway.

First, let's begin with the entire list of possible locks including those that we will not use:

- Electrified Panic Hardware
  - Rim
  - Mortise
  - Vertical Rods
- Electrified Mortise Lock
- Electrified Cylinder Lock
- Electric Strike
- Magnetic Locks
- Three-point Locking Mechanism
- Securitech Locks
- Hi-Tower Lock
- Blumcraft Lock

Now, let's analyze the selection process, step by step:

1. The doors are fire-rated doors, so no electrified strike and better not to use magnetic locks.
   a. Electrified Panic Hardware
      i. Rim
      ii. Mortise
      iii. Vertical Rods
   b. Electrified Mortise Lock
   c. Electrified Cylinder Lock
   d. Electric Strike
   e. Magnetic Locks
   f. Three-point Locking Mechanism
   g. Securitech Locks
   h. Hi-Tower Lock
   i. Blumcraft Lock
2. The doors are hollow metal, so no Blumcraft Lock.
   a. Electrified Panic Hardware
      i. Rim
      ii. Mortise
      iii. Vertical Rods
   b. Electrified Mortise Lock
   c. Electrified Cylinder Lock
   d. Electric Strike
   e. Magnetic Locks
   f. Three-point Locking Mechanism
   g. Securitech Locks
   h. Hi-Tower Lock
   i. Blumcraft Lock
3. The doors are automatically operated so any lock requiring coordination of the two door leaves is out (electrified mortise lock, electrified cylinder lock, 3-point locking mechanism, securitech locks, and hi-tower lock).
   a. Electrified Panic Hardware
      i. Rim
      ii. Mortise
      iii. Vertical Rods
   b. Electrified Mortise Lock
   c. Electrified Cylinder Lock
   d. Electric Strike
   e. Magnetic Locks
   f. Three-point Locking Mechanism

    g. Securitech Locks

    h. Hi-Tower Lock

    i. Blumcraft Lock

4. This leaves electrified panic hardware, so let's finalize the selection. The door is automatically operated, so no rim or mortise version.

    a. Electrified Panic Hardware

        i. Rim

        ii. Mortise

        iii. Vertical Rods

    b. Electrified Mortise Lock

    c. Electrified Cylinder Lock

    d. Electric Strike

    e. Magnetic Locks

    f. Three-point Locking Mechanism

    g. Securitech Locks

    h. Hi-Tower Lock

    i. Blumcraft Lock

5. The selection is electrified panic hardware with vertical rods (EPH-VR; Fig. 14.1) that are interfaced to the automatic operator to unlock the vertical rods before actuating the automatic operator and then reengaging after the doors are closed again by the operator.

Let's try a few more examples.

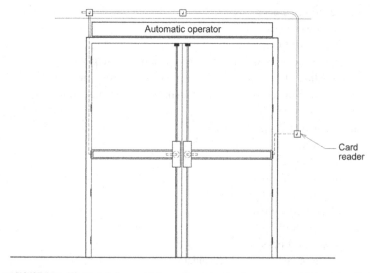

■ **FIGURE 14.1** EPH-VR on hollow metal doors with autooperator. *Image courtesy of Horton Automatics.*

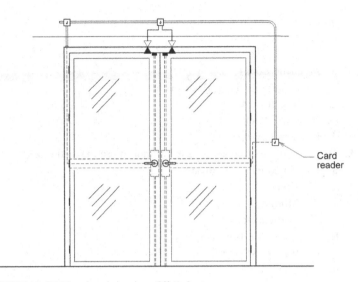

Card
reader

■ **FIGURE 14.2** EPH-VR on framed glass doors. *IRCO-VonDuprin.*

Framed glass door

This time let's look at a set of framed glass doors that leads from a company cafeteria to an outdoor employee patio (Fig. 14.2). The doors are unlocked from the outside by a card reader. These doors are from an area with occupancy of 200, so they are classed as fire exit doors.

1. This is a set of framed glass doors, so no Blumcraft Hardware.
   a. Electrified Panic Hardware
      i. Rim
      ii. Mortise
      iii. Vertical Rods
   b. Electrified Mortise Lock
   c. Electrified Cylinder Lock
   d. Electric Strike
   e. Magnetic Locks
   f. Three-point Locking Mechanism
   g. Securitech Locks
   h. Hi-Tower Lock
   i. Blumcraft Lock
2. Double doors must operate independently, so no coordinator is allowed.
   a. Electrified Panic Hardware
      i. Rim
      ii. Mortise
      iii. Vertical Rods

    b. Electrified Mortise Lock

    c. Electrified Cylinder Lock

    d. Electric Strike

    e. Magnetic Locks

    f. Three-point Locking Mechanism

    g. Securitech Locks

    h. Hi-Tower Lock

    i. Blumcraft Lock

3. These are fire exit doors, so it is best not to use magnetic locks.

    a. Electrified Panic Hardware

        i. Rim

        ii. Mortise

        iii. Vertical Rods

    b. Electrified Mortise Lock

    c. Electrified Cylinder Lock

    d. Electric Strike

    e. Magnetic Locks

    f. Three-point Locking Mechanism

    g. Securitech Locks

    h. Hi-Tower Lock

    i. Blumcraft Lock

4. The choice again is EPH-VR.

Herculite lobby doors

These are 10-ft tall, unframed glass doors in the main lobby of a Class A office building. The owner wants to make a good impression on everyone entering the lobby, but he wants to lock the doors after hours, allowing only those who have an access card to enter.

1. First let's eliminate all locks that do not work with frameless glass doors.

    a. Electrified Panic Hardware

        i. Rim

        ii. Mortise

        iii. Vertical Rods

    b. Electrified Mortise Lock

    c. Electrified Cylinder Lock

    d. Electric Strike

    e. Magnetic Locks

    f. Three-point Locking Mechanism

    g. Securitech Locks

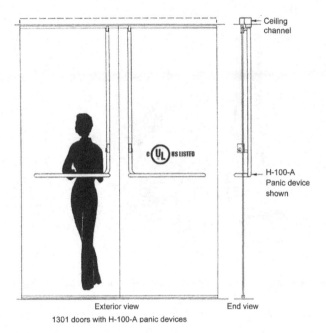

Ceiling channel

H-100-A Panic device shown

Exterior view

End view

1301 doors with H-100-A panic devices

■ **FIGURE 14.3** Blumcraft door hardware. *Photo courtesy of CRL Blumcraft.*

   h. Hi-Tower Lock

   i. Blumcraft Lock

2. This leaves magnetic locks, EPH-VR, and Blumcraft locks. Since the owner wants the doors to make the best impression, we will select the Blumcraft electrified door hardware (Fig. 14.3).

High-rise building stair-tower door

Next, let's look at a stair-tower door in a high-rise building (Fig. 14.4). The stair-tower is pressurized and the door is hollow metal. The door will be electrically locked to permit reentry only to authorized personnel. The door leads from occupancy of less than 50.

1. This is a hollow metal door so let's eliminate all locks that work only on glass doors.

   a. Electrified Panic Hardware

      i. Rim

      ii. Mortise

      iii. Vertical Rods

   b. Electrified Mortise Lock

   c. Electrified Cylinder Lock

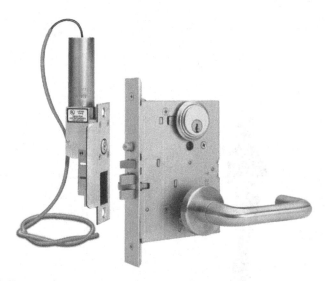

■ **FIGURE 14.4** High-rise stair-tower door with Hi-Tower lock. *Image courtesy of Security Door Controls.*

    d. Electric Strike
    e. Magnetic Locks
    f. Three-point Locking Mechanism
    g. Securitech Locks
    h. Hi-Tower Lock
    i. Blumcraft Lock
2. There is a stair-tower door so it must latch on closure.
    a. Electrified Panic Hardware
        i. Rim (can be dogged, so not acceptable)
        ii. Mortise
        iii. Vertical Rods (can be dogged, so not acceptable)
    b. Electrified Mortise Lock
    c. Electrified Cylinder Lock
    d. Electric Strike
    e. Magnetic Locks
    f. Three-point Locking Mechanism
    g. Securitech Locks
    h. Hi-Tower Lock
    i. Blumcraft Lock
3. Must be a robust electrified lock.
    a. Electrified Panic Hardware
        i. Rim
        ii. Mortise
        iii. Vertical Rods

    b. Electrified Mortise Lock

    c. Electrified Cylinder Lock

    d. Electric Strike

    e. Magnetic Locks

    f. Three-point Locking Mechanism

    g. Securitech Locks

    h. Hi-Tower Lock

    i. Blumcraft Lock

4. No dead-bolt is allowed.

    a. Electrified Panic Hardware

        i. Rim

        ii. Mortise

    b. Electrified Mortise Lock

    c. Electrified Cylinder Lock

    d. Electric Strike

    e. Magnetic Locks

    f. Three-point Locking Mechanism

    g. Securitech Locks

    h. Hi-Tower Lock

    i. Blumcraft Lock

5. This leaves us with the three choices as listed above. Perhaps we will select the Hi-Tower lock as it requires the least door modification.

Rear-exit door on warehouse with hi-value equipment

Now, let's take a look at a single, rear-exit door on a warehouse. This is a hollow metal door and the warehouse has been broken into a couple of times through weak door hardware. We are looking for something very robust. The door is a fire egress door.

1. Hollow metal door, so let's rule out all locks that will not work with these.

    a. Electrified Panic Hardware

        i. Rim

        ii. Mortise

        iii. Vertical Rods

    b. Electrified Mortise Lock

    c. Electrified Cylinder Lock

    d. Electric Strike

    e. Magnetic Locks

    f. Three-point Locking Mechanism

      g. Securitech Locks

      h. Hi-Tower Lock

      i. Blumcraft Lock

2. The door is a fire egress door, so let's rule out all locks that should not be used with these.

      a. Electrified Panic Hardware

          i. Rim

          ii. Mortise

          iii. Vertical Rods

      b. Electrified Mortise Lock

      c. Electrified Cylinder Lock

      d. Electric Strike

      e. Magnetic Locks

      f. Three-point Locking Mechanism

      g. Securitech Locks

      h. Hi-Tower Lock

      i. Blumcraft Lock

3. We need a very robust lock that can foil even the best attempts at entry.

      a. Electrified Panic Hardware

          i. Rim

          ii. Mortise

          iii. Vertical Rods

      b. Electrified Mortise Lock

      c. Electrified Cylinder Lock

      d. Electric Strike

      e. Magnetic Locks

      f. Three-point Locking Mechanism

      g. Securitech Locks

      h. Hi-Tower Lock

      i. Blumcraft Lock

4. This leaves us with the two choices listed above. The best choice would be panic hardware equipped electrified dead-bolts. Let them try to get past those locks!

Office suite door

In this selection, we have a 12-ft solid core wood door that leads to/from an office suite with occupancy of less than 50. The owner wants to maintain the original look of the door. The door swings out of the suite into the corridor.

1. First, let's remove all locks that would not work with solid core
   doors.
   a. Electrified Panic Hardware
      i. Rim
      ii. Mortise
      iii. Vertical Rods
   b. Electrified Mortise Lock
   c. Electrified Cylinder Lock
   d. Electric Strike
   e. Magnetic Locks
   f. Three-point Locking Mechanism
   g. Securitech Locks
   h. Hi-Tower Lock
   i. Blumcraft Lock
2. Now let's remove all locks that would not maintain the elegant look
   of the finely finished solid core wood door.
   a. Electrified Panic Hardware
      i. Rim
      ii. Mortise
      iii. Vertical Rods
   b. Electrified Mortise Lock
   c. Electrified Cylinder Lock
   d. Electric Strike
   e. Magnetic Locks
   f. Three-point Locking Mechanism
   g. Securitech Locks
   h. Hi-Tower Lock
   i. Blumcraft Lock
3. We need good security on this suite, so let's remove the weak locks.
   a. Electrified Panic Hardware
      i. Rim
      ii. Mortise
      iii. Vertical Rods
   b. Electrified Mortise Lock
   c. Electrified Cylinder Lock
   d. Electric Strike
   e. Magnetic Locks
   f. Three-point Locking Mechanism
   g. Securitech Locks
   h. Hi-Tower Lock
   i. Blumcraft Lock

4. Finally, this is an existing door on an occupied office suite, so the owner cannot remove the door to be cored for an Electrified Mortise Lock, so the final selection is a Hi-Tower Lock, which does not require coring the door.

Double-egress doors—hospital corridor

In hospitals, it is common to see double-egress doors that allow the passage of a gurney or cart in both directions. The doors will not have a center mullion, so there will be a clear expanse of space when both doors are open. These doors will get a lot of abuse as carts go through them, and each will be operated by its own automatic door operator. This is a new hospital.

1. This is new construction, so we can add Total Doors as an option.
   a. Electrified Panic Hardware
      i. Rim
      ii. Mortise
      iii. Vertical Rods
   b. Electrified Mortise Lock
   c. Electrified Cylinder Lock
   d. Electric Strike
   e. Magnetic Locks
   f. Three-point Locking Mechanism
   g. Securitech Locks
   h. Hi-Tower Lock
   i. Blumcraft Lock
   j. Total Doors
2. We need doors that will work with no mullions.
   a. Electrified Panic Hardware
      i. Rim
      ii. Mortise
      iii. Vertical Rods
   b. Electrified Mortise Lock
   c. Electrified Cylinder Lock
   d. Electric Strike
   e. Magnetic Locks
   f. Three-point Locking Mechanism
   g. Securitech Locks
   h. Hi-Tower Lock
   i. Blumcraft Lock
   j. Total Doors

3. The doors are in a fire corridor, so let's eliminate locks that should not be used in this environment.
   a. Electrified Panic Hardware
      i. Rim
      ii. Mortise
      iii. Vertical Rods
   b. Electrified Mortise Lock
   c. Electrified Cylinder Lock
   d. Electric Strike
   e. Magnetic Locks
   f. Three-point Locking Mechanism
   g. Securitech Locks
   h. Hi-Tower Lock
   i. Blumcraft Lock
   j. Total Doors
4. Since the doors will be in a very high abuse environment, let's use Total Doors, which are especially well suited to this need (Fig. 14.5).

Inswinging office door

This is a simple application. The office includes storage of important files and has occupancy of six workers. The door is a solid core wood door in a middle-management environment. It is an existing door in an aluminum frame, so frame strength is an issue. The door is existing, in use, and is currently equipped with a mechanical cylinder lock.

■ **FIGURE 14.5** Double-egress Total Doors. *IRCO-VonDuprin.*

1. Let's eliminate all locks that are not appropriate to this type of door (inswinging solid core wood door).
   a. Electrified Panic Hardware
      i. Rim
      ii. Mortise
      iii. Vertical Rods
   b. Electrified Mortise Lock
   c. Electrified Cylinder Lock
   d. Electric Strike
   e. Magnetic Locks
   f. Three-point Locking Mechanism
   g. Securitech Locks
   h. Hi-Tower Lock
   i. Blumcraft Lock
2. Let's consider that the frame is weak so let's not modify the frame if possible.
   a. Electrified Panic Hardware
      i. Rim
      ii. Mortise
      iii. Vertical Rods
   b. Electrified Mortise Lock
   c. Electrified Cylinder Lock
   d. Electric Strike
   e. Magnetic Locks
   f. Three-point Locking Mechanism
   g. Securitech Locks
   h. Hi-Tower Lock
   i. Blumcraft Lock
3. We want a strong lock as the information assets in this room are significant, so let's eliminate the electrified cylinder lock.
   a. Electrified Panic Hardware
      i. Rim
      ii. Mortise
      iii. Vertical Rods
   b. Electrified Mortise Lock
   c. Electrified Cylinder Lock
   d. Electric Strike
   e. Magnetic Locks
   f. Three-point Locking Mechanism
   g. Securitech Locks
   h. Hi-Tower Lock
   i. Blumcraft Lock

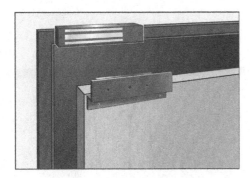

■ **FIGURE 14.6** Inswinging office door with magnetic lock. *Image courtesy of Security Door Controls.*

4. We don't have time to ship the door out to be modified to accommodate an electrified mortise lock, so we will select a magnetic lock with an infrared request-to-exit sensor and a pneumatic exit switch (Fig. 14.6).

Revolving door—emergency egress side door

This is a revolving door on a fourth floor corridor that leads directly from an elevator lobby to an adjacent parking structure. We need to ensure that visitors do not enter at this level and we want to ensure that we account for every person who comes and goes into the adjacent trading room. A revolving door has been recommended to deal with the accounting of people coming and going, but in this circumstance, fire code requires an adjacent exit door. The adjacent door is a framed glass door. The owner wants the adjacent door locked at all times, which is not allowed under fire code.

1. First let's look at which locks we have used in previous lists that fit both the fire code and the Owner's wishes.
   a. Electrified Panic Hardware
      i. Rim
      ii. Mortise
      iii. Vertical Rods
   b. Electrified Mortise Lock
   c. Electrified Cylinder Lock
   d. Electric Strike
   e. Magnetic Locks
   f. Three-point Locking Mechanism
   g. Securitech Locks

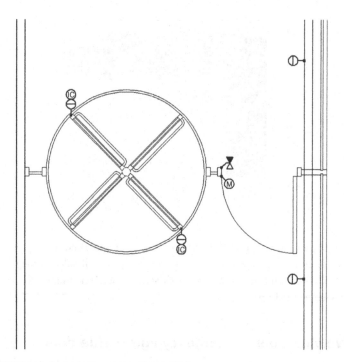

■ **FIGURE 14.7** Revolving door with delayed egress hardware.

    h. Hi-Tower Lock
    i. Blumcraft Lock
2. Oops! We can't use any of these. But if we look at fire code, we may find that a delayed egress lock is permitted. It complies with fire code and allows the door to remain locked at all times except in an emergency. The delayed egress hardware must be equipped with an interface to the fire alarm system (Fig. 14.7).

CHAPTER SUMMARY

1. For standard application rules in selecting locks, here are the priorities I suggest:
    a. As always, life safety first.
    b. For Fire Doors on Corridors and Occupancies 50 and over—Electrified Panic Hardware
    c. For Fire Doors off Corridors and Occupancies less than 50—Electrified Mortise Locks
    d. For Nonfire Doors

    i. Electrified Mortise Locks

    ii. Magnetic Locks with safety provisions

2. When selecting locks, the variables to consider when choosing door-locking hardware include:

    a. Is the door fire rated?

    b. What kind of door is it?

    c. What is the level of traffic through the door?

    d. Is the door standard or reversed swing? (A standard door opens toward you from the outside while a reverse swing door opens toward you from the inside.)

Q&A

1. *What is the first consideration for selecting the right lockset for a door?*

    a. For Fire Doors off corridors and occupancies less than 50—Electrified Mortise Locks

    b. For Fire Doors on corridors and occupancies 50 and over—Electrified Mortise Locks

    c. For Nonfire Doors: Electrified Mortise Locks and Magnetic Locks with safety provisions

    d. Life safety first

2. *Which of the following are criteria you should consider when selecting the right lock for a door?*

    a. Is the door fire rated?

    b. What kind of door is it?

    c. Both a and b

    d. Neither a nor b

3. *Which of the following are criteria you should consider when selecting the right lock for a door?*

    a. What is the level of traffic through the door?

    b. Is the door standard or reversed swing?

    c. Both a and b

    d. Neither a nor b

4. *A standard door swing*

    a. Opens away from you

    b. Opens toward you

    c. Swings to and fro

    d. Swings on hinges

5. *A reversed door swing*

    a. Opens away from you

    b. Opens toward you

    c. Swings to and fro

    d. Swings on hinges

6. *What kind of lock is best for a pair of double doors leading to an information technology server room, off a fire corridor?*

    a. Electrified Mortise Lock

    b. Blumcraft Hardware

     c. Hi-Tower Lock

     d. Electrified Panic Hardware with Vertical Rods

7. *What kind of lock is best for a set of framed glass doors that leads from a company cafeteria to an outdoor employee patio?*

     a. Electrified Mortise Lock

     b. Blumcraft Hardware

     c. Hi-Tower Lock

     d. Electrified Panic Hardware with Vertical Rods

8. *What kind of lock is best to use on 10-foot tall, unframed glass doors in the main lobby of a Class A office building?*

     a. Electrified Mortise Lock

     b. Blumcraft Hardware

     c. Hi-Tower Lock

     d. Electrified Panic Hardware with Vertical Rods

9. *What kind of lock is best to use on an existing hollow metal door in a pressurized stair tower within a high-rise building?*

     a. Electrified Mortise Lock

     b. Blumcraft Hardware

     c. Hi-Tower Lock

     d. Electrified Panic Hardware with Vertical Rods

10. *What kind of lock is best to use on a single, rear-exit door on a high-value warehouse that has been broken into several times in the past. The door is a fire egress door.*

     a. Magnetic Lock

     b. Panic Hardware equipped Multipoint Dead-Bolt Securitech Lock

     c. Electric Strike

     d. Electrified Mortise Lock

11. *What kind of lock is best to use on a 12-foot solid core wood door that leads to/from an occupied office suite with an occupancy of less than 50, where the owner wants a robust lock and to maintain the original look of the door and the owner does not want to core the doors?*

     a. Magnetic Lock

     b. Hi-Tower Lock

     c. Electrified Mortise Lock

     d. Electric Strike

12. *What kind of lock is best to use on double-egress hospital doors with no center mullion?*

     a. Electrified Panic Hardware with Vertical Rods

     b. Electrified Panic Hardware with Horizontal Rods

     c. Electrified Panic Hardware with Rim Hardware

     d. Total Doors

13. *What kind of lock is best to use on a solid core inswinging office door with an aluminum frame and occupancy of six workers where the lock is protecting important assets and the lock must be installed immediately?*

     a. Electric Strike

     b. Electrified Mortise Lock

     c. Hi-Tower Lock

     d. Magnetic Lock

14. *What kind of lock should be placed on a single-leaf fire egress door located right next to a revolving door, where we wish to guide people through the revolving door except in an emergency?*
    a. Securitech Lock
    b. Blumcraft Lock
    c. Delayed Egress Lock
    d. No lock is legal on this door

Answers: (1) d; (2) c; (3) c; (4) b; (5) a; (6) d; (7) d; (8) b; (9) c; (10) b; (11) b; (12) d; (13) d; (14) c.

Chapter 15

Specialized Portal Control Devices and Applications

CHAPTER OBJECTIVES

1. Learn the basics in the chapter overview
2. Get to know about specialized portals for pedestrians
3. Learn about automatic doors
4. Learn about electronic turnstiles
5. Learn about man-traps
6. Learn about full-verification portals
7. Understand specialized portals for vehicles
8. Learn about high-security portals
9. Learn about sally ports
10. Pass a quiz on specialized portal control devices and applications

CHAPTER OVERVIEW

On rare occasions, you will encounter some specialized portals for either pedestrians or for vehicles. Unlike conventional portals where an authorized user simply presents their credential to a credential reader and the portal opens automatically, passage through specialized portals can be more involved from a procedural and/or technology point of view.

We will examine a variety of specialized portals, beginning with the simplest—automatic doors and electronic turnstiles. In the case of automatic doors and electronic turnstiles, the complexity is not procedural, only technological. For man-traps and full-verification portals all have procedural as well as technological complexities.

Specialized portals for vehicles include high-security portals, which are procedurally simple, and sally ports, which are both technologically and procedurally more complicated.

Electronic Access Control. DOI: http://dx.doi.org/10.1016/B978-0-12-805465-9.00015-4

SPECIALIZED PORTALS FOR PEDESTRIANS
Automatic doors

Automatic doors come in a variety of configurations, but they all share one thing in common—they all have an automatic operator. Automatic doors are not complicated. The basic automatic door has a door, an automatic operator, and sensors with logic circuitry to initiate the automatic operator when someone approaches the door. Most older folks got their first introduction to automatic doors at a grocery supermarket.

Although basic automatic doors are simple, they become more complicated when coupled with electronic access control systems. Why? Because the door must lock until the automatic operator kicks into action. For the automatic operator to open the door, a mechanism must first unlock the door. This must be repeated in reverse order when the door is closed by the automatic operator. For doors with a higher security requirement, the coordination with the automatic operator can become significant. A typical automatic door interface for a high-security door can include:

- Receiving a door unlock signal from the access control panel
- Initiating a time delay sequence
- Sequence Step 1: Electrically retract panic hardware vertical rods
- Sequence Step 2: Sense that vertical rods are retracted
- Sequence Step 3: Energize the automatic operator
- Sequence Step 4: Monitor door opening for presence of a person in the opening
- Sequence Step 5: After no presence (all people have passed) initiate automatic door closing by the door operator
- Sequence Step 6: Look for door closure (door position switches)
- Sequence Step 7: Upon sensing of door closure, release the panic hardware vertical rods (relock the door)
- Sequence Step 8: Monitor vertical rod position to ensure that doors are latched

Most automatic operator manufacturers offer control circuitry for this purpose (Fig. 15.1).

Automatically operated access controlled doors may include:

- Single or double sliding doors
- Single or double swing doors
- Bifold or fourfold doors

■ **FIGURE 15.1** Automatic door with access control. *Image courtesy of Horton Automatics.*

Man-traps

A Man-trap is a sequence of two doors, enclosing a vestibule (Fig. 15.2). Each door is electrically locked and the two doors are interlocked such that only one can open at a time. This makes certain that the enclosed space always has one door closed, thus ensuring security for the enclosed space.

Man-traps require an interlocking circuit and special care to handle emergencies. Remember the basic mantra: life safety above everything. The interlocking circuit is as critical for life safety as is the life safety circuit itself. To that end, every man-trap must be equipped with both a fire alarm interface and should also have an emergency man-trap override switch within the secure space near the inner door. This may be in the form of a large red mushroom-shaped push button or as a special switch such as a fire pull-box (usually colored blue).

The basic man-trap interlock circuit interfaces to:

■ Door position switches on both doors
■ Electric locks (usually magnetic) on both doors
■ Entrance credential readers on both outer and inner doors (may include card reader, keypad, and/or biometric reader on outer door and usually only one credential reader on inner door)
■ Request-to-exit push button switches or credential readers
■ Redundant lock power supplies

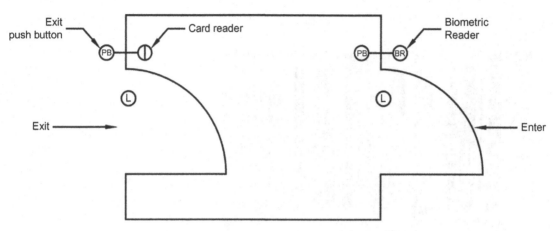

■ **FIGURE 15.2** Man-trap.

The interlock operates as follows:

- Both doors closed (quiescent condition)
- Entrance process:
  - System begins in quiescent condition
  - Authorized user presents credential(s) to outer credential reader(s)
  - Card is authorized by access control system
  - Access control system notifies interlock circuit to release outer door
  - Interlock circuit locks power to the inner door, ensuring that it cannot open while outer door is open
  - Interlock circuit unlocks the outer door
  - Authorized user enters the man-trap
  - Outer door closes behind authorized user who is now enclosed in the man-trap
  - Authorized user presents credential to inner door credential reader
  - Card is authorized by access control system
  - Access control system notifies interlock circuit to release inner door
  - Interlock circuit locks power to the outer door, ensuring that it cannot open while inner door is open
  - Inner door opens and authorized user enters secure area
  - Inner door closes behind authorized user
  - System returns to quiescent state

- Exit procedures:
  - System begins in quiescent condition
  - Authorized user approaches inner door to exit and presses exit push button
  - Access control system notifies interlock circuit to release inner door
  - Interlock circuit locks power to the outer door, ensuring that it cannot open while the inner door is open
  - Interlock circuit unlocks the inner door
  - Authorized user enters the man-trap
  - Inner door closes behind authorized user who is now enclosed in the man-trap
  - Authorized user presses outer door exit push button
  - Access control system notifies interlock circuit to release outer door
  - Interlock circuit locks power to the inner door, ensuring that it cannot open while outer door is open
  - Outer door opens and authorized user exits the secure area
  - Outer door closes behind authorized user
  - System returns to quiescent state
- Emergency operation:
  - System is in any condition
  - Emergency is initiated by the emergency pull station, emergency push button, or fire alarm interface, any of which will notify the interface circuit
  - Regardless of any other condition, the interface circuit drops power to both inner and outer doors, releasing them both for exit
  - All occupants within the secure space can exit freely.

Man-traps require fire code variances in every jurisdiction. Several electrified lock manufacturers manufacture off-the-shelf man-trap interface circuits that have been approved by various fire authorities. Although custom-made relay and programmable logic circuit devices have been used, from a dependability and liability standpoint most designers utilize an interface circuit that is manufactured for this specific purpose.

Full-verification portals

Full-verification Portals add a little something extra to man-traps (Fig. 15.3). The idea of a full-verification portal is that it adds visual verification to the process of admitting authorized users through a man-trap.

■ **FIGURE 15.3** Full-verification portal. *Image courtesy of Hirsch Electronics.*

As the user's credential is authorized by the electronic access control system, it initiates a display of the photo of the authorized user on a video monitor at a security post located at the man-trap. The person in the man-trap is compared against the photo of the authorized user on the computer monitor at the man-trap security post.

The security officer then releases the inner man-trap door using a door control panel within the security post at the man-trap.

This adds a high level of certainty to the operation of the man-trap for very high-security facilities.

Electronic turnstiles

"Positive access control" is when you want to be sure that every person entering a controlled or secured space is authorized to enter. One of the best ways to ensure positive access control is to use a turnstile that operates on the principle of one authorization, one grant of access.

■ **FIGURE 15.4** Paddle-type electronic turnstiles. *Image courtesy of Smarter Security Systems.*

While we are all familiar with the old-fashioned tripod style stadium turnstiles, electronic turnstiles are designed for commercial building environments. The most famous electronic turnstiles are those at the CIA headquarters building, which have appeared in countless movies and news photos. Electronic turnstiles come in three main varieties: simple electronic turnstiles, paddle-type turnstiles (Fig. 15.4), and glass-wing turnstiles.

All three types of turnstiles utilize a tunnel comprising a pair of stainless steel panels enclosing a short passageway (usually 1 to 1−1/2 m or 36−60 in.) in length. There is a card reader at each end on the right-hand side. In the case of the paddle-type and glass-wing type electronic turnstiles, the user is presented by a physical barrier barring their path. For the simple electronic turnstile, there is no physical barrier. An authorized user approaches the turnstile and presents their card. There is usually a red indicator that changes to green upon card authorization signaling the user to enter the turnstile.

With the paddle-type or glass-wing type electronic turnstile (Fig. 15.5), these elements open to allow passage. As the user transits through the turnstile, he/she breaks a series of infrared beams at ankle and thigh-high levels that track his/her progress through the turnstile.

If a person on the opposite side tries to enter from the wrong direction while the paddles or glass wings are open, sensors on that end of the turnstile sense the presence of an unauthorized person and a local alarm

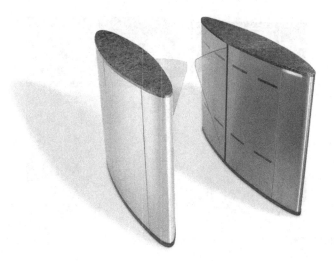

■ **FIGURE 15.5** Glass-wing type electronic turnstile. *Image courtesy of Smarter Security Systems.*

is sounded and the paddles or glass wings close to bar entry of the unauthorized person.

If a person attempts to enter the turnstile without authorization, the infrared beams sense this and sound a local alarm.

If an unauthorized person attempts to follow an authorized person through the turnstile while it is still open, the infrared beams sense this and the paddles/glass wings close immediately behind the authorized user stopping the unauthorized user. The local alarm is also sounded.

Care must be taken in the selection of the electronic turnstile to be certain that it will comply with the business culture of the organization and that it will operate fast enough to handle the peak traffic flow. A traffic flow study is critical to determine how many electronic turnstile lanes are required to maintain peak traffic flow.

Electronic turnstiles should be used in conjunction with a visitor management system (see Chapter 21: Related security systems) to ensure that visitors as well as employees have credentials to pass through the electronic turnstiles.

■ **FIGURE 15.6** Antitailgate alarm. *Image courtesy of Smarter Security Systems.*

Antitailgate alarm

A variation on the electronic turnstile is the antitailgate system, which uses a single column of infrared beams to sound a local alarm if two people pass through a door on a single card authorization (Fig. 15.6).

The antitailgate alarm interfaces to the access control system to acknowledge each card read by the card reader. Typically the access control panel door lock relay is set to zero seconds (i.e., it pulses but does not hold itself open), and this signals a circuit controlling the door lock. Upon that signal, the antitailgating system control circuit releases its own door lock relay, which holds the door lock unlocked for a longer time period, typically five to ten seconds. The antitailgate alarm is suppressed for one authorized user. The authorized user can turn the lever and enter the door. This time delay also allows the door to be open while a second authorized user presents his/her card allowing a second authorized user to enter without triggering the antitailgate alarm. Any number of authorized users can present their access cards and enter without triggering the antitailgating alarm. However, the first unauthorized person to try to "tailgate" behind an authorized person causes the alarm to sound.

SPECIALIZED PORTALS FOR VEHICLES

High-security barrier gates

High-security requires physical security. High-security barrier gates are used to prevent crash-through entries into secure facilities such as ports and nuclear facilities (Fig. 15.7). Crash-rated gates are rated by the US Department of State (DoS) and the US Department of Defense (DoD). The current standard is SD-STD-02.01, Vehicle Crash Testing of Perimeter Barriers and Gates, Revision A, dated March 2003. Crash-rated gates are available in four ratings:

- K4 barriers will stop a 15,000 pound (6810 kg) vehicle at 30 mph (45 kph)
- K8 barriers will stop a 15,000 pound vehicle traveling at 40 mph (60 kph)
- K12 barriers will stop a 15,000 pound vehicle traveling at 50 mph (80 kph)
- K54 barriers will stop a 65,000 pound (29,483 kg) vehicle traveling at 50 mph

Crash-rated gates actually carry two ratings, K/L, where the "K" indicates the DoS/DoD certified barrier's maximum vehicle impact speed and the "L" indicates the test vehicle's distance of penetration upon impact with the barrier (Fig. 15.8).

■ **FIGURE 15.7** K8 Lift-arm barrier gate. *Image courtesy of Delta Scientific.*

■ **FIGURE 15.8** K12 Phalanx Gate. *Image courtesy of Delta Scientific.*

- L1 = 20–50 ft (6–15 m)
- L2 = 3–20 ft (1–6 m)
- L3 = 3 (1 m) or less

When combined, a crash-rated gate is listed in both dimensions, such as K12/L3, which would stop a 15,000 pound vehicle traveling at 50 mph in 3 ft or less penetration (typically that would be the windshield flying out of the vehicle). To be DoS certified at any rating, the penetration of the cargo bed of a truck must not exceed 1 m beyond the preimpact inside edge of the barrier. Thus, most crash-rated gates are L3 gates (Figs. 15.9 and 15.10).

Crash-rated gates are available in a variety of designs including:

- Lift-arm barrier gates
- Sliding gates
- Web fabric gates
- Phalanx gates
- Rising bollards
- And a few other miscellaneous uncommon types

Sally ports

Sally ports are a secure, controlled entry-way for vehicles having two gates, similar to a man-trap in operation (Fig. 15.11). The entrance is

■ **FIGURE 15.9** Web fabric K12 gate. *Courtesy of Smith Wesson Security Solutions.*

■ **FIGURE 15.10** K12 sliding gate. *Image courtesy of Delta Scientific.*

typically part of an outer perimeter such as a secure fence line or prison perimeter wall, and the interior gate is also fortified against easy entry. The purpose of a sally port is to deter, defend, and delay against unwanted entry (or exit). The name sally port comes from two words: port from the Latin word *portus* meaning door and sally from the Latin *salire* (to jump) or sortie, a military maneuver designed to delay and

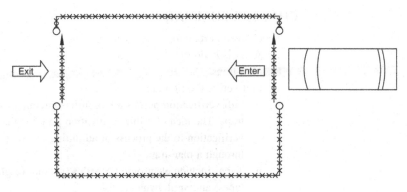

■ **FIGURE 15.11** Sally port.

harass an opposing force. A sally port is a delaying area in which an opposing force can be trapped to permit the friendly force to overwhelm and stop the offender. Gates on a sally port are usually wider and higher than normal vehicle gates to provide for greater security.

Sally ports are usually operated remotely by a security officer from a separate secure space where they can supervise the entry and exit. Only one sally port door is ever opened at a time to ensure that no one can sneak in (to a military compound) or out (for a prison). The remote security officer will have some method to verify the credentials of the people requesting passage through the sally port and can supervise the passage in its entirety from their remote location, typically using video cameras or another guard with a radio.

Sally ports require caution to prevent a person from being trapped during an emergency. Emergencies could include a fire at the facility or a security event related to the sally port itself. The size of the middle space must be adequate for the largest vehicles that could ever transit the port. The middle space is typically much wider than a normal vehicle lane to permit a thorough search of the vehicle seeking passage. It is common to require all occupants of vehicles to stand down and be searched before being authorized to proceed. Vehicles are also searched completely including inside, under hoods, within trunks, and under the vehicle. Accordingly, it is common for sally ports to have multiple entries and exits all controlled to allow for the safety of any occupants during any type of emergency. If all is not as it should be, the sally port is locked down and unauthorized users are taken into custody. Sally ports are also used in maximum security prisons to control passage from one area to another of a quantity of prisoners, similar to a man-trap, but for a much larger number of pedestrians, and completely under the control of a remote security officer.

CHAPTER SUMMARY

1. Specialized portals for pedestrians include:
   a. Automatic doors
   b. Man-traps
   c. Full-verification portals
      i. Full-verification portals add a little something extra to man-traps. The idea of a full-verification portal is that it adds visual verification to the process of admitting authorized users through a man-trap.
      ii. Types include electronic turnstiles (paddle- or glass-wing types) and antitailgate alarm
2. Specialized portals for vehicles include:
   a. High-security barrier gates: The current standard is SD-STD-02.01, Vehicle Crash Testing of Perimeter Barriers and Gates, Revision A, dated March 2003. Crash-rated gates are available in four ratings:
      i. K4 barriers will stop a 15,000 pound (6810 kg) vehicle at 30 mph (45 kph)
      ii. K8 barriers will stop a 15,000 pound vehicle traveling at 40 mph (60 kph)
      iii. K12 barriers will stop a 15,000 pound vehicle traveling at 50 mph (80 kph)
      iv. K54 barriers will stop a 65,000 pound (29,483 kg) vehicle traveling at 50 mph
   b. Crash-rated gates actually carry two ratings, K/L, where the "K" indicates the DoS/DoD certified barrier's maximum vehicle impact speed, and the "L" indicates the test vehicle's distance of penetration upon impact with the barrier.
      i. L1 = 20−50 ft (6−15 m)
      ii. L2 = 3−20 ft (1−6 m)
      iii. L3 = 3 (1 m) or less
   c. Common types of high-security barrier gates include:
      i. K8 Lift-arm Barrier Gates
      ii. K12 Phalanx Gate
      iii. K12 Web Fabric Gate
      iv. K12 Sliding Gate
   d. Sally ports are a secure, controlled entry-way for vehicles having two gates, similar to a man-trap in operation.

Q&A

1. *Automatic doors all share one thing in common. They all have:*
   a. Sliding panels
   b. Swinging panels
   c. Pneumatic operators
   d. Automatic operators
2. *Automatic doors become more complicated when coupled with:*
   a. Sliding Panels
   b. Swinging Panels
   c. Access Control Systems
   d. Pneumatic Operators
3. *Automatic doors must be kept locked until:*
   a. The pneumatic operator kicks into action
   b. The hydraulic operator kicks into action
   c. The automatic operator kicks into action
   d. The user passes through the door
4. *For the automatic operator to open the door a:*
   a. Mechanism must first unlock the door
   b. Hydraulic operator must first unlock the door
   c. Pneumatic operator must first unlock the door
   d. Security guard must first unlock the door
5. *Automatically operated access controlled doors may include:*
   a. Single or double sliding doors
   b. Single or double swing doors
   c. Bifold or fourfold doors
   d. All of the above
6. *A man-trap is a*
   a. Large steel cage with a trap door
   b. Sequence of two doors, enclosing a vestibule
   c. Sequence of three doors, including a trap door
   d. Room with an interrogator, a bright light, and a chair
7. *Man-traps require:*
   a. An interlocking circuit and at least four access control readers
   b. An interlocking circuit and special care to handle emergencies
   c. An interlocking circuit and a fire suppression system
   d. Panic Bars on both doors
8. *Every man-trap must be equipped with:*
   a. A fire alarm interface
   b. An emergency man-trap override switch within the secure space near the inner door
   c. Both a and b
   d. Neither a nor b

9. *Full-verification Portals add:*
   a. Visual verification to the process of admitting authorized users through a man-trap
   b. Video verification using automated software to verify the identity of the user
   c. An additional card reader to exit the man-trap
   d. None of the above

10. *Electronic turnstile types include:*
    a. Tripod turnstiles
    b. Paddle turnstiles
    c. Glass-wing turnstiles
    d. All of the above

11. *Electronic Turnstiles provide:*
    a. Negative access control
    b. Positive access control
    c. Additive access control
    d. Multiplying access control

12. *Antitailgate alarms always have:*
    a. Infrared beams across a door, a logic circuit, and an audio alarm
    b. Paddles
    c. Glass wings
    d. Excellent acceptance by the users

13. *High-security vehicle barrier gates are used to:*
    a. Allow large trucks to enter securely
    b. Prevent employees from using secure entries
    c. Prevent crash-through entries into secure facilities
    d. Establish the "look" of a secure facility

14. *Crash-rated gates are available in four ratings:*
    a. K2, K4, K6, and K8
    b. K4, K8, K12, and K24
    c. K4, K8, K12, and K48
    d. K4, K8, K12, and K54

15. *Crash-rated gates actually carry two ratings, K/L, where the "K" indicates the DoS/DoD certified barrier's maximum vehicle impact speed, and the "L" indicates the test vehicle's distance of penetration upon impact with the barrier. Which rating below is not correct?*
    a. L1 = 20–50 ft (6–15 m)
    b. L2 = 3–20 ft (1–6 m)
    c. L3 = 3 (1 m) or less
    d. L4 = 1 (0.3 m) or less

16. *Which type of gate is not available for crash-rated gates?*
    a. Lift-arm barrier
    b. Sliding
    c. Rising finger
    d. Web fabric

17. *Sally Ports are:*

   a. A secure, controlled entry-way for vehicles having two gates, similar to a man-trap in operation

   b. A secure, uncontrolled entry-way for vehicles and pedestrians having at least three gates

   c. A secure entry-way for vehicles and pedestrians that uses only automated gates

   d. A secure, controlled entry-way for vehicles only that uses only automated gates

Answers: (1) d; (2) c; (3) c; (4) a; (5) d; (6) b; (7) b; (8) c; (9) a; (10) d; (11) b; (12) a; (13) c; (14) d; (15) d; (16) c; (17) a.

Industry History That can Predict the Future

CHAPTER OBJECTIVES

1. Get to know the basics in the chapter overview
2. Learn about the first generation
3. Understand the second generation
4. Learn about the short-lived third generation
5. Learn how the industry matured in the fourth generation
6. Learn why the future is with the fifth generation systems
7. Answer questions about how access control history affects design, installation, and maintenance

CHAPTER OVERVIEW

There have been five generations of alarm/access control systems. With each movement from one generation to the next major changes in architecture and capabilities occurred. If one looks through the lens of history, one can see into the future through the pathway the industry has taken.

From the first generation to the fifth, there have been major increases in capabilities. Although the fifth generation has yet to realize its full potential (and for many manufacturers this may never happen), the promise of current technology is amazing. By the end of this chapter you will understand fully the capability of the current generation's technology.

Much of the information in this chapter is derived from Chapter 3, How electronic access control systems work, of my book Integrated Security Systems Design. For those readers who are interested in systems design, this book is a good reference manual.

Electronic Access Control. DOI: http://dx.doi.org/10.1016/B978-0-12-805465-9.00016-6

A LITTLE BACKGROUND

In my capacity as a consultant, I am approached frequently by manufacturers who want to explain their wares. As I question them on the capabilities of their systems I am intrigued that most manufacturers are not even aware that their technologies have crossed a generational border, nor are they aware of what promise their new platforms can bring. Many manufacturers have adopted a "me too" approach to system development and do not seem to understand that in their efforts to "keep up" with the development of competing systems, they now possess a system capable of far more than it delivers.

As we go through this chapter, you will begin to see how to unlock the hidden capabilities of many systems on the market. Then you will be able to make your system do things even their manufacturers did not know they could.

Does history repeat itself? Does history draw a vaguely linear line into the future? Can we learn anything at all about our career if we know the history of how it became what it is? The answer to all of those is a qualified yes.

The history of the access control industry has been one of continual consolidation into ever larger and larger companies held in fewer and fewer hands. To some extent this has been good for the industry, for integrators, and for clients. Has the technology gotten better? Most certainly it has. Does it do more? Yes it does. Is there that one system that can do everything? No, not yet. There are far fewer brands and models of alarm/access control systems on the market today.

Alarm and access control systems are in most cases part of a larger integrated security system that includes digital video, security intercom, 2-way radio, photo identification, visitor access system, and often other systems as well.

The alarm/access control system was, for many years, the heart of all integrated systems. This is not the case today. Alarm/access control system manufacturers have abdicated their leadership role to the digital video industry, which is also not yet fully aware of what their systems can do, especially when coupled with a powerful alarm/access control system.

The access control industry has evolved, in my opinion, along a line that has served the short-term interests of the access control system manufacturers well, but not the needs of the integrators and their clients. In the end, the access control system industry, which should have been the focus of all system integration, has abdicated that role to the digital video industry to the detriment of its clients, its integrators, and, ultimately, the access control system manufacturers themselves. The market always vindicates and often that happens with a vengeance.

In the end, the market share will go to the segment of the industry that provides the infrastructure for all other segments to operate under. That *was* the access control industry. But increasingly, I believe it *will be* the digital video industry.

FIRST GENERATION

In the beginning, there were guards. Or rather, I should say, in the beginning there were night watchmen. It was up to the receptionist and office staff to control access, such as it was. In most cases, it was possible for a total stranger to walk right up to the organization's president's office with no one stopping to ask anything. Those were simpler times with simpler crimes.

However, as night fell and all employees went home it was left to a night watchman to conduct his "rounds." He would walk around and through the building(s), looking for anything that was out of place. If he found an intruder, he would challenge him and sound a whistle that hopefully would fall on the ears of a police foot patrol.

The first McCulloh Loop telegraph-type alarm system was installed in Boston in 1882 (Fig. 16.1). Later systems worked by sending a

■ **FIGURE 16.1** McCulloh Loop system.

20 milliamp current down a loop of wire and then monitoring the current on the wire. If there were any changes in the current, an alarm meter would change state or a pen would move on a paper tape, signaling a change in state in the alarm circuit. These were widely used in police call boxes and fire pull stations. The first magnetic stripe access control cards appeared in the 1960s. Early intercom systems date back to the 1877, 1 year after Alexander Graham Bell patented the telephone ("From the Call Box to the Computer," an ITI Whitepaper on the Changing Face of Technology in Law Enforcement, Information Technologies, Inc.), and were first used in police call boxes. In 1961, the London police began using closed-circuit television (CCTV) to monitor activities in train stations (UrbanEye.net, "CCTV in London," Michael McCahill & Clive Norris, Working Paper No. 6.). All of these were discrete, individual systems. For example, there were no camera switchers, but each camera reported to an individual video monitor. Videotaping did not occur due to the expense. Alarm recording was done by hand notes and maintaining the McCulloh Loop paper tape.

The first generation of access control systems is still in use today, and is most often seen on hotel room doors (Fig. 16.2). These are stand-alone

■ **FIGURE 16.2** First generation access control reader.

card readers, each one controlling a single door. Although unconnected to any other card reader, and therefore not part of a network, first generation access control systems were a major step forward.

In the first generation of security technology, intercoms and CCTV were a rarity. CCTV systems usually comprised only a few cameras with each camera reporting to its own video monitor. There is a resurgence in first generation access control systems, primarily using a biometric reader or keypad built into an electrified door lock. These are marketed to upscale residential and light commercial users.

SECOND GENERATION

The second generation of access control systems networked eight card readers together to a dedicated computer that was about the size of a huge early electronic desk calculator (Fig. 16.3) (Examples of second generation access control systems include the Rusco 500, the early Cardkey 1200, and numerous Secom systems.). There were a pair of keypads (A–F and 0–9), a nixie tube display, and a 3-in. paper tape. When a person presented a card to the front door of a facility, you would hear the paper tape chatter and the nixie tubes would display something like 1CO3-AA. One would then refer to a book that would indicate that card CO3 was granted access to door 1.

The second generation of alarm systems replaced the impossible to read meters and paper tapes with colored lamps and (at last) an audible alarm (Fig. 16.4) (An example of a second generation alarm system is the Flair alarm panel.). Each alarm had three colored lamps: green for secure, red

■ **FIGURE 16.3** Second generation access control system.

■ **FIGURE 16.4** Second generation alarm panel.

■ **FIGURE 16.5** Third generation alarm/access control system terminal. http://www.old-computers.com/museum/computer.asp?c = 410&st = 1.

for alarm, and yellow when bypassed. There was a switch to bypass the alarms. The second generation began around 1945 and continues today, though it is rarely used now.

During this time CCTV systems were still little used, but intercom systems were becoming only a little more than obscure.

THIRD GENERATION

The third generation of alarm/access control systems began in 1968 and continued until about 1978. For the first time, third generation systems combined alarm and access control into one system (Fig. 16.5). Up to 64 card readers and up to 256 alarm points were wired home-run style to a Digital Equipment Company DEC-PDP-8 or IBM Series 1 mini-computer. These were often the size of a four-drawer filing cabinet. These computers often used core memory, a beehive terminal, and a line printer. A basic 16-card reader system could cost over a hundred thousand dollars. During this time, CCTV began to be used by corporations and there were a few instances of intercom systems being used to remotely help facilitate access control.

Most commands were operated by a combination of keystrokes that the operator had to be trained to remember. Early Cardkey systems used so-called "dot commands," wherein actions to be carried out by the system were initiated by the operator using a two-letter command preceded by a period (dot).

These were a far cry from later mouse-operated graphical user interface systems.

FOURTH GENERATION

In 1971, Intel introduced the first 4-bit microprocessor, the 4044, designed by Intel designer Ted Hoff for a Japanese calculator company (Patent #3,821,715, IEEE Global History Network, Ted Hoff: Biography, http://www.ideafinder.com/history/inventors/hoff.htm.). The processor boasted more than 2300 transistors and more switches than ENIAC. The ENIAC filled an entire room and required a dedicated air conditioning system just for the computer (Fig. 16.6). The Intel 4044 was an entire central processing unit on a single chip about one third the size of a conventional PostIT™ note. The Moss Technology 6502, Zilog Z-80 and Intel 8088 8-bit microprocessors soon followed. These were the basis for a new breed of alarm and access control system technology called distributed controller systems.

Up until 1974, each alarm and access control field device was wired individually back to the mini-computer, where a jumble of wires fed into custom-made circuit boards. All that wire was costly and often cost-prohibitive for most organizations. In 1974, one of the first distributed controller microcomputer-based alarm and access control systems was christened (By Cardkey). For the first time it brought the ability to multiplex alarms and card readers into controller panels and network panels together into a distributed system. The system comprised a series of individual alarm/access control panels, wired together on a proprietary network, all connected to a central mini-computer. This was a radical

■ **FIGURE 16.6** ENIAC computer. *US Army Photo.*

change. Finally, the cost of wiring, which was a major cost of early systems, was dramatically reduced, although the first fourth generation systems still terminated all of those controllers into a central mini-computer, computer terminal, and line printer (Early Cardkey 2000 system). The computers often had what I like to call a "user-surly" interface. When a person presented a card at the front door, the terminal still dutifully displayed something like "1CO3-AA" as the line printer chattered off the same message. The console officer looked that number up in a book. The system still had a long way to go.

Corporations begin using intercoms and CCTV more extensively, although prices were still prohibitive for most users, with basic cameras costing as much as $1200 each (RCA Vidicon camera typical pricing as of 1980.). Major advances in CCTV occurred during this period as well. For the first time, it was possible to use just a couple of video monitors to view many cameras, as the cameras were finally terminated into sequential switchers that switched the monitor view from camera to camera in a sequence (One of the first sequential switchers was made by Vicon.).

As the industry progressed, costs dropped dramatically and the systems became much more user-friendly. CCTV systems saw major advances. First, the advent of consumer videocassette recorders helped drop the price of video storage to practical levels. Toward the early 1990s, video was multiplexed by splitting the 30 frames per second that the recorder used between various cameras so that each camera was recorded two or more times per second onto a single tape. This economized video storage even further. In the alarm and access control system industry, during this time, mini-computers gave way to PCs and server-based networked systems.

The laggard was intercom technology, for which there was no single industry standard, little compatibility between manufacturers, and it was not generally friendly to networking between buildings or sites. Intercom usage was minimal in most systems, and usually reserved to assist access at remote gates and doors.

By the mid-1990s, however, the manufacturers realized that consultants and organizations wanted to integrate their various systems and began to integrate alarm, access control and CCTV, and intercoms into a truly integrated system. These systems could detect an intrusion, automatically call up an appropriate video camera to view the alarm scene, and, if an intercom was nearby, they could sometimes queue the intercom for response. They would also display the video and a map showing the area of the alarm to help the console operator understand more about what he was seeing on the video monitor.

These systems were able to detect and respond to emerging security events in real time. But the system interfaces were highly proprietary and clumsy. Achievement of an interface between any two brands and models of systems would usually not apply to any other brands or models. Each time integration was desired, it was always new turf.

Stalled progress

Many leading consultants believed that true systems integration would be performed by alarm and access control systems (Surprisingly, one of the earliest full integration platforms came from a small company in Branson, Missouri, named Orion Technologies (now Aegis Systems), who intended to create a universal front-end system that could control any combination of Alarm/Access Control Systems, Video Systems, and Intercom Systems. Other major video manufacturers quickly followed suit, leaving alarm/access control system manufacturers as systems to be integrated instead of as the primary integration platform.). But they were wrong. Significant progress stalled for over a decade from the mid-1980s until after 2000. There was a fundamental reason for this that goes back to the foundation of the industry. In the early days of fourth generation microprocessor-based systems, 64 kilobytes (KB) of random-access memory (RAM) cost over $1000 (Typical price for static RAM circa 1980.). With memory this expensive, the founders of the industry solved the problem by putting the "personality" of the systems into Erasable Programmable Read-Only-Memory (EPROMs). This meant that the systems functions were totally defined by the contents of their EPROMs. For the most part, their functions were not very field-programmable. By the end of the 1990s, some of the leading systems had a seemingly infinite parts list, with a part for every function. This was because the industry had failed to change with the times. By the year 2010, it was possible to buy 4 gigabytes (GB) of memory for about $100.00. That means that one could buy over 60,000 times the amount of memory for nearly one-tenth of the price of the 64 KB of RAM from 1980. That is like getting a fleet of 747 airplanes for the price of a pizza! As the computer industry surged forward in capabilities and dropped like a rock in price, some consultants and integrators considered that the access control system industry was desperately trying to maintain its margins and was not generally implementing advances in system architecture. Accordingly, at the same time that microcomputer costs were dropping dramatically, there were arguably no significant reductions in access control system costs. The access control system industry was also hanging onto a long-term strategy of keeping their architecture as proprietary as possible while clients were increasingly asking consultants to design nonproprietary, open-architecture systems. This strategy was doomed to fail. And fail it did, taking well-known access control system manufacturers with it, (most notably Schlage™, formerly a dominant access control system manufacturer, now almost completely forgotten).

Fourth generation alarm and access control system architecture functions were defined by their physical attributes. They did what they did because of the programming attributes of the EPROM. Their functions were controlled by their physical environment. Key strategic principle: It is important to understand that good security is achieved by the ability to control the environment. Any device that cannot control its environment outside of what the product development designer imagined that environment to be is inherently limited in its ability to serve the needs of the client in controlling appropriate behavior. If you cannot control the environment, the environment controls you.

The fourth generation of alarm and access control systems was doomed by its failure to adapt to emerging trends of network technology and trying to hold tenaciously to controller-based systems. Enter the fifth generation.

FIFTH GENERATION

The problem with alarm and access control systems was that their manufacturers thought they were making alarm and access control systems (you read that right). What they were actually making were programmable logic controllers (PLCs), which were equipped with an alarm and access control system database. This is an important distinction because, in their failure to understand this, they clung to EPROM architecture. While EPROMs solved the industry's early problems of high memory costs, as memory costs plummeted the industry confined their systems to the functions that its designers imagined for each box and implemented into the EPROM in the controller box. But corporate and government clients all had special needs that could only be met with changes in the logic structure of the controller. For example, if a client needed a local alarm to sound at a door if held open too long during the daytime, but wanted that door to alarm at the console immediately after hours, the client was sold an adapter box to perform that function for each door. Most manufacturers responded to such needs by building an endless list of adapter boards and boxes to make their system perform each unique need. For those manufacturers who did not make an endless assortment, there were third-party manufacturers who did.

This strategy served the manufacturing community well, but not the clients because of the additional hardware and custom equipment required to perform perfectly logical functions needed by the average client. For example, if a client wanted to have confirmation of perimeter alarms by two detection systems (requiring both to trigger to cause an alarm, but using each individually to cause an alert), most fourth generation systems required each detection system to have its own inputs (providing the alert: the "or" function), then the system would reflect the status of those inputs on output relays that would then be hardwired together to create "and" functions and those hard-wired outputs would be wired to additional inputs as "and" function alarms. This may not sound like a challenge, but for perimeter systems comprising fifty or more detection zones in each detection system, the cost of hardwire and wiring often ran thousands of additional dollars over what a PLC-based system would cost.

But a few industry experts understood that there is no function desired by any client that cannot be met by a simpler, not a more complicated, system architecture. Every imaginable function can be met by a simple combination of inputs, outputs, memory, logic cells, counters, and timers. Like dancing, where endless variety can be achieved with only a few basic steps, so too can an endless variety of functions be achieved with only these few basic logic objects. That structure defines PLC architecture. A few manufacturers introduced PLC-based alarm and access control systems, notably those who made building automation systems (BAS) that were already based on PLC architecture. But these products never made a significant impact in the marketplace because they continued to focus their marketing efforts on their BAS product lines instead of their access control system products.

The key to fifth generation alarm and access control systems is that they are all based entirely on software-based functions (mostly Second Query Language, SQL) and reference databases, whereas earlier systems were based on functions that were defined in hardware. From the late 1980s through the 1990s, however, the industry was driven to integrate multiple buildings and sites together. The industry struggled with a variety of unsuccessful system architectures until it gradually adopted a network infrastructure already in place in most businesses—the Local Area Network (LAN), Municipal Area Network (MAN), and Wide Area Network (WAN) Ethernet architecture—already in use in their computer networks. As the manufacturers began adapting their systems to Ethernet architectures, a convergence of security systems began to occur, bringing together alarm, access control, CCTV, and voice communication systems into a single integrated system.

That convergence continues today. In its next phase, distributed alarm and access control system controllers may totally disappear as CCTV manufacturers realize that they can totally integrate alarm and access and intercom functions into their CCTV Ethernet architecture. This resulted in a system comprised entirely of "edge devices" (cameras, intercoms, door access hardware, etc.), and a server/workstation for the user interface. These are interfaced together on the LAN.

Today's digital video cameras are already appearing with digital signal processing (DSP) chips and up to 64 megabytes (MB) of RAM. Soon, each individual door's access devices will be served by a single tiny controller in a junction box above the door. That tiny circuit board will comprise inputs, outputs, a card reader port, door lock output, and an exit device port, all served by a DSP chip and several megabytes of memory. A database of all users for that door will be contained in the DSP's memory board. A single centrally located controller may manage all of the microcontrollers in the entire building.

As these "edge devices" begin to connect directly to the LAN without an intervening controller, costs will drop dramatically. Although the industry seems terrified of this development, it will result in a substantially higher use of the systems. The industry need not fear this development. In the same way that CCTV manufactures feared the price drop below $1000 per camera, only to find that camera sales soared, so too will alarm and access control and intercom sales soar as the introduction of microcontrollers causes prices to fall. When these devices are combined with PLC programmability, which will allow an infinite variety of applications without an endless stock of hardware, the industry will finally have arrived. Before the end of this decade, the industry will very likely be served by only "edge devices" and software connected together on a Security LAN.

AVOIDING OBSOLESCENCE

Obsolescence is a funny thing. When we specify obsolete equipment, it does not seem so at the time. How can we tell what is going to be obsolete? How long is a good enough life cycle? What is the migration path to the future? There are signs.

Planned obsolescence

At one time, a major access control manufacturer had in their current line five (count 'em 5) different access control systems. Each was designed to serve a different market growth segment from a small system that could support only about 32 readers to a large system that could serve multiple sites (sort of) and could handle thousands of alarms, card readers, and so forth. None of those systems could migrate to the next. If a foolish consultant were to specify a smaller system from this manufacturer, and the client outgrew that system in say, only 2 years, then the client would have to abandon much of the capital investment and purchase a new system. That is built-in obsolescence. That company (Cardkey™) does not exist anymore. It was bought by a big corporate integrator. It is their problem now. (Oh yes, and now it is their clients' problem, too).

Unplanned obsolescence

The industry is undergoing a sea change as this book is written. Much of today's access control and video technology will not exist in any form before this decade ends. How do I know? From market forces, that's how. Some have argued that the industry has long been based on serving the manufacturers, not the clients. If that is true, information technology-based systems will largely end that. It is possible that someday the industry will manufacture edge devices and software and nothing else— no digital video recorders (DVRs) and no access control panels, nothing but small microcontrollers and digital cameras that connect field devices directly to and powered by the Ethernet, oh yes, and software. It is a good bet that all of those bent metal boxes that manufacturers love to sell will be obsolete soon. This is about to become a very thin market with very few manufacturers (but with lots of software vendors early on).

What the future holds

When I wrote *Integrated Security Systems Design* in 2006, I predicted that the alarm/access control system industry would universally adopt something that none of the manufacturers offered at that time. I introduced the idea of microcontrollers to the industry. These are small single- or double-door controllers that sit above a door, controlling that door only and having enough memory to store all of the users for that door and all of the functions needed at that door. Each microcontroller would communicate via TCP/IP to its host. This has now come to be. Today, many manufacturers offer microcontrollers.

I also predicted that microcontrollers would include a small four-channel digital switch used to connect one or two digital video cameras, an intercom, and another microcontroller down the hallway. So far no one has introduced that, but I still think it will happen because the market wants it.

You can create any function imaginable if you combine inputs, outputs, card reader interface, counters, timers, and access to the status and functions of all the system's software. I recently had a client request that the security console officers have no access to highly sensitive central bank counting room cameras when counting is under way and then regain access to those cameras when counting hours ended. We were able to accommodate an automated password function for the security console officers that automatically disabled and enabled access to the counting room cameras based on the hours of operation of the counting room. That is a powerful function.

I had another client request that when a computer programmer arrived late at night to work on an end-of-month database problem that lighting could be enabled to the server room and workstations area so that the programmer (often a lone female) would not have to transit all that distance in the dark. I was able to specify a system that could automatically light a pathway to and from her car to her workspace and also turn on the air conditioning to her office and set the coffee maker to brew a cup to meet her when she arrived, all based on the status of her parking structure access card as a programmer. Similar conveniences were also provided to late-working executives, building engineers, and so forth.

Information is power when it is coupled with a really good logic cell.

CHAPTER SUMMARY

1. First generation alarms used McCulloh Loop telegraph-type alarm circuits.
2. First generation access control systems were single-door assemblies and are still available as hotel room door locks.
3. The second generation of access control systems networked eight card readers together to a dedicated computer that was about the size of a huge early electronic desk calculator.
4. The second generation of alarm systems replaced the meters and paper tapes of the first generation alarm systems with colored lamps and (at last) an audible alarm.
5. The third generation of alarm/access control systems began in 1968 and continued until about 1978. For the first time, third generation systems combined alarm and access control into one system.
6. Up to 64 card readers and up to 256 alarm points were wired home-run style to a PDP-8 or IBM Series 1 mini-computer. These often used core memory, a beehive terminal, and a line printer.
7. In 1971, Intel introduced the first 4-bit microprocessor, the 4044. This became the basis for a new kind of alarm and access control system technology called distributed controller systems. This was the beginning of forth generation access control systems.
8. Fifth generation alarm and access control systems are all based entirely on software-based functions (mostly SQL language) and reference databases, whereas earlier systems were based on functions that were defined in hardware.

Q&A

1. *Alarm/access control systems are in most cases:*
   a. Part of the digital video System
   b. Used mostly to control visitors
   c. Part of a larger integrated Security System
   d. Separate systems on separate computers
2. *First generation alarm systems were:*
   a. McCallan line systems
   b. McCulloh Loop systems
   c. McDonalds alarm systems
   d. McWilliams alarm systems

3. *First generation access control systems can:*
   a. Provide access to no more than eight doors
   b. Provide access only to magnetic stripe cards
   c. Still be found on old buildings
   d. Still be found on hotel room doors

4. *Second generation alarm systems:*
   a. Replaced audio annunciators with colored lamps
   b. Replaced meters and paper tape with colored lamps
   c. Replaced most guards
   d. Replaced most guard dogs

5. *Second generation access control systems:*
   a. Networked 8 card readers together to a computer the size of a 4-drawer filing cabinet
   b. Networked 8 card readers together to a computer the size of a large early electronic calculator
   c. Networked 24 card readers together along a continuous coax cable
   d. Networked 24 card readers together along a common circuit at distances not to exceed 100 m

6. *Third generation alarm/access control systems:*
   a. Wired card readers and alarms to a computer the size of a 4-drawer filing cabinet
   b. Used plug-in dynamic random-access memory
   c. Used dot matrix printers
   d. Used color televisions as computer monitors

7. *Fourth generation alarm/access control systems:*
   a. Used discrete transistors for processing
   b. Used Nixie Tubes for displays
   c. Used the first microprocessors as CPUs
   d. Used highly educated guards to monitor them

8. *Fourth generation alarm/access control systems were the first to use the concept of:*
   a. Distributed processing
   b. Distributed card readers
   c. Distributed consoles and monitors
   d. Distributed power supplies

9. *By the mid-1990s, organizations wanted to integrate:*
   a. Their employees racially
   b. Their remote offices into many fewer locations
   c. Alarm, access control, CCTV, and intercoms into a single system
   d. Alarm, access control, business computer network, and irrigation systems into a single system

10. *Fourth generation alarm/access control system architecture functions were defined by:*
    a. Their logical attributes
    b. Their physical attributes
    c. Their programmers
    d. Their operators

11. *Fifth generation alarm/access control systems are based on:*
    a. Software-based functions (mostly SQL language)
    b. Hardware-based functions (mostly EPROMs)
    c. ANSI standards
    d. Archimedes principle
12. *In the past, several alarm/access control system manufacturers used _____ to ensure good sales as organizations grew.*
    a. A highly trained sales force
    b. A highly trained technical force
    c. Highly trained monkeys
    d. Planned obsolescence
13. *Microcontrollers:*
    a. Are small, single- or double-door controllers that communicate via TCP/IP
    b. Are small egg-shaped elements that control most functions in the system
    c. Are small battery-powered elements that control only the door lock
    d. Are small battery-powered elements that communicate via RS-485
14. *In the future, microcontrollers may include:*
    a. A video camera
    b. An intercom
    c. A four-port digital switch
    d. An eight-port digital switch

Answers: (1) c; (2) b; (3) d; (4) b; (5) b; (6) a; (7) c; (8) a; (9) c; (10) b; (11) a; (12) d; (13) a; (14) c.

17

Access Control Panels and Networks

CHAPTER OBJECTIVES

1. Get to know the basics in the chapter overview

2. Learn about panel components

3. Learn about panel functions

4. Learn about panel form factors

5. Learn about typical panel locations

6. Understand local and network cabling

7. Understand redundancy and reliability factors

8. Answer questions about access control panels and networks

CHAPTER OVERVIEW

Access control panels perform the heavy lifting in alarm/access control systems. Every single function that access control systems perform in the real world occurs first in the "mind" of an access control panel.

All access control panels share certain attributes and components. Alarm input and output control components may be internal or external to the basic access control panel. There are a variety of access control panel form factors that can be used in a variety of locations as system needs dictate.

In the past a wide variety of different types of wiring schemes and wiring protocols existed, but today most systems use one or two of only a few. Finally, we will discuss reliability and redundancy issues—how to get the most out of the hardware.

ACCESS CONTROL PANEL ATTRIBUTES AND COMPONENTS

You will remember that the second generation of access control systems was the first to wire all access control devices centrally. The purpose of

Electronic Access Control. DOI: http://dx.doi.org/10.1016/B978-0-12-805465-9.00017-8

this was to achieve both centralized control and reporting. Both second and third generation access control systems required that all devices be wired from wherever they were in the building all the way back to the location of the central processing unit (CPU). In the second generation this comprised a large object that looked like a giant desk calculator, and in the third generation this comprised a mini-computer the size of a four-drawer filing cabinet.

Early second generation systems had little if any storage. Their purpose was control, not recordkeeping, of just a very few doors. Third generation systems operated up to 64 doors. But both second and third generation systems suffered from the same limitation—the number of doors that could be wired back to the CPU. Although this presented a practical limitation in terms of wiring difficulty, it also presented a barrier because of the cost. Even in those times, copper and the labor to run it was expensive. Additionally, the reality of Ohm's Law meant that there was a practical limitation to the distance that doors could be placed away from the CPU.

The whole idea of fourth generation access control systems was to find a way to increase the capacity of the systems (numbers of doors connected) and decrease the amount of wiring necessary for each door, and increase the distance between the door and the central computer. I have done budgets that saw as much as 34% of the entire security system budget going to cable and conduit. Although necessary for the operation of the system, cable and conduit contribute nothing to functionality, so anything that can reduce their costs can drive down the overall cost of the entire security system without losing any functionality. Fourth generation access control systems were the first to "farm out" the access control decisions and operation of remote field devices (locks, etc.). To access control panels instead of connecting all of those devices centrally to a single computer that made all decisions centrally.

The constant themes of security system development have been and continue to be:

- More functions
- Easier to use
- Lower cost

With that in mind, let's look at what we have to work with in an alarm/access control system to achieve those goals. An alarm/access control system comprises the following elements:

- Field elements
  - Credential reader
  - Door lock

- ❏ Door position switch
- ❏ Request-to-exit sensor
- ■ Alarm/access control system panels
  - ❏ Access control panels
  - ❏ Alarm input boards
  - ❏ Output relay boards
- ■ Server(s)
- ■ Workstation(s)
- ■ Communications infrastructure
  - ❏ Communication boards (protocol specific—RS-485, etc.)
  - ❏ Digital switches
  - ❏ Cables and conduits
- ■ Software
  - ❏ Operating systems
  - ❏ Software

Of all these, the best places to achieve our three goals are:

- ■ More functions—system software and hardware interfaces
- ■ Easier to use—system software
- ■ Lower cost
  - ❏ System architecture (alarm/access control system panels)
  - ❏ Anything to reduce the need for cables (panel design)
  - ❏ We can also expect the cost of servers and workstations to continue to decline

We look at access control panel form factors next in this chapter, and you will see how panel design can dramatically affect cable and conduit costs.

So back to access control panels. Let's look at how their components and attributes contribute to their functions. Once this is understood, we can begin manipulating them to do more at lower costs.

Let's call the term "access control panel," a generic phrase that can include:

- ■ An electronics panel that can interface with and control access control system field devices (credential reader, electrified lock, door position switch, and request-to-exit devices)
- ■ An electronics panel that can interface with alarm devices (alarm input board)
- ■ An electronics panel that can control electrical devices (output control panel)

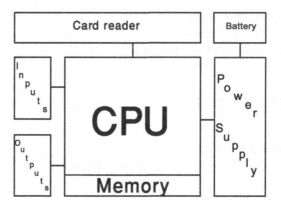

■ **FIGURE 17.1** Basic access control panel components.

The panel must communicate with the access control system. It must be able to make access control decisions locally based on information stored about the user presenting his/her credential and then after operating the access portal, and it must report the event to the master access control system database archive.

So the alarm/access control system panel includes all the elements shown in Fig. 17.1.

COMMUNICATIONS BOARD

The communications board provides the communications to and from the system server(s). It is wired to the servers using a digital protocol. The most common digital protocols used to connect access control panels to their servers include:

- Ethernet (TCP/IP)
- RS-485 (4-Wire)
- RS-485 (2-Wire)
- RS-232
- Protocol-B (Older Cardkey Protocol)

In addition to allowing communications with the server(s), the communications board may also allow downstream communications to other access control panels and/or a local connection to a diagnostic laptop computer. It is common to see one access control panel that connects using TCP/IP that connects to the server and then that same panel connects to a number of other access control panels using RS-485. This reduces demands on digital switches and extends the distance over which communications can take place.

There are several different types of cabling and networking used in the security industry. These are discussed later in this chapter (local and network cabling).

Power supply and battery

Each access control panel is equipped with a low-voltage power supply that converts mains ac power to low-voltage DC power. Access control panels are often also equipped with a back-up battery to ensure consistent power in case of a sudden voltage drop or other disturbance. The battery should be sufficient to power the panel and its locks for up to 4 hours.

Central processing unit

The CPU is the "brain" of the access control panel. This device performs the following functions:

- Receives downloaded data from server(s)
  - Time zones
  - Portal configurations
  - Database of authorized users for portals under the control of the access control panel including:
    - Users
    - Portals authorized
    - Time zones authorized
- Makes access control decisions
  - Receives request from the credential reader
  - Queries the on-board memory for the authorization for this user for that access portal
  - Makes the access control decision (grant/deny)
  - If the decision is to grant access:
    - Unlocks the door
    - Suppresses the door alarm
    - Relocks the door on closure
    - Resets the door alarm on door closure
- Receives alarm status information from the alarm inputs
- Directs the output control to activate a relay
- Communicates all event data to the on-board memory (first in-first out)
- Communicates all event data to the server(s)

Erasable programmable read-only memory

The Erasable Programmable Read-Only Memory (EPROM) is a data circuit that retains its memory even when power is off. Security system manufacturers use EPROMs to hold both the operating system and the program for the access control panel. The CPU performs the logical functions, but it gets its instructions from the EPROM.

The difference between fourth and fifth generation access control systems is the extent to which the EPROM determines what functions are performed and how they are performed. Fourth generation systems get all their instructions from EPROMs. Whereas fifth generation systems get some of their instructions from EPROMS and the balance of their instructions from the contents of Structured Query Language (SQL) databases (or their equivalent in another language than SQL).

This is an important distinction because fourth generation systems can perform ONLY the functions described by the EPROM, but fifth generations systems can perform virtually any function based upon the contents of the SQL database, plus the status of inputs and outputs, counters, and timers. Thus, virtually any imaginable function can be performed by a fifth generation access control system.

Random access memory

The access control panel needs random access memory (RAM) to store its decision resource information (time zones, access levels, user information, etc.) and its event history. In early fourth generation access control systems, this memory was limited due to cost. Today it is common to see systems that sport comparatively enormous memory.

The larger memory of fifth generation systems allows for SQL functions that earlier systems were incapable of performing, making those very capable machines.

Input/output interfaces

As capable as all of the previously mentioned components are, they are of no use if they cannot connect to devices in the real world. Most access control panels have the following input/output connections:

- Between 1 to 16 credential reader inputs
- Between 1 to 16 door position switch inputs
- Between 1 to 16 door lock control relays
- Between 1 to 16 request-to-exit sensors

- Between zero to 8 alarm input points
- Between zero to 8 output relay points

Increasingly, Inputs and outputs of all kinds are relegated off the main access control panel electronics "motherboard" and onto so-called "daughter" boards. Often the only difference between an access control panel that can control 2 readers and one that can control 16 is the connectors on the motherboard and the programming of its EPROM.

Additionally, there is a trend toward software rather than hardware interfaces to other systems. Long ago if one wanted to integrate an access control system to a video or intercom system, e.g., to trigger a camera in response to an alarm, it was always done by connecting a relay dry contact from the access control system to an input on the video system, or by connecting a dry contact on an intercom call button to an alarm input on the access control system. Today, this kind of interface is mostly accomplished by a software interface using a "software developers kit." Accordingly, several manufacturers have discontinued older "multiconnection" input and output boards in favor of smaller 8 input or output boards. At this time, there is a healthy black market for the older 32- and 48-point boards, which were far more economical.

Access control manufacturers are always looking for new ways to make a buck and as software is replacing hardware, some are relying on making the rarer hardware interfaces more costly.

Regardless, hardware interfaces have their benefits. The foremost problem with interfacing two systems from different manufacturers using software is that the interface is only as good as the latest software version from both manufacturers. There have been cases where software has been integrated in one version only to be abandoned in later versions, forever compelling the owners of the earlier version to hold onto their now outdated software versions or else lose the integration. Hardware interfaces do not have this problem. As long as the new software version still talks to the old hardware, you are golden. Manufacturers tend to support their own older hardware much longer than the software interfaces of other manufacturers.

ACCESS CONTROL PANEL FORM FACTORS

The architecture of the basic panel shown in the previous section can take many forms (Fig. 17.2). The most common include:

- 2, 4, 8, or 16 door connections
- Additional inputs and outputs on the main access control panel

■ **FIGURE 17.2** Typical alarm/access control panels. *Image courtesy of DSX Access Systems, Inc.*

- Input and Output boards moved off the main access control panel board as separate boards attaching to the access control panel or directly on the communications path.

If we remember that our goal is to increase functionality, ease of use, and reduce costs, how then can system access control panel form factors contribute to these goals?

Let's look at each goal:

- *Increase functionality:* This derives from the design of the CPU, the amount of memory available, and the functions defined by the EPROM.
- *Improve user experience:* Sorry, this is almost entirely a software function. However, there are a couple of brands of access control systems that label their access control panels so well that as-built documentation is virtually unnecessary in the field (Fig. 17.3). Those are a pleasure for installers and maintenance technicians to work with. (Thanks DSX!)
- *Reduce Costs:* Here is where system architecture can make a very large difference.

Virtually unlimited functionality can be achieved by allowing the integrator to access the status and function of every single access control system attribute through SQL commands and data fields.

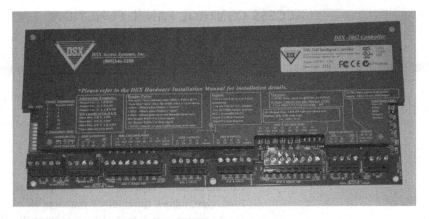

■ **FIGURE 17.3** Self-documenting 8 reader access control panel. *Image courtesy of DSX Access Systems, Inc.*

The access control industry is still in the "toddler" phase of fifth generation technology. A surprising number of access control system manufacturers have no idea what their access control systems are capable of when properly programmed. Quite understandably, these are the manufacturers that limit what system attributes and functions their SQL programming can access, and generally these systems also make it complicated to access their SQL programming at all.

This will all change as smart manufacturers begin to see profits from allowing their system integrators and end users to access the full capabilities of their systems. This process may take a decade.

When I wrote *Integrated Security Systems Design* in 2006, I predicted that a new type of access control panel would be introduced into the market. I called it a "microcontroller." I described a tiny access control panel that could fit into an electrical box above a door, designed to control just one door. I indicated that microcontrollers would be connected via TCP/IP to their servers. When I first introduced this idea, I was universally declared an idiot. Today, most major manufacturers have microcontrollers in their product line.

The Microcontroller has not fully matured. I predicted that the microcontrollers would not only operate a single door, but would also include a couple of alarm inputs and output control relays, and most important, include a 4-port 100Base-T Digital Switch that provided 2 ports for local devices (camera and intercom) and one each of uplink/downlink ports. Several microcontrollers include the alarm input and output control points, but none so far includes the digital switch. This will eventually change, because the market wants it.

ACCESS CONTROL PANEL FUNCTIONS

Access control panels have a number of basic functions that we have come to expect from anything defined as an access control panel; however, they can also provide many more functions.

Every access control panel *must* do the following to *be* an access control panel:

- Receive downloaded data from server(s)
  - Time zones
  - Portal configurations
    - Credential reader type/time zone for operation
    - Lock type, unlock time, relock function, etc.
    - Door position switch (normally open/normally closed)
    - Request-to-exit sensor (normally open/normally closed)
    - Portal operation time zone (otherwise unlocked or locked without any access possible)
  - Database of authorized users for portals under the control of the access control panel
    - List of authorized users (their credential number)
    - Portals authorized on this access control panel
    - Time zones authorized for each user on each portal on this access control panel
- Make access control decisions
  - Receive request from the credential reader
  - Query the on-board memory for the authorization for this user for that access portal
  - Make the access control decision (grant/deny)
  - If the decision is to grant access:
    - Unlock the door
    - Suppress the door alarm
    - Relock the door on closure
    - Reset the door alarm on door closure
- Receive alarm status information from the alarm inputs
- Direct the output control to activate a relay
- Communicate all event data to the on-board memory (first in-first out)
- Communicate all event data to the server(s)

Many access control panels also do the following:

- Communicate with other access control panels
- Communicate with outboard alarm input boards

- Communicate with outboard output relay boards
- Interlink with other access control panels to create a global antipassback function

Fifth generation access control panels can also:

- Allow access to all system functions and attributes through SQL or similar language.

True, fully developed fifth generation systems will exist in both the physical and virtual world. That is, for every device or attribute in the physical world (card reader, alarm input, output control, etc.) there will be an equivalent device in software that allows access to and from the physical device. Each device has attributes that can be accessed for status (open/closed, secure, alarm, bypass and trouble, card reader digital data stream, etc.), and each device has functions that can be executed (unlock, sound buzzer, light a lamp, etc.). In addition to all the devices in the physical world, the access control panel also has purely logical devices such as counters, timers and clocks.

Fully developed fifth generation systems blur the line between physical and logical worlds. Every alarm exists in both worlds and every output control exists in both worlds, so one can have both a physical alarm display on a workstation map and create a virtual alarm that can cause events to actuate, counters, timers and in logical combinations with other system device's status can execute very complicated logical actions. For example, on the first valid card read at the parking structure vehicle entry of a corporate building in the morning, the system will:

- Store a video image of the driver, the car, and the license plate in the access control system database along with the card reading event
- Set an antipassback event to prevent an authorized user from passing back their card to the next driver in line who is unauthorized to enter the employee parking area
- Set the employee door to unlock on reading the same authorized user's credential
- Turn on building core lighting
- Turn off building signage lighting (if also daylight)
- If a working day:
  - Turn on building HVAC
  - Turn on power to vending machines in employee lounges
  - Change PABX answering message from night to working day message

- If a weekend day or holiday:
  - ❑ Turn on power to vending machines in employee lounge nearest to this authorized user's work area
  - ❑ Turn on HVAC zone for this authorized user's work area
- If night:
  - ❑ Turn on path lights from parking structure to work area for this particular authorized user
  - ❑ Turn on HVAC zone for this authorized user's work area
  - ❑ Turn on power to vending machines in employee lounge nearest to this authorized user's work area
- If same authorized user leaves the parking structure:
  - ❑ Scan database to see if any other authorized users remain in the building
  - ❑ If no other authorized users during night, weekend, or holiday day:
    - − Sweep off HVAC
    - − Sweep off power to all vending machines
  - ❑ If other authorized users still in the building:
    - − Turn off HVAC for unused zones
    - − Turn off vending machine power to unused zones
  - ❑ Close out the antipassback event

In other words, virtually anything you can imagine, the system can do. Although fifth generation systems can possess truly awesome power, few do today because few manufacturers provide access to the full complement of system devices, attributes, and functions. This too will change.

ACCESS CONTROL PANEL LOCATIONS

Remembering that one of the key goals of access control system design is to minimize the cost of system installations, it is worthwhile to note that where you locate access control system panels greatly affects overall system cost.

Imagine for a moment that we have an access control system in a 10-story building with 100 access-controlled doors and that all of the access control panels are located in the same room with the server.

As you can imagine, the cost for cabling from access control panels to servers would be tiny, while the cost for cabling and conduit from access control panels to all 100 doors spread across a 10-story building would likely dwarf the cost of the access control system. The decision on placement of access control system panels has cost consequences.

As access control systems moved from older proprietary wiring schemes to TCP/IP Ethernet communication, this has opened up new opportunities for cost savings (more on this in the next section).

Generally, conventional old-style access control panels should be located in a secure location such as a telecom room. The room should have a locked door, and I recommend that it also have a card reader and door position switch on the door. I often design access control panels into rooms that include a camera facing the door with video motion detection to confirm the door opening. Access control panels should be located as near as possible to the doors they serve to minimize cable and conduit costs.

Each access control panel should be accompanied by a lock power supply. These must all be served by reliable power, preferably power from an uninterruptable power supply (UPS) that is backed up by an emergency generator. The access control panel should also be equipped with 4-hour battery back-up. All panels should be key-locked and equipped with a tamper switch.

There is a fine balance between reducing the number of credential reader inputs to minimize cabling and conduit costs, and understanding that access control panels that can accommodate fewer readers often cost more than those with a higher reader capacity. I suggest that a per foot (or meter) panel to portal cable/conduit cost should be calculated and then costs should be estimated for cabling and conduit based on maximum reader density panels, centrally located on each floor (or section) of a building. Then look to see if the additional cost of lower reader density panels is outweighed by the lower cost of cable and conduit if access control panels are distributed throughout the floor.

When distributed, security of the panels and lock power supplies may be maintained by placing the panels above the ceiling, accessible through removable ceiling tiles or ceiling access panels. Wherever the panels are located, they *should always* be located on the "secure" side of the wall, behind an access-controlled door, and *never* above the ceiling in a public or semipublic hallway.

Microcontrollers with integral digital switches offer the maximum potential cost reduction as they become commodities. With a typical microcontroller located in a junction box (j-box) just above the controlled door, microcontrollers provide the minimum possible wiring to each door.

Adding to that a digital switch-equipped microcontroller also reduces wiring costs for a nearby digital camera and digital intercom. The wiring to a microcontroller then becomes only two wires (Ethernet and power), with the added possibility for power over Ethernet, depending on the configuration. Microcontrollers should always be placed in a j-box over the secure side of the door, never on the unsecure side.

Note: One manufacturer, who I otherwise respect highly, makes a microcontroller that has an integrated card reader, thus placing the microcontroller in a j-box at waist height on the unsecure side of the door. These should be used with extreme caution as one can remove the card reader and have immediate access to lock power, and thus the ability to enter the door from the unsecure side with only hand-tools. It goes without saying that these should never, under any circumstances, be used on exterior doors.

LOCAL AND NETWORK CABLING

In the early days of access control systems, each manufacturer used its own wiring scheme, its own operating system, and its own communications protocol. This provided excellent security, since a "hacker" could not easily break into the system, but upon doing so, he would find data that he could not understand. This was especially true for those higher echelon systems that encrypted their data. However, this also made for completely proprietary systems that were impossible to interface with other security systems except by using alarm inputs, dry contacts, and RS-232 connections between computers of different systems.

Gradually, the security industry gravitated toward more universal communications protocols in an effort to minimize development and manufacturing costs. The most common communications protocols for access control systems became:

- RS-232
- RS-485 (2- or 4-Wire)
- RS-422
- And finally, TCP/IP
- *RS-232 (Recommended Standard 232):* This is the name for a series of serial binary single-ended data and control system signals connecting between Data Terminal Equipment (DTE) and Data Circuit-Terminating Equipment (DCE). Until recently it was the most common wiring scheme for computer serial ports (replaced by Universal Serial Bus, USB). It is still used extensively to communicate between access control panels and a local diagnostic laptop computer, and sometimes it is still used to communicate to a

system server, although this has been mostly replaced by TCP/IP Ethernet connections.

- *RS-485 (also known as TIA/EIA-485):* This is a standard that defines the electrical characteristics of drivers and receivers used in balanced digital systems. It is capable of point-to-point or point-to-multipoint wiring configurations. RS-485 is especially attractive because it can support communications across long distances as compared to Lantronics, RS-232, and Ethernet. The RS-485 standard only specifies the electrical characteristics of the transmitter and receiver. It does not specify the digital communications protocol. This factor made it highly attractive to access control system manufacturers who wanted to maintain their own proprietary communications protocols. RS-485 offers data transmission speeds of between 100 KB/second at 1200 m and up to 35 MB/second up to 10 m. Because it uses a differential balanced line over twisted pair copper, it can easily communicate over long distances up to 4000 ft, or just over 1200 m. RS-485 can operate as half-duplex (talk or listen but not both at the same time) over 2 wires and can be configured for full-duplex operation (talk and listen simultaneously) over 4 wires. RS-485 can operate as point-to-point (between two devices) or as multipoint (daisy-chaining several devices along a single RS-485 line). RS-422 can be configured to operate in a star-communications fashion by using a repeater (star coupler), which isolates each leg of communications from the others.
- *RS-422 (also known as ANSI/TIA/EIA-422-B):* This is an ANSI standard. RS-422 can operate as point-to-point or multidrop, whereas RS-485 operates as multipoint. Unlike RS-485, RS-422 does not allow for multiple transmitters, only multiple receivers. RS-422 allows for higher data rates than RS-485, but its architecture characteristics have limited its acceptance as a primary wiring protocol in access control systems. Like RS-485, RS-422 does not specify a communications protocol, allowing manufacturers to use their own proprietary protocols. RS-422 is often used as an RS-232 extender.
- *Ethernet TCP/IP:* Today, most access control panels communicate to their server(s) using Ethernet (TCP/IP). This is good for a number of reasons, but chiefly for standardization. The speed with which the security industry has adopted Ethernet connectivity is purely a result of customer demand. As companies moved toward "Enterprise" systems (systems that span across facilities, campuses, cities, states, and even country borders), all other communications systems except for Ethernet are unable to do this.

TCP/IP is an industry standard for both the information technology industry and for the security industry. This also means that an access control system that communicates using Ethernet TCP/IP also uses a common operating system and can "talk" with other systems using the same operating system, such as Microsoft Windows. This also opened up access control panels to communicate using the same protocol as the server and workstations, and of course other related systems like digital video and digital intercom systems. Finally, all of these systems can speak the same language (usually SQL), enabling functionality and interfacing options on a scale never before possible.

Ethernet systems link together across digital switches and routers. Although all Ethernet enabled access control systems communicate using TCP/IP, it is only one of several protocols available on Ethernet.

All TCP/IP systems ensure that each packet is received (it retransmits the same packet as many times as necessary until its receipt is acknowledged). But TCP/IP protocol is a bad idea for digital video and intercom systems. If we lose a pixel in a picture, we do not want to see it inserted into the next frame. Using TCP/IP protocol on audio also results in "grainy" static-filled audio that can become unintelligible.

For those technologies, UDP/IP and RTP/IP are protocols that do not try to resend lost packets. These are called "streaming" protocols because the receiving device receives an uninterrupted stream of data from the sender. TCP/IP is also well adapted to unicast protocol, which is point-to-point in nature (one transmitter and one receiver). In unicast protocol, there is one data stream for each receiving device. If a transmitting device needs to send to four receiving devices, it will send four individual data streams.

For digital video and intercom systems, where there is a need to send to multiple receiving devices, multicast protocol can be used. This is point-to-multipoint communication (one-to-many).

Multicast protocol should not be used on a switch port serving TCP/IP devices, as it will cause communications problems. Accordingly, it is important to establish a "Virtual Local Area Network (VLAN)" on the digital switch network to isolate TCP/IP devices such as access control panels from UDP/IP or RTP/IP devices such as digital video or digital intercom systems. A typical VLAN configuration is as follows:

> VLAN 0—Not used
> VLAN 1—Administrative VLAN for digital switches
> VLAN 2—Digital video VLAN

VLAN 3—Alarm/access control system VLAN

VLAN 4—Digital intercom VLAN

VLAN 2—Addressing scheme

- ❏ 10.128.2.XXX
- ❏ 10.128.2.10—Beginning address for digital video servers
- ❏ 10.128.2.50—Beginning address for digital video workstations
- ❏ 10.128.2.100—Beginning address for digital video cameras and encoders

VLAN 3—Addressing scheme

- ❏ 10.128.3.XXX
- ❏ 10.128.3.10—Beginning address for alarm/access control system servers
- ❏ 10.128.3.50—Beginning address for alarm/access control system workstations
- ❏ 10.128.3.100—Beginning address for alarm/access control system panels

VLAN 4—Addressing scheme

- ❏ 10.128.4.XXX
- ❏ 10.128.4.10—Address for digital intercom matrix switch

Each port on each digital switch must be configured for the correct protocol as well (TCP/IP, UDP/IP, or RTP/IP) and unicast and/or multicast enabled. Do not mix TCP/IP and UDP/RTP on the same port. Unicast and multicast can both be enabled on UDP/RTP ports, but multicast cannot be enabled on TCP/IP ports.

NETWORKING OPTIONS

Today, virtually all access control systems operate on an Ethernet network, the days of proprietary communications wiring schemes (RS-485, 20 mA Current Loop, Protocol-B, etc.) are virtually over. And that's all good. Ethernet makes for a much more flexible and scalable environment. But Ethernet also comes with inherent security risks, which MUST be managed, and there are a number of options on how Ethernet networks can be implemented for security systems, each with their own benefits and challenges. Let's look at options first, and then deal with the risks.

Ethernet implementation options:

- Proprietary security system wired network
- Blended (business and security system) network
- Wireless network

Proprietary security system wired networks (wired networks that only serve the security system and nothing else) are the easiest to manage from a risk standpoint because they are closed networks, with few nodes, few users and requires little maintenance. The simplicity of these can however, also create complacency on the part of the designer, installer and operator so that risks can be introduced into the network at any of these stages. Being off the organization's business network can be both a blessing and a curse. First, the network will never go down when the business network is taken down for routine maintenance. So you can achieve very high "up-time" with this approach. However, if the designer is not a competent network designer (and I dare say that designers of many proprietary security system networks are not competent), the design can be sloppy, resulting in poor communications optimization, (little or poor network redundancy and little attention to Quality of Service), inherent bandwidth limitations (an especially big problem if the system also accommodates a digital security video system), and have built-in network security vulnerabilities. This is especially true if the security network extends outside the skin of the building. Unless the security network is itself completely secured in such cases, one might as well be hanging a network drop on the side of the building!

"Blended networks" are those that share the organization's business and security IT systems on the same IP Network. This brings both advantages and challenges.

Advantages include:

- The network is much more likely to be designed to industry best practices.
- The network is much more likely to be well managed.

Challenges include:

- Business networks are very rarely designed to handle the large bandwidth requirements of digital video systems, and many business network designers are unfamiliar with the special design demands of digital video systems.
- Business network security protocols CAN create operational problems for security system software. Security software is typically UL rated, but only for a designated network environment, and some antivirus software can disrupt or even disable certain security software.
- Politics! IT directors and security directors rarely see eye-to-eye on network issues. As such, the IT director is likely to

win and the security director may not get the performance the
system is actually capable of, or indeed the ability to scale
the system as necessary, since the IT director, not the
security director's budget will be impacted by adding security
system devices.

Wireless (Wi-Fi) Networks:

Wireless network elements are extremely commonplace today, and these
flexible network options DEFINITELY bring security risks that MUST
be managed. And there are other issues as well:

- Wi-Fi networks can only accommodate a limited number of devices
within a given geographic region (typically no more than 4 channels
together in the same space)
- This can be mitigated by hard-wiring security system device to a
local gathering point and then connecting a number of local devices
to a Wi-Fi router or access point. Doing this, dozens of devices can
work together in the same space.
- Wi-Fi networks MUST use advanced security protocols to assure the
security of the attached edge devices (card readers, intercoms, video
cameras). Typical security processes include:
  - Private rather than public VLAN
  - Virtual private network
  - MAC Address tabling
  - Port disabling of unused ports, and on device disattachment
  - Disabling any unused protocols
  - Disabling bi-directional communications on video cameras

As the industry progresses, I expect to see IT and physical security merging. This is in the long-term best interests of both industries. One cannot
have IT Security without physical security, and one cannot have physical
security without IT security. The sooner both industries understand this,
the better for everyone.

REDUNDANCY AND RELIABILITY FACTORS

Many access control systems are "mission critical systems"; i.e., the operation of the access control system is critical to the mission of the organization. Accordingly, reliability is a primary factor in the design and
maintenance of access control systems. The key to reliability is good
design, good installation, good wiring, good power, and a good
data infrastructure.

Good wiring and installation

I am occasionally asked to visit an existing access control system installation that is not working properly. Invariably, I am consistently stunned at poor wiring practices when I tour these installations. Good wiring practices do more than look good, they help ensure reliable system operation.

Good design

What makes a good design, and why is it so important? A good design is one that ensures that the system will be reliable, expandable, and flexible. Reliable design includes enclosing all exposed system cables in conduit and ensuring that power is reliable (see the section "Good Power").

System panels must be in secured locations that are environmentally appropriate (no janitor's closets). The environment should ideally be temperature and humidity controlled to within the rated temperature and humidity operating range of the access control panel. The closer that is to the extremes, the less ideal it is. The temperature should be largely within the middle of the range, so it is best not to mount access control panels outdoors in very hot or cold climates. Although an environment with air conditioning would be ideal, it is not always necessary if the area generally maintains a stable temperature and humidity.

Be careful to select an access control system that provides extensive expansion capabilities, well beyond the logical immediate needs of the owner. Many systems have been replaced due to unforeseen expansion of a system into an Enterprise class system when it was found that the original system could not communicate across a TCP/IP network.

Similarly, the system should have capabilities far beyond those needed immediately. This ensures that no matter what needs develop, the system can accommodate them.

Good power

After poor wiring, power quality and reliability are probably the biggest problems with unreliable alarm/access control systems. On too many systems, designers and installers assume that "power is the owner's responsibility." I routinely see access control panel A/C power plugs simply plugged into a convenient nearby power receptacle from who knows what source. Oh yeah, and that plug is in fact plugged into a three-way tap on the top half of the receptacle, and there is another one on the lower half of the receptacle that is serving a nearby window air conditioner, a small refrigerator, and a small electrical heater. This is a formula for a highly unreliable access control system.

Alarm/access control system power should be provided from a dedicated clean power source, on its own circuit breaker, and not on a circuit with any "noise-producing" equipment such as air conditioners, refrigerators, heaters, or fan units.

Power for the security system should be backed up by a UPS (one for the whole system per building), and that should be backed up by an emergency generator.

Each access control panel should be equipped with a 4-hour back-up battery, which should be sufficient to power the panel and locks for 4 hours.

Good data infrastructure

For systems that communicate using TCP/IP, this is often also equally overlooked. Reliable digital communications require quality digital switches. I strongly advise you not to skimp on switches. Digital switches should be a major brand of commercial grade switch such as Cisco, 3Com, HP, or similar. Use "computer store brand" switches at your peril. There is a vast difference in capabilities between a $2000 switch and a $20 switch.

Especially important is to *never under any circumstances use a hub instead of a switch* in an alarm/access control system network. Although the security system may not include a digital video system in the beginning, that element could be added later on and when there is a "BANG," your system will go down! Hubs are completely incompatible with digital video multicast protocol (Internet Group Management Protocol, IGMP). The use of a hub in a digital system that includes any IGMP devices such as digital video cameras or encoders *will cause the entire system to stop communicating*. This is because the TCP/IP devices see the multicast signals on their hub port and attempt to answer the IGMP group membership request. Repeatedly! Again and Again! Eventually resulting in what is effectively a Denial of Service attack from within the system, effectively bringing the entire system to its knees until there is no effective communications. Don't use hubs at all and you will be safe from this issue.

As part of a mission critical system, the switch should have a redundant power supply and lots of buffering, and it should be remotely configurable across the network. The switch should be capable of supporting multiple VLANs, Virtual Private Networks (VPNs), and protocols (TCP/IP,

UDP/IP, RTP/IP, unicast, and multicast to each individual port). I strongly recommend that you do not mix brands of switches on the same network. This leads to configuration complications and reliability problems. Except for switches in the same physical room, digital switches should uplink via Fiber to avoid data interference.

Redundancy

Everything fails. And redundancy helps ensure reliability when things fail. The following items should be configured with redundant systems or devices:

- A/C Power:
  - Provide power from two sources to each server and digital switch
  - Provide power to the whole system from a UPS
  - Back-up the UPS with an emergency generator
  - Back-up access control panel power with 4-hour back-up batteries
- Digital communications:
  - Configure the uplink/downlink on digital switches as a fiber-optic loop such that if communication is lost between any two digital switches, data can flow in the opposite direction around the loop
- Servers:
  - Use a redundant mirrored fail-over server (operates continuously mirroring the primary server and takes over if the primary server fails)
  - At a minimum, use a fail-over server (takes over only if the primary server fails)
- Heartbeat and Watchdog Timer:
  - Configure the digital communications system with a heartbeat and a watchdog timer to ensure continuous communications (the heartbeat is checked constantly by the watchdog to ensure that the digital infrastructure is communicating and the Watchdog timer resets digital switches if communications fail and notifies the operator of the reset (event and success/failure)). This requires digital infrastructure monitoring software.

CHAPTER SUMMARY

1. The constant themes of security system development have been and continue to be
   a. More functions
   b. Easier to use
   c. Lower cost

2. Fourth generation access control systems were the first to "farm out" the access control decisions and operation of remote field devices (locks, etc.) to access control panels instead of connecting all those devices centrally to a single computer that made all decisions centrally.

3. Access control panel is a generic phrase that can include:
   a. An electronics panel that can both interface with and control access control system field devices (credential reader, electrified lock, door position switch, and request-to-exit devices)
   b. An electronics panel that can interface with alarm devices (alarm input board)
   c. An electronics panel that can control electrical devices (output control panel)

4. Basic access control panel components include:
   a. A communications board
   b. A power supply and battery
   c. A central processing unit (CPU)
   d. An EPROM
   e. Random access memory (RAM)
   f. Input/output interfaces

5. Access control panel form factors include:
   a. 2, 4, 8, or 16 door connections
   b. Additional inputs and outputs on the main access control panel
   c. Input and output boards moved off the main access control panel board as separate boards attaching to the access control panel or directly on the communications path

6. Access control panel functions include:
   a. Receiving downloaded data from the server(s)
   b. Making access control decisions
   c. Receiving alarm status information from the alarm inputs
   d. Directing the output control to activate a relay
   e. Communicating all event data to the on-board memory (first in-first out)
   f. Communicating all event data to the server(s)

7. Many access control panels also do the following:
   a. Communicate with other access control panels
   b. Communicate with outboard alarm input boards
   c. Communicate with outboard output relay boards
   d. Interlink with other access control panels to create a global antipassback function

8. Fifth generation access control panels can also:
   a. Allow access to all system functions and attributes through SQL or similar language

9. Access control panel location selection has a very big effect on cabling costs
10. Common local and network connection communication protocols for access control systems include:
    a. RS-232
    b. RS-485 (2- or 4-Wire)
    c. RS-422
    d. TCP/IP
11. Multicast protocol should not be used on a switch port serving TCP/IP devices as it will cause communications problems.
12. A typical VLAN configuration is as follows:
    a. VLAN 0—Not used
    b. VLAN 1—Administrative VLAN for digital switches
    c. VLAN 2—digital video VLAN
    d. VLAN 3—Alarm/access control system VLAN
    e. VLAN 4—Digital intercom VLAN
13. The key to reliability is good design, good installation, good wiring, good power, and a good data infrastructure.
14. Good wiring practices do more than look good; they help ensure reliable system operation.
15. A good design is one that ensures that the system will be reliable, expandable, and flexible.
16. Reliable design includes enclosing all exposed system cables in conduit and ensuring that power is reliable.
17. After poor wiring, power quality and reliability are probably the biggest problems with unreliable alarm/access control systems.
18. Reliable digital communications requires quality digital switches.
19. Everything fails. Redundancy helps ensure reliability when things fail.

Q&A

1. *The purpose of wiring all access control devices centrally was to:*
   a. Achieve both centralized control and reporting
   b. Achieve both centralized command and control
   c. Achieve both centralized management and distribution
   d. Achieve both centralized delivery and maintenance
2. *The constant theme of security system development has been and continues to be:*
   a. More functions
   b. Easier to use
   c. Lower cost
   d. All of the above

3. *An alarm/access control system comprises:*
   a. Field elements and system panels
   b. Servers and workstations
   c. Communications infrastructure and software
   d. All of the above
4. *An Access Control Panel must communicate with the:*
   a. Digital video system
   b. Access control system
   c. Security intercom system
   d. Building automation system
5. *The access control panel must make:*
   a. Access control decisions
   b. Alarm decisions
   c. Employee plan decisions
   d. Visitor plan decisions
6. *Access control panels include:*
   a. A communications board and EPROM
   b. A power supply and random access memory (RAM)
   c. A central processing unit and input/output interfaces
   d. All of the above
7. *Access control panel functions must:*
   a. Receive downloaded data from servers
   b. Make access control decisions
   c. Receive alarm status information from alarm inputs
   d. All of the above
8. *Access control panel functions must:*
   a. Direct the output control to activate a relay
   b. Communicate all event data to the on-board memory
   c. Communicate all event data to the server(s)
   d. All of the above
9. *Access control panel functions must:*
   a. Direct the output control to turn on the workstation
   b. Communicate all event data to the building automation system
   c. Both a and b
   d. Neither a nor b
10. *Access control panel functions may:*
    a. Communicate with other access control panels
    b. Communicate with outboard alarm input panels
    c. Both a and b
    d. Neither a nor b
11. *Access control panel functions may:*
    a. Interlink with other access control panels to create a global antipassback function
    b. Communicate with outboard output relay panels
    c. Both a and b
    d. Neither a nor b

12. *Access control panel functions may:*
    a. Interlink with the CCTV system to monitor intercoms
    b. Interlink with the Intercom system to monitor fire alarms
    c. Both a and b
    d. Neither a nor b

13. *Fifth generation access control panels can also:*
    a. Allow access to all system functions and attributes through SQL or similar language
    b. Allow access to all system functions and attributes through Pascal, Fortran, or similar language
    c. Allow access to all CCTV system functions and attributes through dry contact interfaces
    d. None of the above

14. *Fully developed fifth generation systems:*
    a. Blur the line between physical and logical worlds
    b. Blur the line between Alarm and CCTV Systems
    c. Blur the line between operators and guards
    d. None of the above

15. *TCP/IP Ethernet communication*
    a. Has made alarm/access control systems more costly
    b. Has opened up new opportunities for cost savings
    c. Has allowed PC computers to operate card readers through USB connections
    d. None of the above

16. *The most common communications protocols for access control systems included:*
    a. RS-232, RS485, RS-422, and TCP/IP
    b. RS-232 and RS485
    c. RS-422 and TCP/IP
    d. None of the above

17. *Because of _____ it is important to establish a VLAN on the digital switch network to isolate devices' access control panels from digital video or intercom systems.*
    a. Unicast Protocol
    b. Multicast Protocol
    c. UUNet Protocol
    d. International Protocol

18. *Redundancy helps assure reliability when:*
    a. Earthquakes strike
    b. Tsunamis strike
    c. Employees strike
    d. Things fail

Answers: (1) a; (2) d; (3) d; (4) b; (5) a; (6) d; (7) d; (8) d; (9) d; (10) c; (11) c; (12) d; (13) a; (14) a; (15) b; (16) a; (17) b; (18) d.

18

Access Control System Servers and Workstations

CHAPTER OBJECTIVES

1. Get to know the basics in the chapter overview

2. Discover server/workstation functions

3. Learn all about panel and global decision processes

4. Learn the elements of access control system scale

5. Understand access control system networking

6. Learn about legacy access control systems

7. Answer questions about access control system servers and workstations

CHAPTER OVERVIEW

In this chapter we will discuss servers and workstations—the heart of the access control system. While access control panels are the workhorses of the system, servers and workstations are the beating heart of the system.

Servers store all of the system configurations and historical data, manage communications throughout the system, and serve the workstations with real-time data and reports. Servers also control so-called "global" system decisions or functions that span across multiple access control panels. We will also discuss the elements of system scale, such as how to scale a small system into a large one. Using these methods, a system can grow from a single access control panel into an international system with thousands of credential readers (just the thing for a budding dot.com company).

Network design is perhaps the most misunderstood aspect of alarm/access control system design. In this chapter, we will fully explore security system network design. Finally, it is important to note that security system

Electronic Access Control. DOI: http://dx.doi.org/10.1016/B978-0-12-805465-9.00018-X

designers, installers, and maintenance technicians will face many existing installations using older "legacy" systems. We will explore how those systems differ from current offerings and how to interface newer systems with their older cousins.

SERVER/WORKSTATION FUNCTIONS

Servers perform the following functions:

- Store all of the system configurations
- Store all of the system's historical event data
- Manage communications throughout the entire system
- Serve Workstations with real-time data and reports

The alarm/access control system server is the all-seeing, all-knowing entity that completely commands all other activities within the system. absolutely nothing happens in an alarm/access control system that the server does not know about and keep notes on.

Store system configurations

When you first unpack, install, and hook up an alarm/access control system it is as dumb as a rock. Okay, it is a rock with electricity going through it, but it is still dumber than every politician in the world. Yes, it is really that dumb. But just like Albert Einstein as a baby, it is going to get a whole lot smarter.

After loading the operating system and alarm/access control system program, and any other necessary software, you will hook up the server to the network and begin programming device hardware configurations for the entire alarm/access control system.

Every device has a number of configurations that are required for the system to work properly. The server manages all of these and distributes the configurations out to field devices such as access control panels.

Common configurations typically include (these may vary by system brand):

- Access control panel configurations:
  - Network, dial-up, and serial communications
  - Cluster configurations
  - Distributed management configurations
  - Firmware upgrades to all access control panels
  - Seamless integration with the host server

- Access control module configurations (may be either part of the access control panel or may be a separate board):
  - Door alarm input configurations
  - Door lock output configurations
  - Card reader configurations
  - Optional boards (if applicable)
- Cluster configurations:
  - Often access control panels are configured for network communications into user-defined groups (logical clusters).
  - Cluster master configuration (configure one access control panel in the logical group as the master).
  - Cluster member configuration (configure all other members of a cluster under their master).
  - The primary communications path must be configured (master access control panel to host server).
  - The connection type for master to host communications (TCP/IP over Ethernet, serial connection such as RS-232 or dial-up modem). Cluster members will connect to their master using TCP/IP over Ethernet only.
  - The secondary communications path—masters are often programmed with a redundant path in case the first path fails.
- Distributed Cluster Management:
  - This function allows the cluster master and members to share communications and commands between them even when the connection to the host server is lost, thus allowing continued uninterrupted operation even when the host connection is temporarily lost.
  - In this case, the master (and sometimes the members) will maintain their own event history until connection between the master and host server is reestablished, at which time all event data will be uploaded to the host server through the master.
  - Types of distributed cluster management:
    - Event control
    - System activity such as cluster output control at one member from an event in a member controller
    - Global antipassback by cluster

Store the system's historical event data

Servers store all system historical event data. Everything that happens out in the field or at a console workstation is recorded into a historical log file (or files).

Typical historical data stored will include a record where the first field is the year/month/day and that will be followed by the type of historical event, followed by the specific change of state or command. Historical data may include:

- Access control events
  - Access granted
  - Access denied
  - Card/credential not recognized
- Alarm events
  - Secure
  - Alarm
  - Bypass
  - Trouble
- Output control events
  - Open/closed/momentary change of state
  - What commanded the change of state (door lock release from authorized card or request-to-exit sensor)
- Antipassback events
  - Initiate antipassback event when a card enters an antipassback zone
  - Close out antipassback event when the card exits the antipassback zone
  - Invalid use of card at entry reader to antipassback zone by a card that is already within the zone (access denied).
- Scheduled events
  - Unlock door for daytime entry
  - Lock door for nighttime entry
  - Turn off vending machines after hours
  - Turn on vending machines at opening or when the department is occupied after hours
  - Etc.
- Operator logs
  - Operator logs onto or off of an alarm/access control system workstation
  - Operator log may include:
    - Year/month/day
    - Operator authorization level (administrator, supervisor, operator, etc.)
    - Operator name
    - Operator event

❑ Operator events
  – Logon/logoff
  – Area of program accessed/viewed
  – Program element commanded (report, unlock a door or gate, etc.)
  – System configuration change (add/modify/delete record or hardware attribute)

Manage communications throughout the entire system

Servers also manage all digital communications throughout the entire alarm/access control system. The server:

- Sends instructions to the access control panels
- Sends instructions to output relay panels
- Receives event data from access control panels
- Receives event data from alarm input panels
- Sends and receives communications to/from other related systems
  ❑ Digital video system
  ❑ Digital intercom system
  ❑ Building automation system
  ❑ Elevator controllers
  ❑ Lighting controllers
  ❑ Etc.

It is common to see both a primary host server and a back-up host server installed on a system to minimize the possibility of a server failure. By the way, i strongly recommend this unless the client does not need the access control system to function. I am not being flippant. A redundant server is always recommended.

There are two operating modes for back-up servers:

- Fail-over host server
- Redundant host server

A fail-over server is a server programmed exactly like the primary host server, but is "standing by," constantly waiting in the wings for the primary host server to somehow fail. Failure can happen when the primary host server is taken down for maintenance or due to a malfunction. The fail-over server receives programming and configuration updates constantly from the primary host server so it is always prepared for the emergency. When the emergency passes and the primary host server comes back online, the fail-over server relinquishes its role back to the primary server. The fail-over server may not maintain current historical archives.

A redundant host server is a server that performs all of the functions of the fail-over server, but is also updated constantly and online just like the primary host server. The only difference between the two in normal times is that all commands are being distributed to the system through the primary host server. That role changes should the primary host server go down for any reason.

Serve workstations with real-time data and reports

The primary access control host server also serves the system workstations with all of the data they request. Workstations interact continuously with their server, constantly receiving data from and sending data to the server.

Data received by workstations may include:

- Access control events information screens
  - Year/month/day/time—building/door authorized entry/denied access
  - Scrolling chronological events
  - Specific event information
- Alarm event information screens
  - Year/month/day/time—location of alarm, secure/alarm/bypass/trouble
  - Scrolling chronological events
  - Specific event information
- Map displays
- Guard tour information
- Hardware status
- Nonhardware status
- Reports
- List of connected servers
- Workstation views content (choice of many screens to view)
- Help screens
- User images related to access events
- Camera snapshot images related to security events
- Live video window (some systems interface with a digital video system to display video related to an event)
- Message dialog boxes (error messages)
- Third-party applications

Data sent to server by workstations may include:

- Login/logout
- Access control commands (manual actions such as unlock a door)
- System-wide threat level
- Options selections
- Third-party applications

Servers also serve reports to workstations and printers. Typical reports may include:

- Roll-call reports (all personnel in the system)
- Connected devices
- Connected servers
- Connected device status
- Journal reports
  - Access control activity report
  - Alarm activity report
  - Manual event activity report
  - Automated event activity report
  - Antipassback report
  - Etc.

DECISION PROCESSES

Although basic system decisions (grant/deny access) are made by the access control panels, it is the primary host server that programs these and distributes the programming appropriately to all the access control panels.

One panel may serve 10 doors within a department. It may have 10 readers connected to it and a total of 400 authorized users for those readers. Another panel may serve 8 turnstiles from the employee lot of a large factory. This panel may have 16 readers connected to it (one in/one out for each turnstile) and have a daily throughput of 12,000 authorized users.

It is the primary host server that decides what access programming to send to each panel and which authorized users to place in the panel's authorized user database. Not all panels receive the entire user database. Typically, they only receive those that their readers are authorized to process.

Additionally, decisions based on schedules (day of week, time of day, etc.) are also downloaded on an "as needed" basis to each access control panel.

But what happens if a card is presented to a card reader that has not been programmed into the system? The access control panel may be programmed to reject the card if the card does not include the facility code of the facility in question.

But sometimes large corporations use a common facility code across several facilities, so the card may be valid for the facility, but not programmed into the access control panel. When this occurs, the access control panel recognizes the card as being from a valid facility, but does not know what to do with the card. It is not void, and it is not accepted. Here is what happens. The access control panel simply queries the primary host server and asks it what to do with the card. The primary host server looks at the total database (all related sites) and recognizes this card as valid across three of the eight facilities. It sends a "grant access" command to the access control panel and the door unlocks.

Most decisions are made inside the access control panel, but some decisions are reserved for the primary host server.

SYSTEM SCALABILITY

Scalability is the ability of a system to grow gradually in size and capabilities, without giant price cliffs to climb along the way.

Every large system began as a small one sometime in the past. They do not start large, they grow larger ... and larger. Accordingly, it is important to design and install every access control system as though it will become a large system someday, even if it is small today.

UNSCALABLE SYSTEMS

Most access control system manufacturers have woken up to this fact and are now making scalable systems. A scalable system is one that does not require the abandonment of any equipment in order to grow in scale. The organization may have to purchase a larger license, but they do not have to throw capital investment away to expand their system from 64 to 65 card readers or from 128 to 129 card readers as was often the case in the past. Quite literally, a number of manufacturers required that when a client needed to grow their systems from 128 to 129 card readers, they had to replace all of the access control system panels and software with another, larger version, all for a modest cost of about $50,000. Yikes!!!

Wait! It was worse than you think. When alarm/access control system manufacturers finally began listening to their customers who were screaming for scalable systems, their solution was to create scalable hardware coupled up with nonscalable software. How did they do that? By selling software that was limited to 64, 128, 256, 512, or 1028 card readers. After selling the client on a "scalable" system, when the client needed to grow from 128 to 129 card readers, he discovered that he had to buy the next higher capacity of software, at a modest cost of about $50,000! Yikes again! That is $50,000 for almost the exact same software with a key enabled to grow to 256 readers instead of 128. This was common in the industry, and it was especially true of larger, more capable systems.

Basic scalability

Finally, along came a company who understood how outraged clients were and they offered the first truly scalable system. The cost of the software was mostly built into the hardware cost so that one basically never needed to upgrade the software, only add hardware to it to grow its scale.

That was real competition for the other major access control system manufacturers and little by little true scalability grew across the entire marketplace. Today it is difficult to find a nonscalable system. This approach also saw the introduction of the first multisite systems.

Multisite systems

Up until this time, most access control systems were designed to serve only one facility. the system may have been installed at multiple buildings on a campus, but all were managed from a single primary host computer (at this time, redundant servers were a rarity). This required that all new employees, in order to receive an access card, had to go to a single security badging center to have their Photo ID made and to have their data entered into the facility's access control system. This limitation made operation across multiple sites virtually impossible. Accordingly, it was common to see large corporations with many different brands and models of alarm/access control systems serving their entire enterprise. This created incompatible cards, so management and other employees who worked at multiple sites had to carry multiple access cards, one for each site they visited.

System-wide card compatibility

The first step toward Enterprise-scalable systems was system-wide card compatibility; i.e., the ability to utilize a single access card across the entire enterprise. Due to client demand, access control system manufacturers began using the Wiegand interface as a standard, allowing for different facility codes for each facility on a common card format. Thus, management could hold only one card that was good across the entire enterprise.

Enterprise-wide system

The next step along the way to true Enterprise scalability was the implementation of a single, common brand/model of alarm/access control system across the entire enterprise. This allowed large corporations to take advantage of buying power and provided for uniform training and maintenance. This phase was pushed along by the consolidation of many small independent integrators into large national integrators, who gave large corporations and government entities buying leverage to get all their facilities "under the tent."

Until to this time, each facility had its own primary host server.

Master host

As enterprise-class organizations began to pay attention to improving cost control on their security units (due to bean counters) and to improve uniformity of corporate security policies across the enterprise (mostly due to litigation), a demand developed for the ability to establish common security and access control policies across the enterprise. This illuminated the need to develop a means to control the application of those policies.

One of the best ways to do that was to put access control policies under the control of a single master host server. In the earliest implementation of this, the application was developed for a single master host server, "talking" to administrative workstations outfitted at each remote site. These all communicated across telephone modems, constantly passing data up and down the line. This did not work well because of the amount of data being communicated often exceeded the capabilities of the telephone line and the inherent unreliability of modem speeds during weather events.

Until this time, all intersite communication was over modems.

Super-host/subhost

Finally, the system architecture evolved into what we now call a "super-host/subhost" configuration in which each individual facility is equipped with its own primary host server and these all connect to a "super-host" at the corporate headquarters facility.

This called for the development of TCP/IP Ethernet communications between super- and subhosts to facilitate the larger amount of data communicated and to take advantage of the corporate wide area network that already connected their information technology (IT) systems.

It was not long after that TCP/IP connectivity was extended to include communications to access control panels to allow connection of small unstaffed remote sites. Soon after that, TCP/IP Ethernet was used to connect the cluster master access control panel for each individual building on the campus. This was ultimately followed by using TCP/IP Ethernet to connect most if not all access control panels throughout the system, taking advantage of existing Ethernet systems and more uniform connectivity.

ACCESS CONTROL SYSTEM NETWORKING

Access control systems on TCP/IP Ethernet networks on a single system at a single site may involve four main logical elements:

- The core network
- The server network
- The workstation network
- The access control panel network

Additionally, more complex system integrations may involve:

- Integrated security system interfaces
- Multisite network interfaces
- Integration to the business IT network
- VLANs

The core network

The core network typically comprises between one to any number of digital Ethernet switches for an alarm/access control system. Typically, the network may include:

- A single digital Ethernet switch to connect the primary and back-up host servers and any workstations if there are no other TCP/IP devices such as access control panels
- Multiple digital Ethernet switches as follows:
  - Core switch for the servers and workstations
  - Distribution switches connect multiple "Edge" switches to the "Core"
  - Edge switches for the access control panels

There may be one or two Core switches in a system, a larger number of Distribution Switches (perhaps one in each building on a campus), and many Edge switches (e.g., one Edge switch per floor of a building). The core network should include one or more good quality digital switches such as Cisco, HP, 3Com, and so forth (avoid the cheap computer store brands). The switch should be capable of supporting VLANs, VPNs, and both Unicast and Multicast protocols to the individual port. Redundant power supplies on the switch are a plus. Better switches are more reliable, less prone to the aging effects of the environment (temperature and humidity effects), and are more likely to work well when the access control system becomes part of a larger integrated security system including digital video cameras and digital intercoms.

I recommend that all digital switches in the same network are of the same brand (all Cisco, all 3Com, etc.). This facilitates better management of the switches and greater reliability of the network in real-world operations.

Typically Core and Distribution switches may be Layer 3 switches, while Edge switches may be Layer 2 or 3, depending on the needs of the system. (Layer 2 switches perform switch functions, while Layer 3 switches perform both switching and routing functions). Core, Distribution and Edge switches are also typically of differing capacities. Typically, Edge switches have the smallest capacity, Distribution Switches have the cumulative capacity of all connected Edge switches, and Core switches must have the cumulative capacity of all connected Distribution switches. Initial switch capacities should be at least 3—4 times the throughput of the initial device load. It's not a bad idea to make the switch capacities 10 times the initial load, to ensure scalability over time as the technology matures.

The server network

The Servers are the core of the network. When you have a primary and back-up server, they should be connected together over an Ethernet network. These will network together through a "core switch."

The workstation network

Although workstations can sometimes be connected to servers using serial communications (RS-232 or Universal Serial Bus, USB), TCP/IP Ethernet connections are recommended. These will connect to the servers through the core switch.

The access control panel network

Assuming that the system comprises only a single building, the access control panels can connect to the network through an edge switch located near the cluster master access control panel. Other panels can connect to the cluster master through TCP/IP Ethernet or RS-485 for most brands.

Ethernet has speed and connection distance limitations. Common Ethernet speeds include:

- 10Base-T—10 Mb/second (Mbps)
- 100Base-T—100 Mbps
- 1,000Base-T—1 Gb/second (Gbps)
- 10,000Base-T—10 Gbps
- And higher

TCP/IP can connect via copper or fiber. Copper connections have a nominal distance limitation of 270 ft (100 m) for 10Base-T and 100Base-T systems. Copper connections include Category 5, (CAT-5), CAT-5E, and CAT-6 types. For 10Base-T and 100Base-T connections, CAT-5 and 5E connections are acceptable up to 100 m. CAT-6 connections serve 1 Gbps connections up to 100 m. CAT-6 cabling can also provide up to 1500 ft for 100Base-T connections.

Fiber connections include multi- and single-mode types. Multimode fiber is intended for relatively short runs or runs having lower speeds (1 Gbps or less). This is common for any runs over 100 m, such as between buildings. For 10 Gbps connections, always use single-mode fiber between buildings.

Alarm/access control systems typically push relatively few data as compared to digital video systems (the exception is that those systems also send video with alarm information).

Access control networks for access control panels can typically be 100Base-T networks. Connections between edge switches (at the access control panels) and the core switch (at the Servers) can be over 100Base-T copper Ethernet up to 100 m. Distances over that should connect through multimode fiber using SFP connectors on the digital switch.

Integrated security system interfaces

When you connect an alarm/access control system to other security and building systems, it is often best to do so by Ethernet connections. The exception is for connections between systems using dry contact interfaces, such as alarm or door control interfaces between systems.

Whenever connecting multiple systems on the same network, it is best to do so by placing each system on its own VLAN.

VLANs

VLANs allow you to isolate communications between systems, buildings, and sites to better manage the quality of communications when multiple systems share the same physical network. VLANs are accommodated with programming on the Digital Switches and with a VLAN addressing scheme so that you can easily see which VLAN each system and device are located within.

A typical VLAN addressing scheme might be:

> 10.100.1.XX—Digital Switch Administrative VLAN
> 10.100.2.XX—Digital Video System VLAN
> 10.100.3.XX—Alarm/Access Control System VLAN
> 10.100.4.XX—Security Intercom VLAN

VLANs require the core switch to be a routing switch capable of Level 3 commands. Distribution and edge switches must be capable of accepting VLAN programming. Additionally, VLANs can be programmed for each system for each building.

Multisite network interfaces

When the alarm/access control system expands across multiple sites it will be necessary to configure VLANs for each system for each site, and the VLANs may need to be routed through an existing business it network to avoid the usually unbearable cost of a dedicated wide-area security system network.

Integration to the business information technology network

In such cases, the Security System may need to comply with network and routing protocols and addressing schemes of the IT department. For this reason, it is advisable to coordinate beforehand with the IT department director to obtain VLAN protocols and an addressing scheme for the alarm/access control system network that will comply in the future with protocols and addressing schemes already in use by the IT department.

Although the security system may be on its own network now, as it grows to span multiple sites, it will often need to be routed through the IT Network. Making sure that you have VLAN protocols and addressing schemes that already comply with the IT department's standards will ensure the least possible disruption if the two are merged together in the future and no harm is done if they are not merged.

Additionally, it is recommended to place the security system behind a hardware firewall to protect both the alarm/access control system and the business IT system from each other to ensure sustainability and reliability for both systems.

For the ultimate in protection, I recommend that the security system be routed through a VPN, which both completely isolates and encrypts the security system data from the business IT network. VPNs are also a good solution for merging systems if the VLANs are not protocol/network address compatible.

LEGACY ACCESS CONTROL SYSTEMS

From time to time you will come across older access control systems also known as "legacy" systems. For the purposes of this discussion, we will consider any system that connects to its primary host server through any means other than TCP/IP Ethernet to be a legacy system. Many of today's top systems that connect via Ethernet have older installations that do not. At some point these will need to be updated.

After reading the previous section system scalability, you will recognize where each system you encounter is along the development evolution.

Legacy systems can be brought up to date by:

- Replacing their software with the current version
- Creating a cluster master access control panel with TCP/IP Ethernet connectivity
- Creating a cluster master for each site and building
- Ensuring that all existing access control panels are compatible with the new software version
- Reviewing the entire system (all sites/buildings) for adaptation to TCP/IP Ethernet
- Having no need to upgrade legacy access control panels if they are still compatible with the new software version
- Reviewing organization monitoring and management policies to see if a super-host/subhost configuration will help manage the security units better; if so, implement these
- Reviewing related security systems for interoperability to create an integrated security system for each site

CHAPTER SUMMARY

1. Servers perform the following functions:
   a. Store all of the system configurations
   b. Store all of the system's historical event data
   c. Manage communications throughout the entire system
   d. Serve workstations with real-time data and reports
2. Every device has a number of configurations that are required for the system to work properly. The server manages all of these and distributes the configurations out to field devices such as access control panels.
3. Common configurations typically include (these may vary by system brand):
   a. Access control panel configurations
   b. Access control module configurations (may be either part of the access control panel or may be a separate board)
   c. Cluster configurations
   d. Distributed cluster management
4. Servers store all system historical event data.
5. Historical data may include:
   a. Access control events
   b. Alarm events
   c. Output control events
   d. Antipassback events
   e. Scheduled events
   f. Operator logs
6. Servers also manage all digital communications throughout the entire alarm/access control system.
7. The server:
   a. Sends instructions to the access control panels
   b. Sends instructions to output relay panels
   c. Receives event data from access control panels
   d. Receives event data from alarm input panels
   e. Sends and receives communications to/from other related systems
      i. Digital video system
      ii. Digital intercom system
      iii. Building automation system
      iv. Elevator controllers
      v. Lighting controllers
      vi. Etc.
8. It is common to see both a primary host server and a back-up host server installed on a system to minimize the possibility of a server failure.

9. There are two operating modes for back-up servers:
   a. Fail-over host server
   b. Redundant host server
10. A fail-over server is a server that is programmed exactly like the primary host server, but is "standing by," constantly waiting in the wings for the primary host server to somehow fail.
11. A redundant host server is a server that performs all of the functions of the fail-over server, but is also updated constantly and online just like the primary host server.
12. The primary access control host server also serves the system workstations with all of the data they request.
13. Servers also serve reports to workstations and printers.
14. Although basic system decisions (grant/deny access) are made by the access control panels, it is the primary host server that programs these and distributes the programs appropriately to all the access control panels.
15. Scalability is the ability of a system to grow gradually in size and capabilities, without giant price cliffs to climb along the way.
16. Attributes of scalability may include:
    a. Multisite systems
    b. System-wide card compatibility
    c. Enterprise-wide system
    d. Master host
    e. Super-host/subhost operation
17. Access control systems on TCP/IP Ethernet networks at a single system at a single site may involve three main logical elements:
    a. The core network
    b. The server network
    c. The workstation network
    d. The access control panel network
18. Additionally, more complex system integrations may involve:
    a. Integrated security system interfaces
    b. Multisite network interfaces
    c. Integration to the business IT network
    d. VLANs
19. The core network typically comprises between one to any number of digital Ethernet switches for an alarm/access control system.
20. Digital switches should be capable of supporting VLANs, VPNs, and both Unicast and Multicast protocols to the individual port.
21. Digital switches in the same network should all be of the same brand.

22. The Servers are the core of the network. When you have a primary and back-up server, they should be connected together over an Ethernet network. These will network together through a "core switch."

23. Although workstations can sometimes be connected to servers using serial communications (RS-232 or USB), TCP/IP Ethernet connections are recommended. These will connect to the servers through the core switch.

24. Assuming that the system comprises only a single building, the access control panels can connect to the network through an edge switch located near the cluster master access control panel. Other panels can connect to the cluster master through TCP/IP Ethernet or RS-485 for most brands.

25. TCP/IP can connect via copper or fiber.

26. Fiber connections include multi- and single-mode types.

27. Access control networks for access control panels can typically be 100Base-T networks.

28. Whenever connecting multiple systems on the same network, it is best to do so by placing each system on its own VLAN.

29. When the alarm/access control system expands across multiple sites it will not only be necessary to configure VLANs for each system for each site, but the VLANs may need to be Routed through an existing business IT network to avoid the usually unbearable cost of a dedicated wide-area security system network.

30. Although the security system may be on its own network now, as it grows to span multiple sites, it will often need to be routed through the IT network. Making sure that you have VLAN protocols and addressing schemes that already comply with the IT department's standards will ensure the least possible disruption if the two are merged together in the future and does no harm today if they are not merged.

31. Legacy access control systems can be brought up to date by:
    a. Replacing their software with the current version
    b. Creating a cluster master access control panel with TCP/IP Ethernet connectivity
    c. Creating a cluster master for each site and building
    d. Ensuring that all existing access control panels are compatible with the new software version
    e. Reviewing the entire system (all sites/buildings) for adaptation to TCP/IP Ethernet
    f. Having no need to upgrade legacy access control panels if they are still compatible with the new software version

g. Reviewing organization monitoring and management policies to see if a super-host/subhost configuration will help manage the security units better; if so, implement these

h. Reviewing related security systems for interoperability to create an integrated security system for each site.

1. *Servers perform the following functions:*
   a. Store all system configurations
   b. Store all of the system's historical event data
   c. Manage communications throughout the entire system
   d. All of the above
2. *Servers perform the following functions:*
   a. Manage the processing of video-to-video monitors
   b. Serve Workstations with real-time data and reports
   c. Both a and b
   d. Neither a nor b
3. *After loading the operating system and alarm/access control system program and any other necessary software, you will hook up the server to the network and begin:*
   a. Programming system maps
   b. Programming schedules
   c. Programming device hardware configurations for the entire alarm/access control system
   d. None of the above
4. *The server manages all of the configurations required for the system to work properly and:*
   a. Distributes the configurations out to field devices such as access control panels
   b. Distributes instructions to video cameras about where to point and focus
   c. Distributes peanuts and colas along the main center aisle
   d. None of the above
5. *Common configurations made by access control servers typically include:*
   a. Access control panel configurations
   b. Access control module configurations
   c. Cluster configurations
   d. All of the above
6. *Common configurations made by access control servers typically include:*
   a. Distributed cluster management
   b. Network fiber connections
   c. Router and firewall power conditions
   d. All of the above

7. *Access control servers*
   a. Delete data daily
   b. Acknowledge quality programming by dialog boxes
   c. Manage communications throughout the entire access control system
   d. None of the above
8. *Operating modes for back-up servers include:*
   a. Fail-over host server
   b. Redundant host server
   c. Both a and b
   d. Neither a nor b
9. *Although basic system decisions (grant/deny access) are made by the access control panels, it is the primary host server that programs these and:*
   a. Distributes the programming appropriately to all the access control panels
   b. Distributes minute-by-minute changes to those instructions
   c. Distributes live video to the operator
   d. None of the above
10. *Scalability is the ability of a system to grow gradually in size and capabilities*
    a. Without regard to the type of organization
    b. Without giant price cliffs to climb along the way
    c. Without any decisions being made by management
    d. Without any regard to cost
11. *Examples of scalable systems approaches include:*
    a. Multisite systems
    b. system-wide card compatibility
    c. enterprise-wide system
    d. All of the above
12. *Large systems may put access control policies under the control of a single:*
    a. Master host
    b. Document
    c. Monthly newsletter
    d. Workstation
13. *In super-host/subhost systems:*
    a. Each individual facility is equipped with its own primary host server and these all connect to a "super-host" at the corporate headquarters facility
    b. Each individual facility is served by access control panels and these are all controlled directly by a "super-host" at the corporate headquarters facility
    c. Each facility has its own security director who decides daily which facility will serve as the "super-host"
    d. None of the above

14. *Access control systems on TCP/IP Ethernet networks at a single system at a single site may involve*
    a. The core network and the server network
    b. The workstation network and the access control panel network
    c. Both a and b
    d. Neither a nor b
15. *VLANs allow you to:*
    a. Isolate communications between systems
    b. Isolate communications between buildings and sites
    c. Better manage the quality of communications when multiple systems share the same physical network
    d. All of the above

Answers: (1) d; (2) b; (3) c; (4) a; (5) d; (6) a; (7) c; (8) c; (9) a; (10) b; (11) d; (12) a; (13) a; (14) c; (15) d.

The Things That Make Systems Sing

Security System Integration

CHAPTER OBJECTIVES

1. Understand why security systems should be integrated
2. Discover security system integration concepts
3. Understand the benefits of system integration
4. Learn about types of integration
5. Study some examples of security system integration
6. Pass a quiz on security system integration

CHAPTER OVERVIEW

While alarm/access control systems, security video systems, and security intercom systems all are powerful tools to help manage risk for an organization, they become even more powerful when integrated together into a single, comprehensive security system.

Security systems should be integrated in order to help the security program operators minimize the organization's risk. Benefits of security system integration include a more complete awareness of the security conditions across the organization, faster and more effective detection of inappropriate behavior, more rapid and accurate assessment and filtering of actual threats vs. nuisance alarms, the ability to delay aggressors both coming into and on their way out of the organization's facilities, better coordinated responses to security events, and the ability to gather coordinated evidence of security events.

We will also discuss types of system integration and look at some examples of effective system integration.

WHY SECURITY SYSTEMS SHOULD BE INTEGRATED

In order to talk about why security systems should be integrated, we need to first understand the theory of protecting organizations' assets.

Electronic Access Control. DOI: http://dx.doi.org/10.1016/B978-0-12-805465-9.00019-1

Every organization begins with a mission. The organization develops programs in support of its mission. If the organization's mission is banking, depending on the country, they may develop programs to include retail branch banks, a credit card program, loan programs, investment banking program, a real-estate management program, currency, oil or stock trading, and so forth. As they develop programs, the organization will acquire assets in support of its programs. These assets always include:

- People
  - Employees
  - Contractors
  - Vendors
  - Visitors
  - Customers
- Property
  - Real property (land and buildings)
  - Fixtures, furnishings, and equipment
- Proprietary information
  - Vital records
  - Patents, formulas, etc.
  - Customer lists
  - Accounting records, etc.
- The Organization's Business Reputation (The Brand)

These assets have appropriate and inappropriate users. Appropriate users include those who use the assets for the benefit of and with permission by the organization. Inappropriate users are those who seek to use the organization's assets for their own benefit rather than for the benefit of the organization, or in some cases the assets are used against the benefit of the organization.

Inappropriate users can include employees using a social network website on company time, too many purchases and returns from some customers, or something more serious. "Threat actors" are a category of inappropriate users who present a criminal or terroristic threat to the welfare of the organization and they act on that threat. Threat actors include:

- Terrorists
- Violent criminals
- Economic criminals
- Petty criminals

Organizations must protect their assets from threat actors or face serious reductions in their ability to meet their mission. The role of an organization's security program is to improve the likelihood of appropriate use of its assets and reduce the potential for inappropriate use of the organization's assets. They do that by analyzing the risk they face and developing appropriate security countermeasures to balance the risk.

In its simplest form, risk is a combination of the existence of an active threat actor interested in the organization's assets (probability, P), exploitable vulnerabilities (V), and the degree of consequences (C) of that threat scenario being carried out or $R = (P*V*C)$.

A high probability of a scenario coupled with high vulnerabilities that could result in high consequences represents a high risk. A low probability with low vulnerabilities, resulting in low consequences, represents a low risk. All other things being equal, threat scenarios with low probability and high consequences should receive a higher risk score than those with higher probability and low consequences. For this reason I recommend that one consider risk as $R = (P*V)$, prioritized by consequences. While similar, the second simple risk formula results in a more accurate risk assessment.

Once risks are assessed, security countermeasures should be developed. These should always begin with a comprehensive set of security policies and procedures, upon which all other countermeasures are built. This is to ensure that all countermeasures have a practical basis in security policy.

Good security programs include all three types of security countermeasures:

- Hi-tech
- Lo-tech
- No-tech

Hi-tech countermeasures include electronic systems: alarm/access control, digital video, security intercoms, 2-way radio, X-ray and metal screening, and so forth. Lo-tech countermeasures include locks, barriers, lighting, and signage, and no-tech countermeasures include policies and procedures, security staffing, dogs, law enforcement liaison programs, and security awareness programs. These three types of countermeasures should always be used together in a layered approach to reduce risk.

All security countermeasures are intended to:

- Deter unwanted behavior
- Detect inappropriate behavior
- Help assess what has been detected

- Help security staff respond to security events
- Delay intrusions and exits of offenders
- Gather evidence of security events for prosecution and training

Since deterrence varies substantially depending on the commitment of the threat actor, it cannot be accurately calculated, so you should not factor deterrence into the countermeasure balancing formula. Remember, all security programs should be layered such that the most valuable assets are protected by multiple layers of detection, assessment, delay, and response. That is, a threat actor should have to go through multiple rings of detection and barriers to get to an asset and to get that asset back out of the organization's possession, encountering delaying mechanisms along the way in and out. At all times, the security system should be gathering evidence.

Designing an effective electronic security system is a challenging task. But electronic security systems do their job better when their various components are "integrated" into a single, comprehensive system allowing each part of the overall system to "feed" information to and draw from the other parts of the system to enhance functions and effectiveness.

A well-designed security system should filter unnecessary information; present relevant information in a quick, easy-to-understand format; and provide the security console officer and supervisor with quick and relevant options to defend the organization's assets.

INTEGRATION CONCEPTS

Security is not a challenge of technology; it is a challenge of imagination. The challenge of security system Integration is also a challenge of imagination more than technology. When thinking about integrating systems, you must ask yourself the following questions:

- If I eliminated all of the technology and had a highly qualified guard at the location in question, what tasks would I want to perform?
- What information would I need to make decisions?
- What resources would I like to have to be able to carry out those tasks?

Let the integration begin...

- Suppose that you are at a security console and there is an alarm in a weapons storage facility.
  - If I eliminated all of the technology and had a highly qualified guard at the location in question, what tasks would I want to perform?

- Confirm the alarm (an intruder, not a newspaper blowing across the field of view of a camera)
- Determine what the intentions of the intruder were and what weapons or tools he might have or be using
- Know where the intruder is going while a response is on its way
- Delay the intruder's progress
- Intersect and disorient the intruder, stop the intrusion, and apprehend the intruder

❏ What Information would I need to make decisions?
- Confirm that there is in fact an intrusion
- Confirm the description of the intruder
- Confirm the direction, path, and speed of the intruder
- Determine what weapons and tools the intruder is using and to the extent possible the aggressiveness of the intruder

❏ What resources would I like to have to be able to carry out those tasks?
- Alarm system
- Digital video system cameras at the point of entry and along the path of the intruder
- Graphical user interface (GUI) maps showing locations of the alarm and cameras
- Remotely deployable barriers
- Remotely controllable lighting
- Communications to responding guards

■ What resources are available to integrate?
❏ Alarm/access control system
❏ Digital video system
❏ Deployable barriers
❏ Remotely controllable lighting
❏ Two-way radio system
❏ Armed security guards each with a GPS-enabled mobile phone

■ How would we integrate these systems?
❏ Alarm/access control system notifies main console on a GUI Map, displaying the location of the alarm.
❏ Alarm/access control system also notifies digital video system, which displays an array of video cameras inside and outside of the alarmed perimeter door.
❏ The area was also prerecorded so the console operator can also see the area of the alarmed door in the time leading up to the alarm.

❑ The console operator can witness the description of the intruder and note the tools he is using and that he is carrying a sidearm.

❑ Zooming out slightly on the console GUI Map, the console officer can see the locations of roving armed guards on the compound.

❑ The console officer notifies the three nearest guards of the alarm and vectors them to the alarmed door, while giving them a description of the intruder and instructing the guards that he is carrying a sidearm.

❑ During this time, the console operator follows the intruder on the digital video system and GUI map and releases held-open fire doors in the path of the intruder and engages electric locks on those doors, delaying the intruder at each door.

❑ The console officer notifies the police emergency number and police respond to the site of the weapons storage compound.

❑ As the armed guards combine forces at the point of entry door, the console officer relays the location of the intruder.

❑ Armed guards converge on the intruder who places his weapon on the floor when confronted by multiple armed guards.

❑ The intruder is apprehended and handed over to police by the armed guards.

BENEFITS OF SYSTEM INTEGRATION

This section is largely derived from my 2014 book *Integrated Security Systems Design—2nd Edition.*

Operational benefits

Uniform application of security policies: To get consistent results, it is imperative to use consistent processes and procedures. Imagine how chaotic it would be for a multinational corporation to allow each department at each site in each business unit to perform their accounting using their own choice of different software programs and different accounting techniques. It would be very difficult for the organization's management to consolidate all of these different reports into a single cohesive picture of the organization's finances, and that could easily result in corporate losses and intense scrutiny by regulatory bodies and shareholders. So it is also unwise for any organization to allow its business units and individual sites to establish their own individual security policies and procedures, guard-force standards, and so forth, which results in the potential for legal liability where different standards are applied at different business units. Enterprise-class security systems provide the platform for the uniform application of Enterprise security policies across the entire organization. They can also provide visibility into how other company policies are being applied and followed. What follows can be better managed when that information is made available to management in a cohesive way.

Force multipliers: Integrated security systems are force multipliers; i.e., they can expand the reach of a security staff by extending the eyes, ears, and voice of the console officer into the depths of the facility where he could not otherwise reach. The use of video guard tours enhances patrol officers so that many more guard tours can be made than with patrol staff alone. Detection and surveillance systems alert security staff of inappropriate or suspicious behaviors and voice communications systems allow console officers to talk with subjects at a building in another state or nation while their behavior is observed onscreen.

Multiple systems: The integration of alarm, access control, security video, and security voice communications into a single hardware/ software platform permit much more efficient use of security manpower. Enterprise-class security systems are force multipliers. The better the system integration, the better use the organization has of its security force.

Multiple buildings: When security systems span multiple buildings across a campus, the use of a single security system to monitor multiple buildings further expands the force multiplication factor of the system. The more buildings monitored the more value the system has.

Multiple sites: Like multiple buildings on a single site, the monitoring of multiple sites further expands the system's ability to yield value. It is at this point that a true Enterprise-class security system is truly required, because monitoring multiple sites requires the use of network or Internet resources. Monitoring multiple sites can get a little tricky due to network bandwidth. We will discuss how to get the most out of network bandwidth later in Chapter 21 "Related security systems," see the section titled "Security architecture models for campuses and remote sites."

Multiple business units: Some large organizations also have multiple business units. For example, a petrochemical company may have drilling, transportation, refining, terminaling, and retailing. Each of these can benefit by inclusion in an Enterprise-wide security program by consistent application of security policies across the multiple business units.

Improved system performance: Enterprise-class systems also provide significantly improved system performance. The integration of multiple systems at multiple sites into a cohesive user interface allows for simple straightforward command and control. Gigantic systems become manageable.

Improved monitoring: System monitoring is usually dramatically improved over nonenterprise systems. The integration of alarms, access control, and video and voice communications across the platform provide the console officer with coherent and timely information about ongoing events and trends. In elegantly designed systems, when a visitor at a remote site presses an intercom call button and identifies themself as an authorized user who has forgotten their access card, the console officer can pull up the record for that user quickly and confirm both the identity of the person at the intercom as well as their validity for that door. In an elegantly designed system, the system knows the user, the door, and the date and time. As the console officer drops the person's icon onto the door icon, the system either grants or denies access to the door, based on the person's authorization for that door for that time. (This application requires a relatively simple custom script at the time this was written.)

Reduced training: Enterprise-class security systems also require less training. The most basic console operator functions for a truly well designed Enterprise-class system can be learned in just a few minutes (answering alarms, viewing associated video, and answering the intercom). Because the interface is standardized across the Enterprise, cross-training between buildings and facilities is practical, and, operators from one site can provide support for a console officer or guard at another.

Better communications: The system also provides for better communications. Imagine a single software platform that integrates security intercoms, telephones, and cell phones with integral walkie-talkie functions, 2-way radios, and paging all into one easy-to-manage platform. Imagine a console officer who can wear a wireless headset, a wired headset, or use the computer's microphone and speakers, who can trigger the push-to-talk button with a footswitch or a mouse press. Imagine how much better it is when the system queues the intercom automatically when a camera is called. The more the system presents the console operator with the tools to act as though he/she were there at the scene, the better the system serves its security purpose. (This function is performed by GUI systems' interface management software in addition to conventional integrated security system software.)

Cost benefits

Improved labor efficiency: For many of the reasons stated earlier, Enterprise-class security systems enhance labor efficiencies: fewer consoles, fewer guards, redundant monitoring, nighttime live monitoring where it was not cost-effective before, mutual-aid between sites and buildings. All of these factors free up guards to be on patrol and in live communication with the central console.

Reduced maintenance costs: Enterprise-class security systems are generally built on the use of a common technology across the entire platform. Counter-intuitively, they are also generally built on simpler technology than less sophisticated systems. The key to success is often the elegant combination of simple technologies into a highly refined system. This inherent architectural simplicity often also results in lower maintenance costs. While the results are elegant and sophisticated, the underlying technology is actually simpler than in times past. The key is to combine simple Boolean algebra logical functions (and, or, not, counting, timing, etc.) in elegant ways.

Improved system longevity: Security systems are notoriously short-lived. Contemporary security systems are comprised of numerous delicate components that either fail mechanically or are unable to upgrade as the system scales. Thus, when upgrades are necessary, it is often necessary to throw out components that are only a few years old because they are not compatible with newer technologies. This inbuilt obsolescence has a long tradition in security systems and it drives building owners and consultants totally nuts! Most building systems are expected to last 15–20 years. Some building systems including the basic electrical infrastructure are expected to last the life of the building. I find it shocking that most electronic security systems made by major manufacturers and installed by major integrators last less than 7 years. A well-designed Enterprise-class security system should last 10–15 years between major architectural upgrades. This is achievable using the principles taught in this book.

TYPES OF INTEGRATION

What can be integrated and how? Virtually every system can be integrated with others one way or another. Integration opportunities fall into the following broad categories:

- Dry contact integration
- Wet contact integration

- Serial port integration
- TCP/IP integration
- Database integration

Dry contact integration

A Dry Contact is a switch or relay point that is not a source of power. When an alarm panel is connected to a door position switch (DPS), that switch is a dry contact because it is not a source of power. The DPS is "dry." The power source is the sensing input. A resistor is typically used in series with the dry contact to limit current through the sensing input (Fig. 19.1).

When an alarm/access control system provides relay points to signal a video system to display a camera in response to an alarm, that relay is a dry contact to the video system.

Dry contact connections are common between systems to signal a binary change in state in one system, a logical 1 or 0, such as on/off. These are typically used to signal an infrequent and relatively constant change of state such as alarm on, alarm off.

Wet contact integration

A Wet Contact is a connection point that is a source power that is used as a signal to another circuit. When an alarm/access control system lock relay switches power on to a door lock, it does so through a wet contact. In this case, the wet contact is a lock power relay to which a lock power supply has provided power. The lock relay completes the circuit to the lock when the relay closes.

A wet contact can also be the connection point to a transistor, silicon-controlled rectifier (SCR), TRIAC, or other electronic component. A

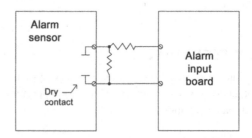

■ **FIGURE 19.1** Dry contact connection.

typical transistor connection can provide either an "open collector or open emitter" connection (Fig. 19.2).

An SCR may provide either an open anode or open cathode connection and a TRIAC may provide a Main Terminal 1 (MT1) or Main Terminal 2 (MT2) connection. An SCR is similar to a transistor except that power can only flow in one direction. When power is applied to its gate it causes power to flow in one direction only through the SCR (Fig. 19.3). SCRs are often used to provide a connection to ground. A TRIAC is like two SCRs back to back with a common terminal in between (like the gate on the SCR; Fig. 19.4).

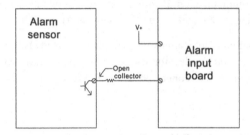

■ **FIGURE 19.2** Open collector and open emitter connection.

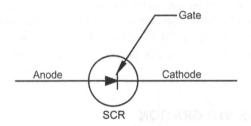

■ **FIGURE 19.3** SCR.

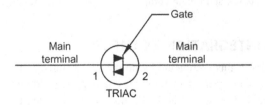

■ **FIGURE 19.4** TRIAC.

■ **FIGURE 19.5** Serial data communication.

SERIAL DATA INTEGRATION

Many systems use serial data to communicate. Serial data is a data connection in which there is a single flow of data bits presented serially (one after the other) rather than parallel (eight bits with one byte at a time).

Whereas dry and wet contacts communicate a simple change of state, serial data communication is often used to communicate instructions from one system to another, such as a command (Fig. 19.5).

Common serial communications pass simple commands between systems, e.g., which floor on an elevator was selected after a valid card was presented. These may be connected via RS-232 or universal serial bus connections.

TCP/IP INTEGRATION

Increasingly, Ethernet TCP/IP connections are used to communicate large amounts of data or data across large distances. In particular, as Enterprise-class systems are being integrated (across many campuses, buildings, etc.) TCP/IP is the preferred communications method because there is usually already a business information technology (IT) connection between the sites that can be used to transmit security system integration information as well.

DATABASE INTEGRATION

When TCP/IP integration is used, it can transmit parallel data, database records and fields, and complicated Structured Query Language (SQL) commands. SQL is coupled with an SQL database at both ends to synchronize databases and create compound command instructions.

SYSTEM INTEGRATION EXAMPLES

The examples in this section are by no means a complete picture of what you can do with systems integration, but hopefully this will give you several ideas of how systems integration can be used to improve effectiveness and reduce the costs of a security program.

Basic system integration

- Systems integrated:
  - ❑ Alarm/access control system
  - ❑ Digital video system
  - ❑ Security intercom system
- Functions achieved:
  - ❑ See and acknowledge alarms on a single map-based GUI
  - ❑ See and answer intercom calls on the map-based GUI
  - ❑ Open vehicle gates and remote doors in response to intercom calls from those locations directly from the map-based GUI
- Benefits derived:
  - ❑ Less technology to interact with for basic functions for all three systems; only one GUI for all three systems
  - ❑ Less training needed for console guards
  - ❑ Faster response to alarm and access control incidents

More advanced system integration

- Systems integrated:
  - ❑ Alarm/access control system
  - ❑ Digital video system
  - ❑ Security intercom system
  - ❑ Building automation system
  - ❑ Stairwell pressurization system
- Functions achieved:
  - ❑ Daytime (work schedule):
    - — Fire stairwell landing next to employee parking lot is also an exit passageway from ground floor corridor to lobby
    - — Smokers gather outside stairwell door for smoke break, smoker forgets access card so he props the door open, causing air to leak from pressurized fire stairwell
    - — DPS on alarm triggers a propped door alarm after 20-second time-out, alarm is audible at door, not transmitted to security console
    - — Smoker ignores alarm at door, 30-second timer counts down
    - — After 30 seconds of alarm at door, alarm is announced on GUI at security console, video camera at door is displayed
    - — Console officer views video at door, sees smokers ignoring alarm
    - — Console officer announces to smokers to close door through intercom at door, stating that this is a pressurized fire stairwell and door must be kept closed, advises smokers to

                      carry access cards, remotely unlocks door to allow smokers entry via intercom after smoking

- ❑ Nighttime schedule:
  - — Door is broken into by intruder, when door opens...
  - — Local alarm sounds, alarm is sent immediately to security console (no delay), GUI shows location of alarm
  - — Up to eight cameras surrounding area of door (inside, outside, and in stairwell) are displayed, showing intruder, who appears to be armed
  - — Console officer follows intruder as he makes his way up stairwell to sixth floor and breaks into sixth floor stairwell door
  - — Console officer notifies police who respond to site
  - — Console officer advises police over telephone of location of intruder
  - — Police capture and arrest intruder with help of console officer's guidance
- ■ Benefits derived:
  - ❑ Case 1:
    - — Stairwell pressurization is not violated
    - — Console officer is not notified if door is closed by nearby persons upon sounding of local alarm at door
    - — Console officer is notified only after alarm is ignored
    - — Smokers are advised to carry access cards
    - — Console officer can remotely unlock door for authorized users
  - ❑ Case 2:
    - — Immediate notification of intrusion to console officer
    - — Video system provides all relevant cameras on alarm
    - — Console officer can follow intrusion on video system
    - — Console officer is not compelled to confront armed intruder
    - — Police receive assistance from console officer as to location of intruder, assisting in his arrest

Advanced system integration

- ■ Systems integrated:
  - ❑ Alarm/access control systems
  - ❑ Digital video systems
  - ❑ Fire alarm systems
  - ❑ Emergency gas pump shut-off system
  - ❑ Two-way audio system
  - ❑ Business IT system

- Functions achieved:
  - Remote monitoring of all branches of a chain of convenience stores/gas stations from a single proprietary central station location
  - Remote monitoring of fire alarms/intrusion alarms/duress alarms
  - Case 1—Store robbery:
    - Robber approaches a store cashier with gun, aggressively demands money—robber appears very threatening to cashier
    - Cashier triggers silent alarm while opening cash register
    - Alarm is sent via business IT system to proprietary central station
    - Alarm rings at central station on GUI showing location of store on map
    - Video from the store is displayed at central station GUI along with audio from microphone above the cash counter at the store
    - GUI displays location and police/fire/emergency numbers next to location on GUI map
    - Central station operator clicks on police emergency number (which dials police and uses recorded message to announce in-progress robbery and store location), announces that that following audio is one way from the store location and that store cannot hear police, then opens audio panel so that police can monitor audio from the store
    - Central station operator announces to robber over loudspeaker above cash counter that he is under observation and that the robbery is being recorded for prosecution (describes the offender so he knows he is under observation), says that police are en route and urges him to leave the store before he is caught by police
    - Robber takes small amount of cash in drawer (policies state large bills must be emptied every hour) and runs out without further aggression, hoping to get away from police and not be seen on video assaulting the cashier
  - Case 2—Fire:
    - Person pumping gas is smoking cigarette, ignites gas vapors, and fire erupts from gas tank of vehicle
    - Intelligent video system sees fire and rings alarm at the store and alarm is sent via business it system to proprietary central station
    - Alarm rings at central station on GUI showing location of store on map

- Video from the store is displayed at central station GUI along with audio from microphone above the cash counter at the store, one camera shows fire and smoke at gas pumps
- GUI displays location and police/fire/emergency numbers next to location on GUI map
- Central station operator directs store personnel to evacuate store through overhead speaker
- Cashier moves everyone out of the store to safety, in the excitement, he forgets to shut-off power to gas pumps
- Central station operator shuts off gas to gas pumps by hitting emergency pump shut-off button on GUI
- Central station operator clicks on fire emergency number (which dials local fire department and uses recorded message to announce fire at store location), then connects central station operator with fire department emergency number for further description of the fire
- Central station operator instructs all persons on gas plaza to evacuate to a safe distance by using microphone at central station and loudspeakers at gas plaza
- Fire department arrives, puts out fire

- Benefits derived:
  - Case 1: Life possibly saved, minimal cash lost
  - Case 2: Lives possibly saved, minimal loss to fire
  - Case 3: Integrating access control systems in multitenant hi-rise buildings:

Multitenant hi-rise buildings present a special problem for security system designers and operators alike. Multitenant hi-rise buildings typically have many tenants, each of which may have their own alarm/access control system on their own office suite, and all of these tenants will need to use common building features that also has its own alarm/access control system (parking, after-hours entry door, floor-by-floor elevator control, building gymnasium, etc.) That are used in common by all tenants. However, the common building access control system may not be of the same brand and type as the tenant's own system, and both the building and tenant systems must have each employee in their access control system database and that database must be kept up to date as new employees are added and terminated.

There are historically only two solutions for this problem, and neither one is good for the tenant's employees or for the building management. The first solution is for the tenant employees to carry two access cards, one for the building and one for the tenant suite. Ughh!!! The second solution is for both the tenant and the building to utilize a common type of access card for their respective access control systems and populate both the building and the tenant access control systems with access card information for all of the tenants' employees. This can be thousands of access card records. This presents a huge problem for building management companies and tenants alike, as they try to synchronize their access control systems and keep their respective databases up to date. It is a logistical nightmare costing thousands of dollars per year for tenants and building management alike. What is needed is a solution that allows tenants' employees to use only one access card for both the tenant suite and the building's common entries, and to eliminate the need to maintain all cardholders in both access control system databases.

One enterprising security entrepreneur, Mr. George Mallard, P.E. has invented a solution for this problem that is so obvious that everyone who sees it cannot believe that they did not think of it themselves.

Following is a brief on how the system works, along with an illustration:

Purpose:

The Parallel Processing Approach provides a method for multitenant building owners to allow their tenants to grant access for their employees to designated base building card readers without the need to maintain the tenant access control system database on the base building access control system.

In all cases, the tenants' employees will be using access cards issued by and under the control of the tenants, so it is not necessary for them to carry one access card to get in the building and a separate one to get into the tenant space.

Using the parallel processing approach, the tenant's access control system database is always under the tenant's control and their card data need never be in the hands of the base building system. This provides both control and privacy, while still informing the base building system that a specific tenant's valid cardholder has entered the premises.

Thus the base building management knows when a tenant's employee has entered, and through which entry, tenant by tenant. That is, if there are a number of tenants in the building, each card usage will reflect the tenant's name, rather than the individual cardholder. When coupled with a video camera at the card reader, identity can also be rendered if needed. (There is an option to pass the card number back to the base building system, for many brands of access control systems, if that is required.)

Environment:

The parallel processing approach will work with any two access control systems, including any mix of brands and models, as long as they both use card readers that utilize Wiegand wiring. Thus there is no need for tenants to carry two cards, and no need for building management to worry about tenant employees being added or terminated. The system will always work.

The parallel processing approach communicates between the two access control systems over an Ethernet network. The network is typically owned by the base building management, in order to allow employees of multiple tenants to the same base building card readers, such as the front door, parking garage or other common area. The approach can accommodate any number of tenants in a building and once set up is maintenance free, requiring no further database synchronization.

Functions:

To the user, the system is completely transparent. The tenant's employee uses the same card to enter the building as to enter the tenant space. This will be the tenant's own access control card.

To the tenant, there is only one card to issue and only one access control system database to control (i.e., their own access control system database). This database is totally under the control of the tenant at all times and privacy of card data is typically assured, except where the tenant has specifically agreed to allow the base building management to see card numbers on entries. Cards for terminated employees will never be granted access to the base building readers once they are terminated on the tenant's own access control system.

To the base building management, the system is maintenance free once the tenant's access control system has initially been connected and configured, there is no need to maintain a database of tenant's employees on their own access control system. There is no need to get a list of new or terminated tenant employees. New and terminated tenant employees are updated continuously by the tenant on the tenant's own access control system and the base building system will not grant access to a card that is not valid on the tenant's access control system (Fig. 19.6).

Operations (simplified):

- Whenever a tenant's employee uses a valid access card on a base parallel processing connected building card reader, the card is read as invalid on the base building system and the card data is passed through the parallel processing system to the tenant's access control system.

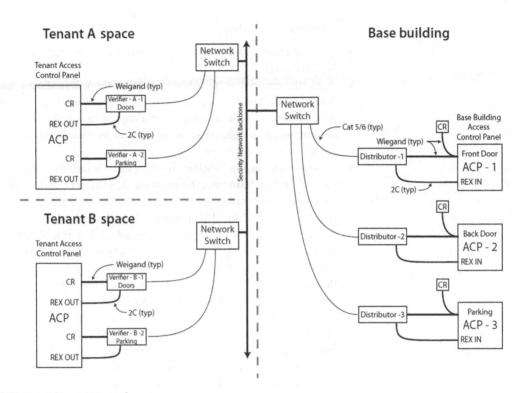

■ FIGURE 19.6 Multitenant hi-rise interface system.

- If the card is read as valid on the tenant's system, the parallel processing system passes a request to exit back to the base building access control panel, thus allowing the tenant's employee to enter.
- At the same time, the parallel processing system passes a token "tenant access card identifier" back to the base building system so that the entry is recorded by the base building access control system as belonging to that specific tenant.

Attributes:

The basic parallel processing system comprises:

- A base building access control panel
- A parallel processing "distributor"
- An Ethernet network
- A parallel processing "verifier"
- A tenant access control panel

Operation of components:

- A tenant presents their tenant access card at a connected base building reader.
- A parallel processing "distributor" acquires the Wiegand card data from the base building access control system and which passes back a "tenant access card identifier" and a request to exit upon presentation of a valid tenant access card at the base building card reader.
- A parallel processing "verifier" receives the Wiegand card data from the parallel processing distributor across the Ethernet network and passes the card data to the tenant access control system access control panel for authorization. Upon authorization, the lock relay on the tenant access control panel is energized and a signal is received at the request to exit input of the parallel processing verifier. It passes this signal back to the parallel processing distributor which in turn signals the request to exit input of the base building access control system panel and also passes the tenant access card identifier back to the base building access control panel. This completes a card transaction. One version of the system also passes a "token" card number to the tenant's access control system, recording a card event at a specific base building card reader, along with the card number. Simultaneously, the base building access control system receives a card event notification specific to the tenant to which the cardholder is employed.
- An Ethernet network connects the parallel processing distributor to the parallel processing verifier.

Larger systems:

- Larger systems may use more base building card readers and may have multiple tenants connected to the system.
- In the most basic configuration, only one tenant card reader connection is required (if all base building doors are in the same access clearance at all base building doors).
- Where multiple access clearances are needed (all employees can enter the front door, some employees can enter the parking structure, and executives only have access to the Gym) then multiple verifiers can be used (one for each clearance).
- Where multiple tenants are participating, at least one verifier is needed for each tenant.

■ One version of the system can also accommodate similar access for elevators, on a floor-by-floor basis.

(For more information about this approach, look up the inventor's name (Mr. George Mallard, PE) on the Internet, along with the words "access control.")

CHAPTER SUMMARY

1. In order to talk about why security systems should be integrated, we need to first understand the theory of protecting an organization's assets.
2. Every organization begins with a mission.
3. The organization develops programs in support of its mission.
4. As they develop programs, the organization will acquire assets in support of its programs.
5. These assets always include:
   a. People
   b. Property
   c. Proprietary information
   d. Business reputation
6. These assets have appropriate and inappropriate users.
7. Appropriate users include those who use the assets for the benefit of and with permission by the organization.
8. Inappropriate users are those who seek to use the organization's assets for their own benefit rather than for the benefit of the organization, or in some cases the assets are used against the benefit of the organization.
9. Threat actors include:
   a. Terrorists
   b. Violent criminals
   c. Economic criminals
   d. Petty criminals
10. Organizations must protect their assets from threat actors or face serious reductions in their ability to meet their mission.
11. In its simplest form, risk is a combination of the existence of an active threat actor interested in the organization's assets (probability, P), exploitable vulnerabilities (V), and the degree of consequences (C) of that threat scenario being carried out or $R = (P*V*C)$.
12. Once risks are assessed, security countermeasures should be developed.

13. These should always begin with a comprehensive set of security policies and procedures, upon which all other countermeasures are built.
14. Good security programs include all three types of security countermeasures:
    a. Hi-tech
    b. Lo-tech
    c. No-tech
15. Hi-tech countermeasures include electronic systems including alarm/access control, digital video, security intercoms, etc.
16. Lo-tech countermeasures include locks, barriers, lighting, signage, etc.
17. No-tech countermeasures include policies and procedures, security staffing, dogs, law enforcement liaison programs, security awareness programs, etc.
18. These three types of countermeasures should always be used together in a layered approach to reduce risk.
19. Since deterrence varies substantially depending on the commitment of the threat actor, it cannot be accurately calculated.
20. A well-designed security system should filter unnecessary information; present relevant information in a quick, easy-to-understand format; and provide the security console officer and supervisor with quick and relevant options to defend the organization's assets.
21. Electronic security systems do their job better when their various components are "integrated" into a single, comprehensive system allowing each part of the overall system to "feed" information to and draw from the other parts of the system to enhance functions and effectiveness.
22. Integration benefits include:
    a. Operational benefits
        i. Uniform application of security policies
        ii. Force multipliers
        iii. Multiple systems
        iv. Multiple buildings
        v. Multiple sites
        vi. Multiple business units
        vii. Improved system performance
        viii. Improved monitoring
        ix. Reduced training
        x. Better communications
    b. Cost benefits
        i. Improved labor efficiency
        ii. Reduced maintenance costs
        iii. Improved system longevity

23. Types of integration include:
   a. Dry contact integration
   b. Wet contact integration
   c. Serial port integration
   d. TCP/IP integration
   e. Database integration
   A parallel processing system can simplify multitenant high-rise design and reduce costs for both tenants and building management.

Q&A

1. *While alarm/access control systems, security video systems, and security intercom systems all are powerful tools to help manage risk for an organization, they become:*
   a. Less powerful in the hands of inexperienced operators
   b. Even more powerful when integrated together into a single, comprehensive security system
   c. Essential methods in the control of terrorism
   d. None of the above
2. *Every organization has four kinds of assets. These include:*
   a. Employees, contractors, vendors, and visitors
   b. People, property, proprietary information, and business reputation
   c. People, fixtures, furnishings, and equipment
   d. Customers, real property, fixtures, and furnishings
3. *Threat actors can include:*
   a. Employees, contractors, vendors, and visitors
   b. Employees, contractors, visitors, and customers
   c. Terrorists, violent criminals, economic criminals, and petty criminals
   d. Terrorists, eco-terrorists, animal-rights terrorists, and bugs bunny
4. *Good security programs include only the following types of countermeasures:*
   a. Hi-tech, lo-tech, and no-tech
   b. Electronics, operations, and investigations
   c. Electronics, operations, and dogs
   d. Electronics, policies, and procedures
5. *All security countermeasures are intended to:*
   a. Deter unwanted behavior
   b. Detect inappropriate behavior
   c. Help assess what has been detected
   d. All of the above
6. *All security countermeasures are intended to:*
   a. Help security staff respond to security events
   b. Delay intrusions and exits of offenders
   c. Gather evidence of security events for prosecution and training
   d. All of the above

7. *Security is not a challenge of technology; it is a challenge of:*
   a. Imagination
   b. Understanding
   c. Apprehension
   d. Arrest

8. *When thinking to integrate systems you must ask yourself:*
   a. If I eliminated all the technology and had a highly qualified guard at the location in question, what tasks would I want to perform?
   b. What information would I need to make decisions?
   c. What resources would I like to have to be able to carry out those tasks?
   d. All of the above

9. *Benefits of system integration include:*
   a. Operational benefits
   b. Cost benefits
   c. Neither a nor b
   d. Both a and b

10. *Operational benefits may include:*
    a. Uniform application of security policies
    b. Force multipliers
    c. Both a and b
    d. Neither a nor b

11. *Cost Benefits may include:*
    a. Improved system longevity
    b. Improved use of dogs
    c. Improved use of drugs
    d. None of the above

12. *Types of integration may include:*
    a. Dry contact integration
    b. Wet contact integration
    c. Serial port integration
    d. All of the above

13. *Types of integration may include:*
    a. TCP/IP integration
    b. Database integration
    c. Both a and b
    d. Neither a nor b

Answers: (1) b; (2) b; (3) c; (4) a; (5) d; (6) d; (7) a; (8) d; (9) d; (10) c; (11) a; (12) d; (13) c.

Chapter 20

Integrated Alarm System Devices

CHAPTER OBJECTIVES

1. Learn the basics in the chapter overview
2. Get to know about alarm concepts
3. Discover types of alarm sensors
4. Learn about alarm system application rules
5. Looking beyond alarm detection
6. Answer questions about integrated alarm concepts and devices

CHAPTER OVERVIEW

In this chapter, we will discuss Alarm Concepts (including how alarms are detected and initiated), Alarm States (including the difference between False Alarms and Nuisance Alarms), Filtering, Alarm Communication and Annunciation, Assessment, Response, and Evidence Gathering. We will examine a variety of types of alarm detection devices, and we will also look beyond Alarm Detection and learn how alarms can be used for Trend Analysis and Vulnerability Analysis.

ALARM CONCEPTS

There is a lot more to an alarm system than most people think. There are many steps between "Bad guy turns up in woods outside our weapons depot" and "Bad guy arrested." This section covers all of the basic concepts involved in alarms from detection to arrest.

Detection and initiation

Alarm detection occurs when the alarm device is initiated by an intruder. Depending on the type of alarm detection device, initiation may be caused by opening a window, door, or gate (equipped with a window/door/gate position switch), breaking a window (acoustic glass break detector), walking or driving over underground seismic or energy field

Electronic Access Control. DOI: http://dx.doi.org/10.1016/B978-0-12-805465-9.00020-8

detectors, walking or driving through a beam (such as infrared or laser beams), or walking through a volumetric energy field (such as infrared or microwave detectors).

Alarm detection devices and their methods of detection are discussed at the end of this chapter.

Filtering and alarm states

Before the alarm is reported to a security console officer as an alarm, it may be filtered to reduce the potential for false or nuisance alarms.

- Alarm states: Alarm devices typically are connected to an alarm input panel and exist in one of four states:
  - Secure: The secure alarm state means that the alarm detection device is connected and communicating properly and that there is nothing to report. The area is secure. Electrically, the alarm circuit is typically closed with a resistor in the circuit to limit the current from the alarm input board. When the secure state is annunciated, the alarm point is typically shown as green (Fig. 20.1).
  - Alarm: When an alarm detection device goes into the alarm state, it is because the alarm detection device has been activated according to its detection technology (a door/window/gate has opened, a motion detector has seen a change in heat or motion, etc.; Fig. 20.2). When the alarm state is annunciated, the alarm point is typically shown flashing red. When the alarm changes from secure to alarm state, there is often also an audible annunciation to draw the security console officer's attention to the alarm.
  - Bypass: There are times when you do not want to receive any alarm even though the alarm detection device may be activated. For example, a motion detection alarm device in a theater hall would be bypassed when the theater is scheduled for activities. In such cases, the alarm detection device is placed into the "bypass" state (Fig. 20.3). Often the alarm is annunciated by a color change as it changes state, but instead of displaying green/red (secure/alarm) it may display green/yellow (secure/sensing but in bypass state). When bypassed, the alarm point shows the change of state but there is not audible alarm.
  - Trouble: On rare occasion, there may be a problem with the alarm detection device wiring. When this happens, it can be sensed by the "trouble" state (Fig. 20.4). Typically the trouble state is

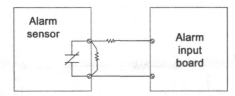

■ **FIGURE 20.1** Secure alarm state.

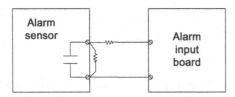

■ **FIGURE 20.2** Alarm state.

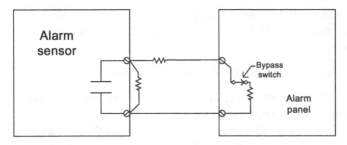

■ **FIGURE 20.3** Bypass state.

annunciated as a flashing yellow light, usually with no audible alarm. Although all alarm detection devices can be wired in a fashion that will not report the "trouble" state when problems arise, they will likely be reported as an alarm. That is the definition of a "False Alarm."

■ Filtering: Alarms may also be filtered to help avoid "nuisance alarms." A nuisance alarm is when an alarm detection device reports a valid detection, but it is an occurrence that has no security implication, such as when a dog leans against a fence.

Every alarm detection device has what are called "Exploit Modes." This is important for system designers, installers, and maintenance technicians

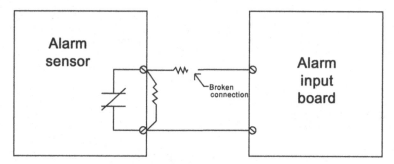

■ **FIGURE 20.4** Trouble state.

to understand. If the exploit mode is used inappropriately, then it will detect things that have nothing to do with security and drive security console officers crazy. Most console operators deal with nuisance alarms by placing those alarm detection devices into "permanent bypass mode." Thereafter they do not cause nuisance alarms, but they also do not report real alarms. This leaves the facility vulnerable to the security event the alarm detection device was installed to detect. The answer is to filter out nuisance alarms.

The most effective way to filter nuisance alarms is to:

- First, understand the exploit modes of all alarm detection devices (more on that later in this chapter).
- Second, use a second alarm detection device for the same area that *does not* possess the same exploit mode as the first alarm detection device; e.g., use an infrared motion detection device along with a pair of infrared photo beams.
- Configure the circuit such that when either one trips, there is no alarm reported but when both circuits trip an alarm is reported.

Communication and annunciation

Most alarms are communicated from the alarm detection device to the alarm panel typically over a two-conductor twisted pair cable (typically UL Listed 20 gauge stranded, individually insulated with an overall sheath; Fig. 20.5).

An alarm panel may be stand-alone (a complete alarm system to which all alarm detection devices connect directly), or it may be an alarm input board, which is a small part of a much larger alarm/access control system

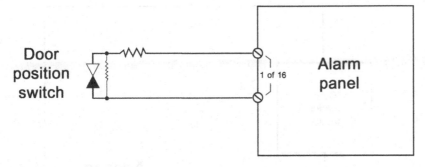

■ **FIGURE 20.5** Simple alarm circuit.

including many other alarm input boards, access control panels, and other electronic devices.

Alarm Input Boards communicate the alarm through the access control panel, to the host server, and then to an alarm/access control workstation where it is often annunciated on a map-based graphical user interface (GUI) that shows the location of the alarm as well as its alarm state (Fig. 20.6).

When an alarm is annunciated, the security console officer's attention is usually drawn to the alarm by an audible alarm. The officer will "silence" the audible alarm and view the point in the alarm state. If this is on a simple lamp-based panel, it will be the flashing red light. If this is on a scrolling screen, it will be the last one on the list, usually flashing red. If this is on a map-based GUI, or map-based annunciating panel it will be the red flashing icon overlaid onto the map showing the location of the alarm event.

Assessment

Once annunciated the alarm must be assessed; i.e., the security console officer must determine if the alarm is a valid "security event" or if the alarm detection device is reporting a "nuisance alarm." This requires that:

- A roving guard is sent to see the conditions at the alarm event location
- The console officer views a video camera showing the alarm event location
- A second alarm detection device in the alarm event location triggers, confirming the first alarm detection device

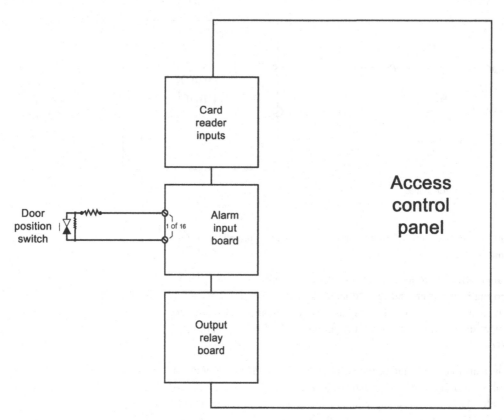

■ **FIGURE 20.6** Advanced alarm communication circuit.

- The console officer listens to an audio source from the alarm event location, confirming inappropriate activity there

The fastest and most accurate way of assessment is to use video to assess the alarm. This provides the console officer with a wealth of additional information including but not necessarily limited to:

- Is the alarm a real security event?
- How many offenders are there?
- What do the offenders look like? (Sex, height, weight, hair, clothing, etc.)
- Are the offenders using tools?
- Are the offenders carrying weapons?
- What direction are the offenders headed?
- Do the offenders appear organized or chaotic?
- How aggressive are the actions of the offenders?

All of this information helps the console officer and supervisor organize an appropriate response plan.

Response

Alerted to intruders, the console officer and supervisor must organize an appropriate response. Typically this may involve sending either police or roving security officers to confront the intruder, thus it is a good thing to know how many intruders there are and whether or not they are armed. For this, the console officer will either use the 2-way radio to notify roving guards or telephone to notify police.

During the pursuit, the console officer should monitor the movements of the intruders using the digital video system. This is best done with video pursuit scripting on digital video software.

The "Video Pursuit Function" is a custom software script that can be overlaid on most digital video systems. The concept was invented by the author to allow console officers to follow subjects easily through a complex facility equipped with a large video system. The video pursuit initiates with the alarm notification and the video system displaying up to nine relevant images, where the center image is the scene of the alarm event and all surrounding images are scenes of "egress paths" away from the alarm scene. Each surrounding image is a mouse button that can be triggered when the console officer "hovers" the mouse over the image. When the intruder exits the alarm scene and goes into the view of a surrounding camera, the console officer clicks on that image and it becomes the new "center focus tile" in a new array of up to cameras. Each movement of the "focus tile" issues a corresponding move on the map-based GUI showing the new location of the subject as he moves. The console officer advises the responding officers of the change in location of the intruder as he moves. This operation repeats itself until the subject is apprehended by police or roving security officers. The video pursuit script can be implemented by competent security system integrators or by the engineering support departments of most digital video system manufacturers.

Evidence

One of the most important functions of alarm systems is to gather evidence of a crime or misbehavior. Alarm/access control systems keep logs of all activity, and one of those logs is the "Alarm Event Log." This is legal evidence in a court of law, civil proceeding, or employment hearing that an event occurred. When the event can be tied to the offender using other evidence such as video, audio, or witness testimony, a case against the offender can be established.

TYPES OF ALARM SENSORS

Alarm sensors can be categorized into several broad areas as follows:

- Outer Perimeter Detection Systems
- Building Perimeter Detection Systems
- Interior Volumetric Detection Systems
- Interior Point Detection Systems
- Intelligent Video Detection Systems

Outer perimeter detection systems

Outer perimeter detection systems are designed to alert whenever a human or vehicle breaches an outer perimeter such as a fence line, gate, or invisible border. Outer perimeter detection systems can also be used along campus boundaries and even national borders.

- *Gate Breach Detection System*
  - Application: Gate breach detectors are typically in the form of magnetic switches that are applied to gates instead of interior doors. They may be configured to work on swing, sliding, or overhead gates. They work by maintaining a magnetic switch in the closed position as long as the switch is held next to the magnet. The magnet is mounted on the gate and the switch is mounted on the adjacent fence. When the gate opens, the switch will change state and send an alarm.
  - Advantages: These switches are generally very reliable. For the highest reliability they should be balanced-bias (wherein the switch exists in a field of several magnets, one on the gate and the others within the switch) so that if anyone tries to place a magnet next to the switch, this act alone will cause an alarm.
  - Disadvantages: As the gate ages, the switch may not align properly with the switch, causing nuisance alarms. Wide gap switches are necessary to accommodate the natural loose fitting of gates.
  - Exploits: Climb over the fence or tunnel below the fence.
- *Seismic Detection Systems*
  - Application: Seismic detectors are essentially rugged waterproof microphones, typically in the form of spikes that are driven into the ground. They are connected by cables (or sometimes radio

frequency (RF)) to a detection circuit that helps to filter out nuisance alarms.

- ❑ Advantages: Very good at detecting vehicles and footsteps on solid ground.
- ❑ Disadvantages: Do not work well in soft ground such as sandy soil and they cannot be used in areas of routine high traffic, such as next to a freeway, as the noise of large vehicles drowns out any nearby footsteps.
- ❑ Exploits: Knowledgeable threat actors use continuous loud sounds such as construction equipment to swamp the detectors so that they cannot hear footsteps.
- ■ *Fiber-Optic Detection Systems*
  - ❑ Application: Fiber-optic cables may be attached to fences to detect climbing or cutting of the fence fabric. Better systems can pinpoint the exact location of the fence breach down to about a couple of meters. Some sensors also provide audio so that the console officer can hear the fence being cut or climbed in order to help verify the alarm.
  - ❑ Advantages: Highly reliable when correctly installed on a good quality fence. They also can detect for very long distances as compared to other types of fence detectors and use fewer electronic modules than most other types, thus reducing their costs on long fence lines. One manufacturer claims to be able to operate a single zone of up to 60 miles (100 km) with a detection accuracy of ± 1.5 m.
  - ❑ Disadvantages: Correct installation is critical and the fence must be kept in very good condition.
  - ❑ Exploits: Exploits include cutting the line in advance of the actual intrusion and making entry before repairs can be conducted, down-line of the cut.
- ■ *Capacitance Detection Systems*
  - ❑ Application: These are typically in the form of two cables placed near each other that set up a capacitive field between the two cables. These can be mounted on a fence, used as a fence topper, or buried underground. Depending on the type and the installation method, these systems can be very good at detecting intruders near the capacitive field. They sense by sensing the water content of human bodies, which disrupts a tuned radio-frequency field between the two cables, changing its frequency. When installed correctly, and in combination with its integral detection filtering equipment, the sensors can detect with relative accuracy the location of an intruder even standing nearby outside the fence.

       ❏ Advantages: These systems are reliable except during rain. They require relatively little maintenance and are much less sensitive to fence fabric maintenance than fiber-optic detection systems.

       ❏ Disadvantages: Unreliable during rain unless carefully monitored and tuned for such. They cannot be used underground in areas of standing water and are less reliable on fences near standing water such as puddles after a rainstorm.

       ❏ Exploits: Knowledgeable threat actors have used heavy rainstorms to mask their entry. The lines can also be cut before an intrusion, with the actual intrusion occurring before repairs can be made. Well-organized security programs place additional patrols or sensor layers when lines are cut to avoid this exploit.

- *Leaky-Coax Cable*
  - ❏ Application: A leaky-coax system comprises a radio-frequency coaxial cable, similar to a television cable, on which the outer braid is designed to be incomplete (about 50% coverage instead of 100%) in order to "leak" their RF signal outside the cable. When the RF signal is interrupted by the presence of a human body with its mass of water, this de-tunes the system to provide for detection. Leaky-coax systems can be buried or placed on fences. Unlike the capacitance system, the leaky-coax system requires only a single cable.
  - ❏ Advantages: Work well on fences and underground. When underground, like the capacitance systems, they are invisible to the offender.
  - ❏ Disadvantages: Do not work well near standing water or in rainstorms.
  - ❏ Exploits: Knowledgeable threat actors have used the slow-move approach (less than 1 ft every 10 minutes) to avoid detection and depending on the area, detection may be avoided during a heavy rainstorm if the detector is fence mounted.
- *Infrared and Laser Detection Systems*
  - ❏ Application: Infrared beams are configured as an infrared light source and a photoelectric detector placed some distance apart. For perimeter use, these are used in vertical arrays so that a person cannot climb over or step under a single beam. To prevent a false light source from being used, they are configured at a specific infrared frequency and are pulsed so that the detector is expecting to see that specific light frequency and pattern, if not,

detection occurs. If the beam is broken, detection occurs; if a false beam is placed on the detector providing an improper frequency of pattern of flashes, detection occurs. Lasers may be used instead of infrared light beams. Detection beams can span short distances or distances up to several hundred meters. All Infrared and Laser Detection Beams are line-of-sight and cannot contour to hills and valleys except by placing additional beams at the vector locations (edges) of hills and valleys. Thus they work best on land with little change in elevation.

❑ Advantages: Very reliable in all weather.

❑ Disadvantages: Detection beams sense anything interrupting the beams including animals, blowing newspaper, leaves, and so forth. This can be mitigated by configuring the beams so that two or more beams must be broken in order to yield detection (alarm filtering).

❑ Exploits: Knowledgeable threat actors cause many false alarms before the actual event so that detection is ignored by the security officer.

■ *Outdoor Passive Infrared (PIR) Detectors*

❑ Application: Although PIR detectors are usually used indoors, there are a few outdoor PIRs available. Outdoor PIRs are equipped with hardened rain-proof enclosures and are designed to work in moderate temperature environments (from somewhat above freezing to about 95°F). They use circuitry that filters the PIR signals to look for human infrared signatures outdoors and avoid common nuisances such as blowing leaves, changes in air temperatures, and so forth.

❑ Advantages: Relatively low cost, work well at short ranges, and their mere presence may serve as a deterrent.

❑ Disadvantages: Do not work at all in high-temperature environments and have difficulty in extreme low-temperature environments because humans bundled up to stay warm so they mask their normal infrared signature.

❑ Exploits: Knowledgeable threat actors can avoid detection by moving very slowly (less than 1 ft per 10 minutes).

■ *Pneumatic Underground Detection Systems*

❑ Application: The most familiar example of a pneumatic detection system is the old-fashioned gas station bell. This comprised a flexible tube (looking sort of like a garden hose) that was stretched across the gas station driveway. When a car drove over the hose, a change in air pressure inside the tube caused the bell to ring. A similar system is available for direct burial and is capable of detecting vehicles or pedestrians across solid ground. The tubes are usually buried just below the surface of the ground with a sensor nearby.

- ❏ Advantages: Accommodate uneven ground and work well in remote areas without quality roads. The detector is invisible as it is buried so detection of the alarm by an intruder is less likely. These are also somewhat less expensive than some other long-line outdoor detectors.
- ❏ Disadvantages: The detection zone is not very precise. Only the entire zone is detected, so it is not possible to pinpoint where along the line the intrusion has occurred. This means that smaller detection zones are necessary. The line has to maintain a constant pressure so the detection circuitry includes pressurization equipment, requiring more power than some other systems. Burrowing animals such as moles have been known to chew through the Pneumatic Hose, causing it to fail. When this happens, the entire line must be searched to find the leak and fix it.
- ❏ Exploits: If a knowledgeable threat actor could find it, he could cut the hose, disabling the entire line until maintenance can occur.
- ■ *Microwave Detection Systems*
  - ❏ *Types:* Available in either monostatic or bistatic versions. Monostatic detectors use a single device that transmits and receives (a transceiver), whereas bistatic detectors use a pair where one is the transmitter and the other is the receiver.
  - ❏ *Monostatic Microwave Detection Systems*
    - – Application: Monostatic detectors fill a volume (detection zone) with microwave energy and detect the echo of that energy. When a person or object enters the detection zone the quiescent echo is changed and that is detected.
    - – Advantages: Quite reliable for short ranges.
    - – Disadvantages: Care must be taken when used near buildings because they can sense through many types of walls. Close proximity to fluorescent lights can cause false alarms. They are sensitive to blowing objects such as leaves, newspapers, wind ripples on a mud puddle, small animals, and so forth.
    - – Exploits: Cause numerous false alarms so that the guard force will ignore the actual intrusion.

- ❏ *Bistatic Microwave Detection Systems*
  - – Application: A detection zone is set up between the transmitter and receiver, with its widest portion in the exact middle of the zone.
  - – Advantages: Accurately detect any movement between the transmitter and receiver.
  - – Disadvantages: Nuisance alarms are a problem. They detect literally everything that goes on between the transmitter and receiver including rain, snow, wind-driven leaves and dust, newspapers, and ripples on a puddle. Everything! Bistatic microwave detectors should always be used with a confirming detector of another technology.
  - – Exploits: Enter during a rainstorm. The unit is effectively blind because it is in constant detection.
- ■ *Ground-Based Radar*
  - ❏ Application: This is a short-range version of long-range marine radar designed to sense vehicles, vessels, and objects close-in (typically less than 2−8 miles).
  - ❏ Advantages: It can track multiple vehicles, moving objects, and people simultaneously and "paint" their location on a Map-based screen.
  - ❏ Disadvantages: It does not work well in close proximity to train tracks and roadways where constant movement can confuse the radar with reflections of other objects. This is one of the most expensive detection technologies.
  - ❏ Exploits: Knowledgeable threat actors can confuse radar with other radar noise on the same frequency.

Building perimeter detection systems

- ■ *Balanced-Biased Door Position Switches*
  - ❏ Application: Like gate position switches, door position switches work by using magnets and a magnetic sensitive switch. A magnet is placed in the top of the door and the switch is placed in the door frame. When the door is closed, the switch rests in a "balanced-bias" magnetic field of several magnets (including the magnet on the door and others within the switch). When the door opens, this unstable magnetic field is disrupted, closing the switch. The switch can also be activated by placing another magnet near the switch, in an effort to bypass the magnet on the door. Thus, the very act of trying to bypass the switch can set it off.

Other versions of this switch are available for overhead and roll-up doors and for windows.

- ❏ Advantages: Very reliable detection.
- ❏ Disadvantages: It is best to conceal the switch and magnet in the frame and top of door. Surface mounting can be used on "back-of-house" doors in rough finish areas.
- ❏ Exploits: Saw through the door, leaving the magnetic switch in place. This is a common exploit for roll-up vehicle doors.

- ■ *Photoelectric Beam Detectors*
  - ❏ Application: Indoor version of the outdoor infrared beam detectors. The beams are pulsed infrared, just like their outdoor cousins, and they sense objects moving between the beams. They are ideal to span a large expanse of doors, and are often used in a warehouse or industrial setting to span up to a dozen vehicle and pedestrian doors.
  - ❏ Advantages: The beam detectors sense entry of an intruder cutting through a warehouse roll-up door, attempting to bypass a door position switch.
  - ❏ Disadvantages: In most installations, beam installation is obvious so they are not often covert.
  - ❏ Exploits: Leave an object blocking the beams, thus putting the alarm zone into the "trouble" state.

- ■ *Glass Break Detectors*
  - ❏ Application: Available in both acoustic and vibration sensitive versions. Older glass break detectors (no longer in use) used foil tape on glass that broke when the glass was broken. Acoustic detectors listen for the distinctive sound of breaking glass. These are usually placed on the ceiling, centrally above several windows. Vibration detectors are small adhesive disks placed on each window which sense the vibration caused when the glass breaks.
  - ❏ Advantages: Very reliable detection of glass breakage.
  - ❏ Disadvantages: Older acoustic detectors were known to detect the sound of vacuum cleaners and sometimes jingling keys.
  - ❏ Exploits: Find another entry point other than through the glass. These detectors will even detect when glass cutters are used.

- ■ *Seismic Detectors*
  - ❏ Application: An indoor version of the outdoor seismic detector, these are used to detect hammering or drilling into walls. They are common on interior vault walls.
  - ❏ Advantages: It is virtually impossible to drill or hammer into a room equipped with seismic detectors.

❑ Disadvantages: Sensitivity must be set to accommodate all kinds of traffic vibrations, so that only a real intrusion is sensed. Unless this is done properly, nuisance alarms are common.

❑ Exploits: Set off a number of false alarms so that the detectors are ignored.

Interior volumetric sensors

■ *Microwave Detection Systems*

❑ Application: Interior microwave sensors are monostatic and are often used to sense activity in stairwells. Because they sense reflected motion, they can sense movement a floor above or below the floor where they are located. I have used these every third floor to sense stairwell motion.

❑ Advantages: Very sensitive to motion, and are particularly sensitive to motion coming toward or going away from the detector. They are less sensitive to detection across their detection zone, but still quite useable for that purpose.

❑ Disadvantages: These detectors are able to see through walls, detecting activity on the other side of a wall. So they must be placed with care, respecting construction materials and normal activity. Microwave sensors can detect plasma moving within fluorescent lights, so they should not be used near them.

❑ Exploits: Create nuisance alarms hoping that the security staff will ignore the alarm. Extremely slow movement can also defeat these detectors.

■ *Infrared Detection Systems*

❑ Application: Infrared detectors create a series of "fingers" and are most sensitive to human movement across their detection zone (the opposite of microwave sensors). The detectors can also create a "curtain" across which motion is detected. They also sense a change in heat across the curtain or fingers in their detection pattern.

❑ Advantages: Provide reliable detection in normal building temperatures and when installed correctly (respecting their limitations).

❑ Disadvantages: Detectors do not work above about 95°F. Heat sources in the detection field such as heating ducts can cause nuisance alarms. These detectors are least sensitive to bodies moving toward the detector. They are designed to detect bodies moving across their detection field.

- ❑ Exploits: Extremely slow movement can sometimes defeat these in high-temperature areas. Knowledgeable threat actors have defeated Infrared detectors by covering their body with a heavy blanket, masking their body heat.
- *Dual-Technology Detection Systems*
  - ❑ Application: Dual-technology detectors combine the best of both infrared and microwave detectors, using both technologies to create a single reliable detector. Both microwave and infrared sensors must trip for the detector to signal an alarm.
  - ❑ Advantages: Sensitive to motion both across and toward the detector. They are far less prone to false alarms than either type of detector alone.
  - ❑ Disadvantages: None.
  - ❑ Exploits: None.
- *Ultrasonic Detection Systems*
  - ❑ Application: Ultrasonic detectors are acoustical microphones and detection circuits that listen for sounds in the ultrasonic frequency range (about 50–100 kHz). Ultrasonic detectors are available in either passive or active versions. Passive detectors listen for sounds in the ultrasonic frequencies whereas Active detectors work like bats, emitting ultrasonic frequencies and listening for changes in the frequency caused by objects reflecting back (a bit like monostatic microwave detectors).
  - ❑ Advantages: None.
  - ❑ Disadvantages: Ultrasonic detectors are less reliable than microwave. Jingling keys, vacuum cleaners, flying insects, and nearby airports and construction zones can all confuse ultrasonic detectors.
  - ❑ Exploits: Many as stated earlier. In one case, an intruder reportedly released a swarm of bees into the room.
- *Thermal Imaging Detection Systems*
  - ❑ Application: Thermal imaging cameras can sense body heat against the background.
  - ❑ Advantages: Provide very reliable detection and also provide an image of the intruder, showing his location in the room or corridor.
  - ❑ Disadvantages: Thermal imaging detectors are very expensive.
  - ❑ Exploits: Cannot penetrate glass, and high ambient temperature near 98°F can mask an intruder.

Interior point detection systems

- *Duress Alarms*

❑ Application: Duress alarms (also called panic alarms) are a detection switch that is activated by a human in response to a security concern. Common forms include switches under a counter or on a wall tripped by hand, a switch on the floor tripped by lifting one's foot up under a paddle, and cash drawer bill traps. Unlike all other alarms discussed here that are designed to be tripped by an unwitting offender, duress alarms rely on an authorized person to intentionally trip the alarm when they are concerned about a security event. Thus these are the highest priority alarms in the system. Typically there is a violent event occurring when a duress alarm is tripped.

❑ Types of Duress Switches:
  - Wall-Mounted Duress Switch: May be placed on a train platform or other public or semipublic area for concerned persons to signal security or police when they are concerned about being attacked.
  - Two-finger Switches: Requires the user to push two buttons simultaneously in order to signal an alarm. The use of two buttons helps ensure that the alarm is intentional and is not being triggered by something bumping against the switch. Commonly used under a desk or cash counter.
  - Shroud Switches: A single switch enclosed by a shroud to keep objects from pushing on the switch.
  - Pull Switches: Requires the user to pull a plunger out from its off position, usually also shrouded to allow fingers to enter between the shroud and the switch but keeping out objects that could accidentally trigger the alarm.
  - Foot Switches: The user places their foot under a paddle and lifts the foot, triggering the switch. These allow the user to have both hands above the desk or counter and are often preferred for their discreet nature.
  - Bill Traps: When the last bill in a cash drawer is pulled out, a clip over the last bill makes electrical contact with a steel plate below the bill, tripping the alarm. Thus the very action of pulling out all of the cash from a drawer sets off a silent alarm.
  - Audio/Video Verification: All duress switches should be used with audio and video to verify the alarm. These tools allow the security console officer to determine the number of aggressors, their actions, their weapons, and how much of a force is needed to quell the event and to protect the innocent people involved.

- ❑ Advantages: An essential tool to allow persons at risk to signal for help.
- ❑ Disadvantages: If not designed or installed correctly, they can cause nuisance alarms.
- ❑ Exploits: The offender may try to defeat the alarm communications so that it will not send an alarm (by cutting a telephone line for example) or try to convince the person under attack not to trigger a duress alarm, usually by threat of force.
- ■ *Explosives Detection Methods and Systems*
  - ❑ *Types:* Explosives detection can be accommodated by a variety of methods and detection technologies including Dogs, X-rays, millimeter wave scanners, spectrographic detection, and visual detection.
  - ❑ *Visual Inspection Approaches*
    - − Application: Visual inspection of packages by trained guards is a common and often effective means of deterrence and detection. A package is opened by hand and inspected for suspicious contents by a trained guard.
    - − Advantages: Can be very thorough. Allows a trained guard time with a subject to analyze their behavior, thus detecting suspicious persons by their behavior as well as by the package itself.
    - − Disadvantages: Most guards are either poorly trained or unconcerned with conducting a thorough search. Most searches are very poor.
    - − Exploits: Of all of the methods of explosives detection, if I were carrying explosives, I would most hope to encounter a visual inspection. Explosives can usually be easily packaged into the container in a manner that evades visual inspection. Also, most visual inspections are profoundly cursory, often missing obvious briefcase pockets and so forth. In almost all inspections I have undergone, my packages were not fully inspected. Also, one could use an accomplice to create a distraction at the time of the hand search.
  - ❑ *Dogs*
    - − Application: Trained dogs are arguably the most reliable explosives detection method. They can detect minute amounts of virtually every explosive compound. Terrorists fear dogs more than any other detection methodology.
    - − Advantages: Dogs are very efficient when properly trained and used. In many countries, people find the use of dogs to

be engaging rather than off-putting. This is not the case in Muslim countries where the use of dogs can be offensive. For some reason, persons carrying explosives also have an aversion to sniffer dogs. Dogs will work for dog food and the love of their handlers.
- Disadvantages: Cannot be used in Muslim countries due to cultural norms. Dogs cannot work effectively for more than about 20 minutes, requiring frequent breaks. Thus multiple dogs are needed for a single checkpoint. When used for multiple checkpoints the cost of training and handlers can become expensive. Dogs' training must be constant for them to be reliable. This is also expensive due to the cost of trainers.
- Exploits: Package explosives hermetically. No other exploits for a well-trained dog.

❏ *Package X-Ray Scanners*
- Application: There are two types: transmission and backscatter. These can image the interior of packages and cargo, including entire vehicles, trailers, and trucks.
- Advantages: Accurately detect most contraband materials including most explosives, drugs, and weapons. Thus, these are among the most versatile explosives detectors.
- Disadvantages: Can miss some substances. The technician must be well trained and continually alert.
- Exploits: Objects can be placed inside the package or container in a manner that is less obvious to a less-well trained operator. For example, a gun can be placed upright against the side of a suitcase instead of flat, where its outline would be obvious.

❏ *Personnel X-Ray Scanners*
- Application: All use backscatter technology. These create a "naked" image of the person scanned showing anatomical features and many kinds of objects placed against the body.
- Advantages: Backscatter imagers can sense weapons, drugs, explosives, and other contraband such as packages of money on the person.
- Disadvantages: Increasingly, these are seen as culturally offensive by all cultures, but especially in Asian and Muslim cultures because they show all anatomy. News stories of TSA agents gleefully selecting attractive women for inspection and horror stories of medical survivors having breast implants or colostomy bags have outraged the public. There continues to

be outrage against these machines, so they are rarely used except at transportation facilities, where their use is detested, but mandatory. Accordingly these are entirely impractical for commercial use.
- Exploits: Certain substances are virtually undetected by these scanners, but many TSA tests have successfully gotten handguns past the detectors.

❑ *Millimeter Wave Scanners*
- Application: These are somewhat similar to backscatter X-ray technology, but work from a distance and provide a visual (not X-ray vision) image of the person, showing the location of the detected objects that the person has over the image of the person. Some versions show a quasi-X-ray image of the person with less definition than backscatter X-rays machines.
- Advantages: Good imaging under ideal environments (no heavy clothing, no significant heat, no rain-wetted clothes, etc.)
- Disadvantages: Rain, heat, and heavy clothes obscure the image.
- Exploits: Enter when there is a heavy rain or wear heavy clothes.

❑ *Spectrographic Detection Systems*
- Application: A variety of electronic tools use spectrum analysis to "sniff" for explosives. Some are configured as wipes that are wiped on a package and then the wipe is placed into the detector, while others are handheld and sniff the article directly.
- Advantages: Very reliable detection, not quite as good as dogs. The very best can be swiped across a car door, steering wheel, or trunk lid and will detect minute traces of common explosives.
- Disadvantages: Only detect explosives leaving residue of vapor.
- Exploits: These are a very good deterrent. Very difficult to beat. However, certain explosive compounds are very difficult to detect, so be certain what you have to detect before relying on these alone.

■ *Radiological Detection Systems*
❑ *Standard Gamma Ray Detectors*
- Application: Handheld and fixed gamma ray detectors sense gamma rays given off by most radiological compounds.

- Advantages: Relatively inexpensive.
- Disadvantages: Can be fooled by common articles such as crockery, bananas, and so forth, which naturally give off gamma rays.
- Exploits: Sufficient lead packaging, although this is very difficult to achieve.
- ❏ *Video Gamma Ray Detectors*
  - Application: These are very rare, but effective. All solid-state video sensors are also sensitive to gamma rays and can thus be used with specialized software to detect gamma rays on a person, in a vehicle, or in a container.
  - Advantages: Can use existing video camera as a sensor. Immediate identification of the suspicious person, vehicle, or object by the video camera.
  - Disadvantages: Can be fooled by common articles such as crockery, bananas, and so forth, which naturally give off gamma rays.
  - Exploits: Sufficient lead packaging, although this is very difficult to achieve.

Intelligent video analytics sensors

- ■ *Types*
  - ❏ Video cameras can be embedded or used with intelligent video software to detect specific or abnormal behaviors that other sensors would not detect.
  - ❏ Common detection algorithms include:
    - Direction of travel (entry into an exit area)
    - Crossing a virtual line (such as a person falling into a subway rail from a platform)
    - Left object detection (briefcase left)
    - Placed object detection (briefcase placed by a bench)
    - Removed object detection (painting on a wall)
    - Lack of movement (person floating in a pool)
    - Loitering behavior of one person among a crowd of people or one person alone in an area like a subway platform
    - Crowd gathering
- ■ *Edge Sensors*
  - ❏ Application: Video cameras equipped with a digital signal processing CPU inside the camera so that the processing is conducted "at the edge" rather than centrally at a server.

- Advantages: Scalable system. Costs of the system are gradual as you add more cameras. These require fewer servers and the cameras use less network bandwidth than some other systems.
- Disadvantages: The camera will process only the algorithms it is programmed to perform and no others.
- Exploits: One must know that a camera is intelligent and what algorithms it is processing.

- *Conventional Central Processing Sensors*
  - Application: A normal digital video stream is decoded by a special server operating designated video motion algorithms.
  - Advantages: Can detect many activities and behaviors that other sensors would not see or notice.
  - Disadvantages: The system will process only the algorithms it is programmed to perform.
  - Exploits: One must know that a video system is intelligent and what algorithms it is processing.

- *Learning Algorithm Systems*
  - Application: Edge Sensors and Conventional Central Processing Sensors both require that a computer programmer design a specific algorithm to detect a specific unwanted behavior. There is another type called Learning Algorithm systems (such as BRS Labs) in which the system is completely ignorant out of the package, but "learns" over time what is normal in the field of view of the camera. These are central processing systems that respond to any out-of-the-ordinary behavior. They will, e.g., equally detect a person carrying an M-16 into an airport or a person carrying a banjo into the airport. Both are uncommon sights to the system.
  - Advantages: These systems are very good at detecting almost any type of unwanted behavior and they get better over time at determining what is unusual but acceptable and what is unusual and unwanted.
  - Disadvantages: No more costly than other central processing intelligent video systems.
  - Exploits: Absolutely none. (Cloak of invisibility, maybe?)

Complex alarm sensing

Combinations of sensors should be used together to minimize nuisance alarms, e.g., outdoor sensors used with video motion detection or beam detectors used with roll-up door position switches. Any combination of sensors that combines the abilities of both to make the exploit of the other impossible to use is a good design. Combinations can be used to verify alarms or as an "and" type logic cell to trigger only when both detectors alarm.

BEYOND ALARM DETECTION

Alarms can be used for more than alerting. They can also be used to detect trends and patterns that would otherwise not be possible and to assist in vulnerability detection and analysis.

Trend analysis

Trend Analysis is the art of determining if there is a trend developing or occurring around a specific type of alarm event. For example, does the same alarm always trigger whenever a train goes by or does the same alarm always trigger when it rains? Is there an instance of crowds gathering too near the edge of a subway platform at afternoon rush hour?

Trend Analysis can be conducted by hand or with specialized software. Common software includes PPM-2000, which can be interfaced with the alarm/access control system and is capable of spotting trends that you otherwise might never notice.

Vulnerability analysis

When a trend spots a common nuisance alarm condition, such as whenever a train goes by or when it is raining or snowing, this points out a vulnerability that should be covered by additional security countermeasures.

Alarm analysis

Alarm/access control systems should also "tag" each alarm event with a digital video clip of the area of the alarm from about 5 seconds before the alarm to about 2 minutes after the initiation of the alarm incident. When the security manager looks through the alarm incident log in the alarm/access control system software, there should be a camera icon next to each alarm event (where a camera can view the area of the alarm). Clicking on that icon should bring up the video clip onto the video screen. This requires integration between the alarm/access control system and the digital video system, and programming of response cameras to each alarm event. These "tagged" alarm video clips should be stored separately from regular archived video so that they are not easily overwritten.

CHAPTER SUMMARY

1. Basic alarm concepts include:
   a. Detection and Alarm Initiation
   b. Alarm States
      i. Secure
      ii. Alarm
      iii. Bypass
      iv. Trouble
   c. Alarm Filtering (usually by using two or more sensors)
   d. Communications and Annunciation
   e. Alarm Assessment
   f. Response
   g. Evidence Gathering
2. Alarm sensors can be categorized into several broad areas:
   a. Outer Perimeter Detection Systems
   b. Building Perimeter Detection Systems
   c. Interior Volumetric Detection Systems
   d. Interior Point Detection Systems
   e. Intelligent Video Analytic Sensors
3. Outer Perimeter Detection Systems include:
   a. Gate Breach Detection System
   b. Seismic Detection Systems
   c. Fiber-optic Detection Systems
   d. Capacitance Detection Systems
   e. Leaky-coax Cable Systems
   f. Infrared and Laser Detection Systems
   g. Outdoor Passive Infrared Detectors
   h. Pneumatic Underground Detection Systems
   i. Microwave Detection Systems
      i. Monostatic Systems
      ii. Bistatic Systems
   j. Ground-based Radar
4. Common Building Perimeter Detection Systems include:
   a. Balanced-biased Door Position Switches
   b. Photoelectric Beam Detectors
   c. Glass Break Detectors
   d. Seismic Detectors
5. Common Interior Volumetric Sensors include:
   a. Microwave Detection Systems
   b. Infrared Detection Systems
   c. Dual-technology Detection Systems

    d. Ultrasonic Detection Systems

    e. Thermal Imaging Systems

6. Common Interior Point Detection Sensors include:

    a. Duress Alarms

    b. Explosive Detection Methods and Systems

        i. Visual Inspection Approaches

        ii. Dogs

        iii. Package X-ray Scanners

        iv. Personnel X-ray Scanners

        v. Millimeter X-ray Scanners

        vi. Spectrographic Detection Systems

        vii. Radiological Detection Systems

            − Standard Gamma Ray Detectors

            − Video Gamma Ray Detectors

7. Types of Intelligent Video Analytics Sensors include:

    a. Dedicated (specific) Algorithm Systems

        i. Edge Sensors (in the camera)

        ii. Conventional Central Processing sensors (in a server)

    b. Learning Algorithm Systems

8. Combinations of sensors should be used together to minimize nuisance alarms, e.g., outdoor sensors used with video motion detection or beam detectors used with roll-up door position switches.

9. Alarms can be used for more than alerting. They can also be used to detect trends and patterns that would otherwise not be possible and to assist in vulnerability detection and analysis.

10. Trend Analysis is the art of determining if there is a trend developing or occurring around a specific type of alarm event.

11. When a trend spots a common nuisance alarm condition, such as whenever a train goes by or when it is raining or snowing, this points out a vulnerability that should be covered by additional security countermeasures.

Q&A

1. *Basic alarm concepts include:*

    a. Detection and Initiation

    b. Filtering and Alarm States

    c. Communications and Annunciation

    d. All of the above

2. *Basic alarm concepts include:*

    a. Assessment

    b. Response

    c. Both a and b

    d. Neither a nor b

3. *Types of alarm sensors include:*
   a. Outdoor Perimeter Detection Systems
   b. Building Perimeter Detection Systems
   c. Interior Volumetric Detection Systems
   d. All of the above
4. *Types of alarm sensors include:*
   a. Interior Point Detection Systems
   b. Intelligent Video Detection Systems
   c. Neither a nor b
   d. Both a and b
5. *Which type of detection system can detect vehicles and footsteps from their sound?*
   a. Seismic Detection System
   b. Capacitance Detection System
   c. Infrared and Laser Detection System
   d. None of the above
6. *Which type of detection system uses radio frequency?*
   a. Seismic Detection System
   b. Infrared and Laser Detection System
   c. Leaky-coax Detection System
   d. None of the above
7. *Which of the following are valid alarm states?*
   a. Secure, Alarm, Bypass, and Enable
   b. Secure, Alarm, Bypass, and Trouble
   c. Unsecure, Alarm, Bypass, and Enable
   d. Unsecure, Alarm, Bypass, and Trouble
8. *Alarm filtering helps avoid*
   a. Nuisance Alarms
   b. False Alarms
   c. Remotely Triggered Alarms
   d. All alarms
9. *Assessment is used to ensure that the alarm is*
   a. Invalid
   b. Valid
   c. Not a Security Event
   d. Reported to the console
10. *Response may typically involve sending either _____ or _____ to confront the intruder.*
    a. Police or Police Dogs
    b. Patrol Car or Military
    c. Police or Roving Security Officers
    d. None of the above
11. *One of the logs in an alarm/access control system that keeps evidence is*
    a. The Evidence Log
    b. The Alarm Event Log
    c. The Alarm Action Log
    d. None of the above

12. *Building perimeter detection systems may include:*
   a. Balanced-biased Door Position Switches
   b. Photoelectric Beam Detectors
   c. Glass Break Detectors
   d. All of the above
13. *Interior volumetric sensors may include:*
   a. Microwave Detection Systems
   b. Infrared Detection Systems
   c. Dual-technology Detection Systems
   d. All of the above
14. *Interior point detection systems may include:*
   a. Duress Alarms
   b. Explosives Detection Methods and Systems
   c. Radiological Detection Systems
   d. All of the above
15. *Intelligent video analytics sensors include:*
   a. Conventional Central Processing Sensors
   b. Learning Algorithm Sensors
   c. Both a and b
   d. Neither a nor b
16. *Alarms can also be used to conduct:*
   a. Trend Analysis
   b. Vulnerability Analysis
   c. Both a and b
   d. Neither a nor b

Answers: (1) d; (2) c; (3) d; (4) d; (5) a; (6) c; (7) b; (8) a; (9) b; (10) c; (11) d; (12) d; (13) d; (14) d; (15) c; (16) c.

Related Security Systems

CHAPTER OBJECTIVES

1. Learn about photo ID systems
2. Learn about visitor management systems
3. Understand the basics of security video
4. Learn about many types of security communications systems
5. Learn about system architectures models for campuses and remote sites
6. Pass a quiz on related security systems

CHAPTER OVERVIEW

Alarm/access control system functions can be expanded and improved by integrating them effectively with related security systems. In this chapter, we will discuss a wide variety of related security systems and how they can all be used together to make a more powerful and effective security program.

Systems discussed include photo ID systems and visitor management systems. We will explore video systems and communications systems in some detail and how to interface them with alarm/access control systems to make a single powerful security system. We will also review a variety of system architecture models for campuses and remote sites and also how security console functions can be improved by system architecture and analysis.

PHOTO ID SYSTEMS

In the early days of access control before electronics, access cards carried all of the information that a guard at a gate needed to know to allow or deny access to the person wearing the card. The card had a facility code (usually a logo of some sort), the user's name and department, and a badge number. Most access cards also displayed a photo of the user so

Electronic Access Control. DOI: http://dx.doi.org/10.1016/B978-0-12-805465-9.00021-X

that the guard could compare the photo and the person wearing it to be certain that the bearer was in fact the person to which the card was issued. Additionally, the cards commonly displayed an array of colors that identified which areas of the facility the bearer was allowed to enter.

As electronic access control systems became more common, Photo ID systems were developed to work with them. Photo ID systems print all of the information on the actual face of the access card except for access authorizations, which of course is inherent in the electronic access card itself.

Photo ID systems include:

- A badging computer that is connected to the access control system database
- A badge printer
- A digital camera, backdrop, and light

Large photo ID operations use multiple computers, printers, and cameras to keep up with a constant stream of new users at factories and other such facilities with high employee hire and turnover rates.

Virtually all alarm/access control systems today have a photo ID module available.

VISITOR MANAGEMENT SYSTEMS

Although access control systems were originally made to provide easy access for authorized employees, visitors also needed access. It is not a good idea to issue permanent photo ID to every visitor.

Different organizations handle visitor access in different ways. One of the best ways to accommodate visitors is by using a visitor management system. A visitor management system is designed to record visitors into a visitor database and issue a temporary visitor access card that can be used at portals equipped with special visitor card readers.

Temporary visitor access cards are typically printed on card stock and include a photo of the visitor, their name, an expiration date, and a card number. The card may also bear a barcode corresponding to the visitor access card number.

Although some facilities only use visitor management systems to keep track of who came through the front door and use the cards as ID cards only for visitors while in the facility, other facilities make full use of visitor cards as access cards, used with the access control system to allow visitors unescorted access to designated controlled areas.

Some visitor management systems track visitor access activity separately from employee access activity, although many access control systems can be programmed to track both. Most visitor management system users also have at least some key entry (and exit) readers that use the visitor access cards as well as employee access cards. Often these readers are placed on lobby turnstiles equipped with both proximity type readers to read employee access cards and barcode readers to read visitor access cards. Visitors are escorted elsewhere throughout the facility. The turnstiles record the visitors going into and out of the facility.

Visitors can be tracked by date or frequency of visits and which employees they came to see as well as other data. The biggest advantage of the visitor access cards is that they are disposable, unlike employee access cards.

SECURITY VIDEO

Without a doubt, after the alarm/access control system, the most important system for a security unit to have and use is a well-designed security video system. Video systems provide the ability to:

- Assess events detected by the alarm/access control system
- Observe all areas of the facility for inappropriate or suspicious activity
- Conduct investigations remotely
- Follow the movements of subjects of security events or investigations
- Produce visual evidence of security events

Today most video systems use analog video cameras and digital storage media; however, this is a snapshot in time. The trend is moving toward digital cameras and digital storage.

Video history you need to know

In the earliest days, security video was an expensive rarity, with even the largest multinational corporations having no more than a few video cameras watching over their most critical assets.

Recorders were unheard of and each camera was wired to an individual video monitor, watched over by a 24-hour guard post.

Today, I have a 16 channel digital video recorder (DVR) and a complement of video cameras watching my home and property. When I received a telephone call in Beirut, Lebanon, from my alarm company in Houston, Texas, telling me that a glass break sensor had triggered on my home in Texas, I asked them to roll the police and then checked my video system to discover that my gardener was mowing the lawn next to the very window that reported the alarm. Although the police found nothing, a check of my home by my wife uncovered a small crack from a stone that was launched by the lawn mower. So how did we come from only a few cameras for multinational firms to private individual assessing alarms from halfway around the world?

The answer is commoditization. I am old enough to remember when the first video cameras were produced for the market with a price of about $1000 each. Integrators exclaimed: "Oh the horror! This will be the end of the security video market! How will we ever be able to make money on security video if cameras are priced under $1000?" Quite the opposite happened. Today security video is the largest segment of the security technology market and small security installers have evolved into large multinational firms doing quite well, thank you.

As prices fell, consumer demand increased. The technology has gone through the following phases:

- 1960s: Individual cameras to individual monitors
- 1970s: Multiple cameras to sequential switchers to fewer monitors
- 1980s: Introduction of quad multiplexers, displaying four cameras on one monitor
- 1980s: Introduction of first analog matrix switches
- 1980s: Introduction of first video recorders for security
- 1990s: Introduction of first 16 channel digital multiplexers
- 1990s: Introduction of first time-lapse video recorders
- 2000s: Introduction of first multiplexing DVRs
- 2000s: Introduction of first digital video cameras
- 2000s: Introduction of first server-based digital archiving solutions
- 2000s: Introduction of first graphical user interface (GUI)
- 2000s: Introduction of exotic security video cameras (infrared, laser illuminated, etc.)
- 2010s: Further development of GUI-based video systems made alarm/ access control systems an attachment to rather than the focus of security monitoring systems

All of this was driven by continuously falling prices, and demand increased. Increasing demand has driven technological development.

Cameras and lenses

Video cameras are an amazing invention. Light falls through a lens onto an imager and is converted into electric signals only to be reassembled as an image on a video monitor or storage device. In the earliest days, this was all done using analog signals. Today, the imagers are digital devices (basically an array of transistors with no cover on the circuit so that light can fall directly onto the transistors). In the case of digital video cameras, this digital imaging device is converted through a standardized digital compression algorithm such as MJPEG, MPEG-4, or H.264, and it is transmitted via TCP/IP to a DVR or server where it is distributed for display on a computer monitor and recorded onto a hard disk. That is the logical development of the industry. But we are not quite there yet.

For some unknown reason, in many (analog) cameras, the digital image is converted to an analog signal where it is transmitted over coaxial wire to the analog input of a DVR where the analog image is converted into a digital image to be displayed and stored digitally. Okay, am I the only one wondering why they convert a digital image to an analog image only to reconvert it back to digital? This is how all technology evolves.

Today, the following types of cameras are available, and in most cases they are available in both analog and digital versions:

- Imager and processor options:
  - Standard resolution color video camera (provides a color image under good lighting conditions)
  - Low-light video camera (works well under low light and standard lighting conditions)
  - Wide dynamic range (WDR) video camera (adjusts scene contrast to accommodate scenes with both light and shadow areas)
- Form factor options:
  - Fixed video camera
  - Fixed dome video camera
  - Pan/tilt/zoom (PTZ) video camera
  - Dome PTZ video camera
- Mounting options:
  - Wall mounting
  - Ceiling mounting
  - Corner mounting
  - Pole mounting
  - Extendable parapet mounting

- Digital options:
  - ❏ Standard resolution cameras
  - ❏ Mega-pixel (MP) video cameras (provides much higher resolution than standard video cameras)
- Lens Options:
  - ❏ Fixed standard lens (fixed focal length)
  - ❏ Vari-focal lens (adjustable focal length to obtain the best scene view with just a single lens)
  - ❏ Zoom lens (remotely adjustable focal length)
  - ❏ Fixed iris (for fixed lighting conditions)
  - ❏ Auto-iris (adjusts automatically to changing lighting conditions)
  - ❏ MP lens (extremely high resolution capabilities to bring the most out of MP Cameras—all MP cameras should be used with MP lenses)

Lighting and light sources

Lighting is essential to good video images. However good lighting is not always available so it pays to know how different cameras will perform under different lighting conditions. First, here is a basic primer on lighting.

Lighting levels

- If we want to measure the amount (level) of visible light (illumination) in an area, we would be interested in knowing its Lux (metric) or footcandles (fcs) (English) level.
- One fc is the amount of light equal to one candle at 1 ft (one lumen on a 1 ft^2 surface).
- Lighting can also be measured in Lux or Lumens per ft^2.
- Multiply fcs by 10.76891 to convert to Lux
- Multiply Lux by 0.09290304 to convert to fcs
- fcs = Lumens per ft^2 (no conversion necessary)
- A typical sunny day can measure between 5000 and 10,000 fc
- An average room may be about 30 fc
- A full moon can provide about 0.2 fc of illumination

Lighting sources

- Different types of lights yield different colors of light. Light is measured in Kelvin (color temperature). Video cameras work best with lighting that is as close as possible to natural overhead daylight at 12:00 noon (5600 K).

- Color temperature affects how warm or cool we perceive the lighting to be. Lower Kelvin temperatures emit warmer light while higher Kelvin temperatures emit cooler light. Reddish colors (like a candle flame) are considered warm lights, whereas bluer colored lights (like moonlight) are considered cooler lights.
- Approximate color temperatures of various lighting sources:
  - 1700 K: match flame
  - 1700 K: low pressure sodium lamps
  - 1850 K: candle flame, sunrise/sunset
  - 2100 K: high pressure sodium light fixture
  - 2700 K: tungsten incandescent light
  - 2900 K: warm white fluorescent light
  - 3000 K: halogen light
  - 3700 K: metal halide light
  - 4100 K: moonlight
  - 4200 K: cool white fluorescent light
  - 5000 K: horizon daylight
  - 5600–6000 K: natural overhead daylight at noon
  - 6500 K: overcast daylight
  - 7000 K: mercury vapor lamp

Important facts about lighting for security video

- Lighting can be either diffuse or direct. Direct lighting provides better contrast, but sometimes there is too much of a good thing.
- Lighting is best for video when it is overhead and diffuse with little or no reflections.
- Lighting problems to avoid include:
  - Strong side lighting, such as from a window
  - Backlighting, placing the subject in a shadow
  - Very low lighting
  - Strong reflections from shiny surfaces
  - Viewing light sources directly

What we see

Our eyes automatically adjust color temperature so that we perceive the same general image under a wide variety of lighting conditions. Color temperature affects our mood more than our actual perception of the light color itself. However, this is not true for color video cameras, which render the color on the monitor screen exactly as the temperature it sees. Thus, one image taken under high pressure sodium lights may appear yellow while another taken under Fluorescent light may appear green.

Auto-white balance

Most color cameras today have an automatic white balance adjustment to assist them in displaying lighting that appears to be more true to the colors we see. While this circuit helps adjust for true color, the results are not always ideal.

Dynamic range

Dynamic range is the camera's ability to differentiate the extremes of light and dark within a scene, especially when one is close to the other. WDR video cameras use processing circuitry to down-contrast light areas and dark areas so that the overall scene becomes more readable. Our eyes do this trick naturally. WDR works best, e.g., when we need to see a subject close to or standing in front of a bright window.

WDR cameras should not be used everywhere. Using WDR in evenly lit scenes results in less overall contrast and a less readable scene.

Display devices

Most display devices today are LCD or LED computer screens. Older display devices included cathode ray tube (CRT) computer screens, plasma screens, and analog CRT video monitors. The best display devices provide fast refresh rates, very high resolution, and very high contrast.

Video recording devices

In the early days of security video, video was recorded on Open Reel 2″ videotape recorders. This was entirely impractical and soon gave way to recording video on VHS video cassettes. This was shortly followed by time-lapse VHS recorders, which have given way to modern digital recording. Today, security video is recorded on either DVRs or network video recording systems.

DVRs (Fig. 21.1) receive either analog or digital video inputs (depending on the brand and model), and for analog inputs, they digitize the video (convert it from analog to a digital signal). Most DVRs also compress the video into a digital format that is more compact than raw video to save storage space. The video is stored on internal or attached hard disk drives (a few also have tape drives for back-up). DVRs may operate alone or networked together as a group. Most DVRs can support a workstation directly with intrinsic software or software that resides on the workstation. Networked DVRs provide the workstation with the ability to view

■ **FIGURE 21.1** Digital video recorder. 2011 Pelco, Inc.

multiple cameras on multiple DVRs. The key element defining DVRs is that cameras typically connect directly to the DVR.

Network video recording systems are available in several versions, but they all share the ability to work only with predigitized sources that are received across the network, and not directly onto the recorder.

Some DVRs can connect with dedicated network video recorders that are designed to work with specific brands of DVRs. Most network recording systems are designed to work with server-based digital video systems. Server-based digital video systems use a quantity rating from 1 to N servers (Fig. 21.2). In a server-based system video cameras are digitally encoded either within the video camera (digital camera) or through a digital encoder. Both the cameras and the servers connect to the security system network. This allows cameras to be placed widely among floors, buildings, and campuses and still to all be centrally recorded on the servers with a minimum of wiring. This would not be possible using DVRs.

Server-based systems also have the advantage of being able to provide fail-over servers so that if one server fails, another takes its role instantly, recording to the same or different recording media. Fail-over servers can also record to redundant media so that even if the media fails, the recording is intact.

Several video recording schemes are available using server-based systems. These include:

- Internal hard disk storage (HDS)
- Direct attached storage (DAS)
- Network attached storage (NAS)
- Storage area network (SAN)

Internal HDS is common on Fail-over servers where the need to record may be short-term (such as hours or a couple of days). Internal HDS is also common on very small systems where there is only a single server and few cameras.

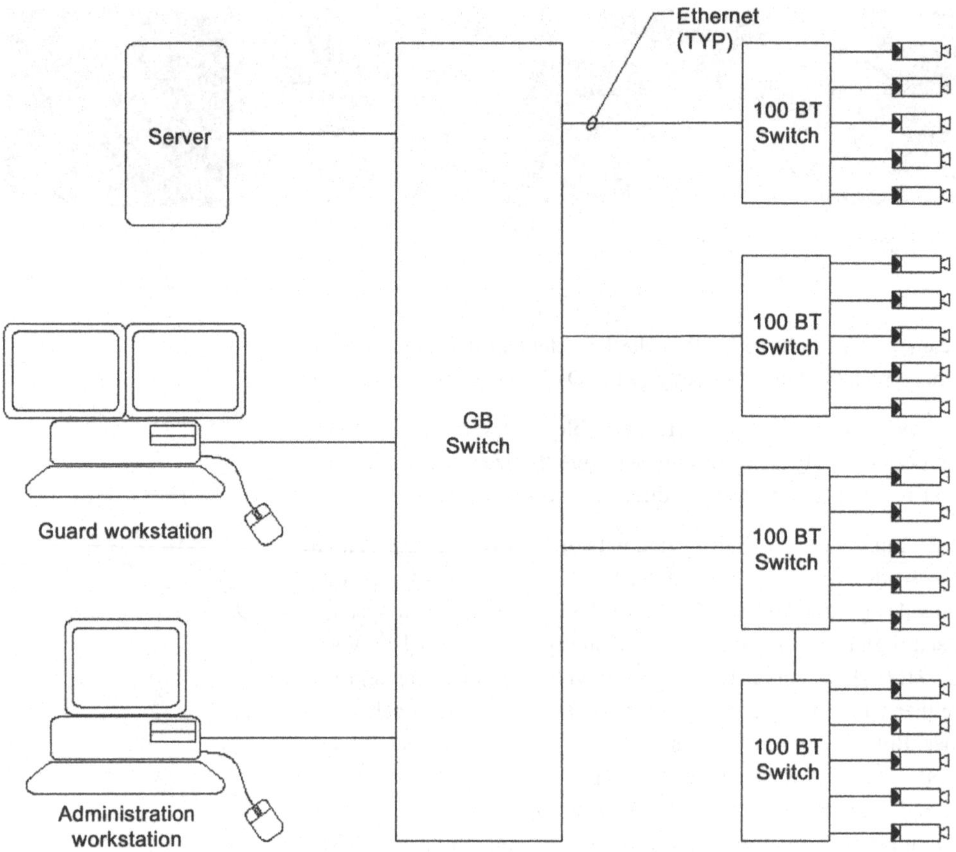

■ **FIGURE 21.2** Basic server architecture.

DAS(Fig. 21.3) is usually configured as an external hard disk that is directly connected to the server (not through a network). This is also common on small systems with a single server. DAS usually provides more storage than HDS. DAS systems typically have from 1 to 8 hard disks in a common enclosure, and these are usually connected through a Firewire or USB-2 port.

NAS(Fig. 21.4) is generally similar to DAS in that it is a box of hard disks (Just a Bunch of Drives, JBOD) except that instead of the box connecting to a single server, it connects to the same network that the servers and cameras or encoders are connected to. NAS has the ability to share its storage across a small number of servers.

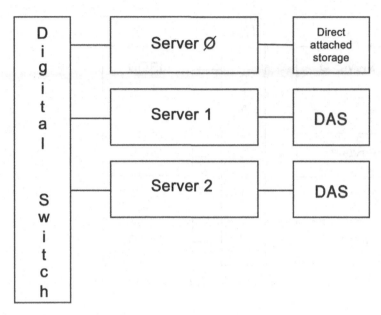

■ FIGURE 21.3 Direct attached storage.

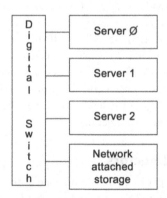

■ FIGURE 21.4 Network attached storage.

A SAN(Fig. 21.5) comprises one or more JBODs that are controlled by a SAN Switch. The SAN switch directs incoming data from video sources (cameras and encoders) to the drives in the JBOD. SANs can store enormous amounts of data. It is not at all unusual to store over 200 Terabytes (TB) of video on SANs.

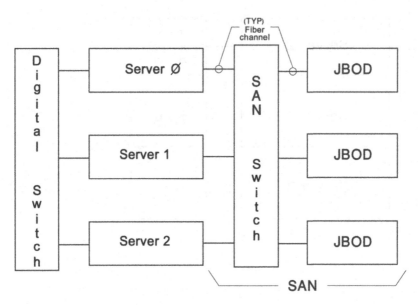

Both NAS and SAN storage configures multiple hard disk drives into RAID arrays that provide more reliability and redundancy than conventional single-disk storage. The most common RAID arrays for security video are RAID-0 and RAID-5. The RAID-0 storage mirror is two or more drives so that the data is completely redundant between the drives. RAID-0 is used for the server operating systems and programs. RAID-5 uses "striping" to span the data across many drives. This process provides a great deal of reliability and redundancy of the data. RAID-5 storage is often used to store video and other data.

Video motion detectors

Of all the different types of alarm sensors, one of the most useful types is video motion detection (VMD). VMD does what it says: it compares subsequent frames of video from a camera comparing one frame to the next looking for any differences between the subsequent frames. When the VMD sees a difference, an alarm is sent to the monitoring workstation. Virtually all VMDs break the image into many small cells and allow the user to identify which cells are to be used for motion detection and which will allow motion without creating an alarm. This is useful, e.g., to detect motion in a storage yard while ignoring motion in an adjacent street. VMDs also let the user adjust sensitivity for each cell, and most have other useful settings to optimize the motion detection. VMDs were originally discrete hardware boxes, but now virtually all are in the form of software algorithms that are built into the DVRs or archive servers.

Video analytics

VMDs are a very basic form of video analytic process. Just as an algorithm can be written to detect motion, so can many other algorithms be written to detect other specific behaviors within the scene view of the video camera. Common video analytics include:

- Object placed in the scene by a person and left behind
- Object removed (painting on wall, etc.)
- Loitering behavior
- Person going in one specific direction (entering through an exit-way)
- Object or person entering a defined detection zone
- Object or person leaving a defined detection zone
- Person, vehicle, or object stopping
- Fence trespassing detection
- Automatic autonomous PTZ tracking of a detected moving object, person, or vehicle

There are two basic types of video analytics: those that have specific algorithms to detect specific behaviors and a newer type that uses no algorithm at all. Instead the second type uses artificial intelligence (AI) to allow the video analytic to "learn" what normal behavior is within the scene view of the camera. Night versus day, busy at morning, lunch and evening rush hours and quiet in between, vehicle traffic moves according to normal traffic rules, and so forth. So when a pedestrian walks in through a vehicle exit gate. That action is detected as abnormal behavior. When a pedestrian walks across the scene carrying an AK-47, this is a profile that the system has never seen before so it yields an alarm. It would do the same if the person were carrying a banjo. When a fight breaks out inside a gymnasium, it yields an alarm. There is no specific algorithm for most of these but the AI system "learns" to detect unwanted behavior that might never have been imagined by the user. AI systems are the future of video analytics (Fig. 21.6).

Video system interfaces

Video systems can interface to other systems in a variety of ways. The most common of these include:

■ **FIGURE 21.6** Artificial intelligence video analytics. The AISight logo is a registered trademark of Behavioral Recognition Systems, Inc. (BRS Labs). *Photograph courtesy of BRS Labs and used with permission.*

- Dry contact interfaces (notify video system to display a camera in response to an alarm or may notify other system for an event in the video system)
- RS-232 interface (receives a script to display a camera in response to an event on another computer or tell another system to perform an action for an event in the video system)
- RS-485 interface (similar results to RS-232)
- TCP/IP interface (essentially any function available; e.g., having the digital video system pass a video frame or clip to the alarm/access control system to file it in the same database as an associated alarm)

SECURITY COMMUNICATIONS

No matter how much people are enamored of electronic security systems, it is the physical and operational security that does the real heavy lifting of security programs. Electronic security systems serve as the eyes and ears of the security operations staff so that they can position guard assets where and when they are needed. It is Security communications systems that provide the link between electronic security systems and the operations staff.

Two-way radio

After an alarm is detected using the alarm system and assessed using the video system, the security console officer will communicate with roving guards using 2-way radios to direct them to respond.

A good 2-way radio system should include:

- One or two 2-way radio base stations at the security console and possibly at the security manager's office
- A transmitter and radio antenna tower
- A sufficient supply of portable radios to allow all officers to carry one
- A sufficient supply of battery chargers for all portable radios
- One or more repeaters to reach distances beyond the reach of the portable radios or to reach into buildings where the signal is weakened by the structure
- Leaky-coax systems to allow the repeaters to transmit clear radio signals far underground where normal radio signals would be blocked

Today it is common to supply each security officer at a post or on patrol with an earpiece and microphone so that discrete messages can be communicated to remote officers and without disturbing those nearby.

Two-way radios are available in two common frequencies—very-high frequency (VHF) and ultra-high frequency, (UHF)—and two common types (analog and digital). Radios are available either openly transmitting or transmitting encrypted communications. Both VHF and UHF radios require licensing in most areas of the world. Most security organizations today prefer digital encrypted radios to ensure that security communications are not publically available.

Telephones

Telephones are the heart of all security communications with the greater organization and the outside world. Virtually all communications with other units of the organization are handled by the internal telephone system and security units routinely reach for the phone to call police, fire, ambulance, and other outside services.

Security consoles should be equipped with two types of phones: PABX (or VOIP) and one or two fixed line phones that connect directly to the telephone company's central office (fixed C.O. line). Normally the console is equipped with one or more PABX or VOIP phone extensions, which provide connection to outside telephone lines and the organization's internal phone intercom system. This is the main telephone system that the console will rely on for most outside communications.

The fixed C.O. lines are used for emergencies. Typically two lines are used and these are programmed such that only one line must be called to reach both (rollover phone number). This is the number given to branch offices for emergencies. It is the number that voice alarm dialers at remote branches call to advise the security console of an alarm (if voice is used). And it is the line that the security console officers will use when the PABS/VOIP lines are all down and not working (and you can bet that *will* happen).

Security intercoms and bullhorns

Most security units use a dedicated security intercom system to communicate with:

- Drivers at vehicle access gates
- People at remote entry doors
- People inside stairwells
- People at loading docks, etc.
- Suspicious people in parking lots, maintenance yards, etc., through intercom bullhorns

Security intercoms provide access to security for people all over the facility. Additionally the system, using intercom bullhorns, allows the security console officers to address intruders and suspicious people outside and inside.

There are three common types of field intercom stations:

- Standard field intercom station (mic/speaker/call button)
- Interview station (mic/speaker and no call button usually placed near a camera)
- Intercom bullhorn (2-way communications at a distance)

Small intercom systems use 2 or 4 wires to communicate from one or more master stations to a number of intercom field stations. These systems use analog audio components and require extensive dedicated wiring. As such, they are relatively inflexible for expansion and functional changes.

Larger analog intercom systems use a centralized or network of matrix switches (one in each building) to connect field intercom stations to a common audio and control buss that runs between buildings and back to the central console.

Newer intercoms use a digital infrastructure (TCP/IP) to connect remote field stations anywhere in the world to intercom master stations also located anywhere in the world. They use the organization's own TCP/IP network for communications. This allows a single security console to monitor intercoms at the organization's facilities located around the world.

Finally, hybrid matrix switchers allow migration of older intercoms to a digital infrastructure. These have an analog matrix switch and a digital intercom all built into one allowing the organization to use its existing analog intercom infrastructure while converting new work to digital, all answered from the same intercom master stations. These systems also allow older analog stations to be answered anywhere in the world.

Public address systems

There are times when Security needs to speak to large numbers of people, such as immediately before or after a natural disaster, such as a hurricane or earthquake, in order to warn people and guide them to safety.

Many facilities are equipped with loudspeakers that play background music or "white noise" to make workspaces more livable. These systems can be used as public address (PA) systems by adding PA amplifiers and a zoned switching circuit to lift the speakers from the primary music or noise source and place them onto the PA source.

Coupled with a microphone and zone selection switches at the console (or using a computer microphone and zone switching system), those speakers that already exist in the workplace can serve a second valuable function.

Nextel phones

Nextel phones are special cell phones that also have a 2-way radio function, allowing one user to speak hands-free with one or many other users. Groups of users can be set up so that the Maintenance or Security Unit can speak to technicians or guards. This is a great way to use one technology for two purposes.

Nextel phones can also be integrated into the security console just like 2-way radios using specialized adaptive hardware.

Voice loggers

For every important communication, there will be some need to report it. The best way to make sure that the report is accurate is to record all verbal communications by using a voice logger. Early voice loggers comprised reel-to-reel audio tape recorders and lots and lots of tape running at very slow speeds (15/16″ per second). These were replaced by slow speed audio cassettes. Modern voice loggers use digital multitrack recording to log and preserve conversations from telephones, 2-way radios, intercoms, PA systems, and even Nextel Phones. All conversations are imprinted with a date/time code from a source that is synchronized with the alarm/access control system and digital video system so that all communications can be heard in the context of events recorded in those systems as well.

Smart phones and tablets

One of the emerging technologies that shows great promise is the use of smart phones and tablet computers to send situational awareness, alarm, video, and even audio to roving guards who are responding to an alarm. These use the public cellular network to pipe high-speed data to the smart phones and tablets. Expect to see more of this in the future.

Consolidated communication systems

At many consoles it is common to see several telephones, a security intercom, 2-way radios, PA systems, and several different elevator intercoms (several for each building on a campus). Console officers are expected to manage all of these different communications devices. Which they do ... until there is an emergency such as an earthquake. Then, as they say, "all hell breaks loose." And quite literally every communications device is ringing for attention. Remember now, the security console officer has two ears, one mouth, and only one brain, so quite naturally they are incapable of answering all these urgent calls, including of course, those urgent calls for medical attention or say, people trapped in elevators. So how can they sort out the critical from the simply curious ("Uh, what just happened")? The answer is a consolidated communications system (CCS).

A CCS manages all of the communications devices and prioritizes the calls into a queue for the console officer to answer. For example, for people trapped in an elevator, a recorded ring tone is heard by the console officer when they push the elevator intercom inside the lift. This is a familiar and comforting sound that gives the console officer perhaps 20 seconds to answer (10 rings), after which the digital recording announces "There has been an emergency in the building that is placing a high demand on the security console. If you have a medical emergency please press the call button twice, otherwise, you are in a queue and please be patient for an answer."

I first designed a CCS in the late 1980s for a campus of high-rise buildings. The design called for 100% custom technology. Today the major digital switch company makes a CCS that can integrate all types of communications hardware off-the-shelf and program it to perform functions like the one I just mentioned. CCSs are useful for large campus facilities and enterprise-class security organizations.

SECURITY ARCHITECTURE MODELS FOR CAMPUSES AND REMOTE SITES

As security systems grow from one building to a campus to multiple campuses, the organization can either treat them as individual separate systems or combine them into one single system under centralized management. This provides great economy of scale, greatly reducing operating costs and improving the probability of a consistent uniform application of corporate security policies.

The key to this is using a TCP/IP digital infrastructure to connect all systems at all facilities at all campuses into a single system.

Each individual building is equipped with a local area network (LAN). Individual systems of each building are allocated to a virtual local area network (VLAN) to segregate the functions and data of the different systems on the network. Each building is also on its own VLAN, so we

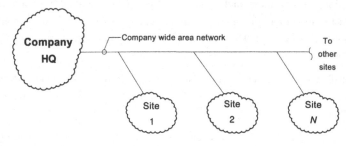

■ **FIGURE 21.7** Enterprise-class network.

have VLANs with sub-VLANs. Different campuses are connected together using Routers and Firewalls to further segregate security data from ordinary business data. Communications between facilities may be over a dedicated business network or across microwave connections, leased data lines, or the Internet. When the Internet is used, security data should be placed onto a Virtual Private Network (VPN), which isolates and encrypts the data to assure its integrity (Fig. 21.7).

COMMAND, CONTROL, AND COMMUNICATIONS CONSOLES

Command, control, and communications (C3) consoles help Security Console Officers manage vast amounts of data in a coordinated fashion. The design of security consoles is as much art as science. I have never found one that could not be improved upon. One must accommodate all of the security technologies and truly understand and appreciate the functions the console officers must perform. These include but may not be limited to:

- Responding to alarms
- Responding to access assistance requests
- Communicating with and dispatching guards
- Conducting video guard tours
    - Looking for security violations
    - Looking for safety violations
    - Looking for security vulnerabilities
- Conducting video surveillance and counter-surveillance
- Reviewing alarm/access control, video, and intercom system condition daily (in conjunction with roving guards) and updating a daily security system maintenance condition log

Consoles should be designed to place the most used equipment closest to the console officers. They should provide equipment redundancy (e.g., additional workstations and phones). They should provide a clear line of sight to video monitors and an adequate number of monitors to maintain vigilance while reviewing alarm video and maintaining surveillance operations.

Console room managers should also be placed in the console room where possible so that they can supervise the operations and maintain awareness of events.

Existing and new security console rooms can be improved by spending several days working side by side with the console officers, categorizing their duties, and observing what makes their day easy and hard. (Believe me, they will have their complaints.)

CHAPTER SUMMARY

1. Photo ID systems include a:
   a. Badging computer that is connected to the access control system database
   b. Badge printer
   c. Digital camera, backdrop, and light
2. Large photo ID operations use multiple computers, printers, and cameras to keep up with a constant stream of new users at factories and other such facilities with high employee hire and turnover rates.
3. One of the best ways to accommodate visitors is by using a visitor management system.
4. Some visitor management systems track visitor access activity separately from employee access activity, although many access control systems can be programmed to track both.
5. Video systems provide the ability to:
   a. Assess events detected by the alarm/access control system
   b. Observe all areas of the facility for inappropriate or suspicious activity
   c. Conduct investigations remotely
   d. Follow the movements of subjects of security events or investigations
   e. Produce visual evidence of security events
6. If we are interested in measuring the amount (level) of visible light (illumination) in an area, we are interested in knowing its Lux (metric) or fcs (English) level. One fc is the amount of light equal to one candle at 1 ft (one lumen on a 1 ft^2 surface).
7. Different types of lights yield different colors of light. Light is measured in Kelvin (color temperature). Video cameras work best with lighting that is as close as possible to natural overhead daylight at 12:00 noon (5600 K).
8. Most color cameras today have an automatic white balance adjustment to assist them in displaying lighting that appears to be more true to the colors we see.
9. Dynamic range is the ability of a camera to differentiate the extremes of light and dark within a scene, especially when one is close to the other.
10. Most display devices today are LCD or LED computer screens.

11. DVRs receive either analog or digital video inputs (depending on the brand and model), and for analog inputs they digitize the video (convert it from analog to a digital signal).

12. Network Video Recording systems are available in several versions, but the thing that they all share is that they work only with predigitized sources that are received across the network, not directly onto the recorder.

13. Several video recording schemes are available using server-based systems. These include:
    a. Internal HDS
    b. DAS
    c. NAS
    d. SAN

14. The most useful type of alarm sensors is VMD.

15. Common video analytics include:
    a. Object placed in the scene by a person and left behind
    b. Object removed (painting on wall, etc.)
    c. Loitering behavior
    d. Person going one specific direction (entering through an exit-way)
    e. Object or person entering a defined detection zone
    f. Object or person leaving a defined detection zone
    g. Person, vehicle, or object stopping
    h. Fence trespassing detection
    i. Automatic autonomous PZT tracking of a detected moving object, person, or vehicle

16. There are two basic types of video analytics: those that have specific algorithms to detect specific behaviors and a newer type that uses no algorithm at all.

17. Video systems can interface to other systems in a variety of ways. The most common of these include:
    a. Dry contact interfaces (notify video system to display a camera in response to an alarm or may notify other system for an event in the video system)
    b. RS-232 interface (receives a script to display a camera in response to an event on another computer or tell another system to perform an action for an event in the video system)
    c. RS-485 interface (similar results to RS-232)
    d. TCP/IP interface (essentially any function available; e.g., having the digital video system pass a video frame or clip to the alarm/access control system to file it in the same database as an alarm with which it is associated)

18. Security communications systems provide the link between electronic security systems and the operations staff.
19. A good 2-way radio system should include:
    a. One or two 2-way radio base stations at the security console and possibly at the security manager's office
    b. A transmitter and radio antenna tower
    c. A sufficient supply of portable radios to allow all officers to carry one
    d. A sufficient supply of battery chargers for all portable radios
    e. One or more repeaters to reach distances beyond the reach of the portable radios or to reach into buildings where the signal is weakened by the structure
    f. Leaky-coax systems to allow the repeaters to transmit clear radio signals far underground where normal radio signals would be blocked
20. Telephones are the heart of all security communications with the greater organization and the outside world.
21. Most Security units use a dedicated security intercom system to communicate with:
    a. Drivers at vehicle access gates
    b. People at remote entry doors
    c. People inside stairwells
    d. People at loading docks, etc.
    e. Suspicious people in parking lots, maintenance yards, etc., through intercom bullhorns
22. PA systems can notify large quantities of people at the same time.
23. Nextel phones are special cell phones that also have a 2-way radio function, allowing one user to speak hands-free with one or many other users.
24. For every important communication, there will be some need to report it. The best way to make sure that the report is accurate is to record all verbal communications with a voice logger.
25. One of the emerging technologies that shows great promise is the use of smart phones and tablet computers to send situational awareness, alarm, video, and even audio to roving guards who are responding to an alarm.
26. A CCS manages all verbal communications into and out of a security console.
27. A TCP/IP digital infrastructure can connect all systems at all facilities at all campuses into a single system.
28. C3 consoles help security console officers manage vast amounts of data in a coordinated fashion.

Q&A

1. *Photo ID systems include:*
   a. A Badging Computer that is connected to the access control system database
   b. A Badge Printer
   c. A Digital Camera, Backdrop and Light
   d. All of the above

2. *Visitor management systems are designed to record visitors into a visitor database and issue a _____ that can be used at portals equipped with special visitor card readers.*
   a. Guide Dog
   b. Temporary Visitor Access Card
   c. Temporary Employee Access Card
   d. Temporary Escort

3. *Video Systems provide the ability to:*
   a. Assess events detected by the Alarm/Access Control System
   b. Identify celebrities when they are on site
   c. Identify intruders inside the IT Server Room
   d. Identify trends in surveillance

4. *Good Lighting is essential to:*
   a. Good moods among employees
   b. Good video images
   c. Good balance between interior and daylight
   d. None of the above

5. *Different types of lights yield different _____.*
   a. Lighting levels
   b. Ambient light to daylight
   c. Colors of light
   d. None of the above

6. *Light color is measured in degrees _____.*
   a. Kelvin
   b. Kevin
   c. Bacon
   d. None of the above

7. *Lighting can be either:*
   a. Diffuse or Diffused
   b. Diffuse or Direct
   c. Direct or Overhead
   d. Overhead or Underhanded

8. *Lighting is best for video when it is overhead and diffuse with little or no _____.*
   a. Direction
   b. Reflection
   c. Announcement
   d. None of the above

9. *Dynamic Range is the ability of a camera to:*
   a. Differentiate levels of color
   b. Differentiate extremes of behavior
   c. Differentiate the extremes of light and dark within a scene, especially when one is close to the other
   d. None of the above

10. *The best display devices provide fast:*
    a. Refresh rates, very high resolution, and very high contrast
    b. Refresh rates, very high ambient light, and very high optimum luminance
    c. Nectar rates, very high lamberts, and optimum maximization
    d. None of the above

11. *Server recording schemes include:*
    a. Internal Hard Disk Storage (HDS), Direct Attached Storage (DAS)
    b. Network Attached Storage (NAS), Storage Area Network (SAN)
    c. Both a and b
    d. Neither a nor b

12. *Video motion detectors are a very basic form of:*
    a. Analog acceptance log
    b. Video analytic process
    c. Arbitrary analysis
    d. None of the above

13. *Video Analytics may use:*
    a. Specific algorithms to detect specific behaviors
    b. Artificial Intelligence (AI) that learns what is normal behavior within the scene view of the camera
    c. Neither a nor b
    d. Both a and b

14. *Video System Interfaces may include:*
    a. Dry Contact Interfaces, RS-232 Interface
    b. RS-485 Interface
    c. TCP/IP Interface
    d. All of the above

15. *2-Way radio is a form of:*
    a. Announcement System
    b. Security Communications
    c. Video Communications System
    d. None of the above

16. *Security communications systems may include:*
    a. Telephones
    b. Security Intercoms and Bullhorns
    c. Public Address Systems
    d. All of the above

17. *Security communications systems may include:*
    a. Nextel Systems
    b. Voice Loggers
    c. Consolidated Communications Systems
    d. All of the above

18. *The key to building an Enterprise-class network is:*
   a. Using RS-485
   b. Using RS-232
   c. Using a TCP/IP Ethernet Network
   d. None of the above

Answers: (1) d; (2) b; (3) a; (4) b; (5) c; (6) a; (7) b; (8) b; (9) c; (10) a; (11) c; (12) b; (13) b; (14) d; (15) b; (16) d; (17) d; (18) c.

Chapter 22

The Merging of Physical and IT Security

CHAPTER OBJECTIVES

1. Understand that there is only one security mission
2. Understand that IT security and physical security share the mission
3. Understand that vulnerabilities exist between IT & physical security
4. Understand that sophisticated threat actors are exploiting those vulnerabilities
5. Learn how to reduce and mitigate those vulnerabilities

CHAPTER OVERVIEW

Virtually every large organization has both a Physical Security Unit and an IT Security Unit. In many organizations, the management of these two separate units frequently treat each other's roles with contempt, each acting as though the other's role is less important and sometimes insignificant. In many organizations, this has led to business culture clashes for budget, and to political struggles within the organization. None of this benefits the organization in any way.

In this chapter we will examine the roles of both IT security and physical security and see that together, they share a common mission, to secure the organization, how their roles are both necessary, and how each relies on the other for its own integrity.

We will also learn that the cultural divide between physical and IT security units has created a chasm of vulnerabilities which have resulted in a joint failure to serve the common mission.

Sophisticated threat actors have begun to understand the vulnerabilities between these two important units and have begun to exploit them to the disadvantage of the organization.

Electronic Access Control. DOI: http://dx.doi.org/10.1016/B978-0-12-805465-9.00022-1

By approaching security as a common mission instead of as two separate missions, both units can see the vulnerabilities and patch them to prevent threat actors from using them to carry out exploits against the organization.

THERE IS ONLY ONE SECURITY MISSION

Let me be clear. The guys in the expensive blue suits that live in the mahogany offices of the corporate headquarters only understand one thing about security. "We need it and it has to work!" THEY understand that there is only "security" or "not security." There is no such thing as physical security and IT security. That is only a technical nuance. There is only "security" or "not security."

IT SECURITY AND PHYSICAL SECURITY SHARE THE SAME MISSION

In the offices of the physical security director and the chief information security officer's offices however, many seem to believe that theirs is the only security mission that matters. That other security unit is irrelevant to them. It's not their mission. It is, however a burning cauldron of competition for the security budget and so it is often dealt with by belittling, condescending and sarcastic comments.

I have seen IT security directors and managers talk about how unsophisticated their counterparts in physical security are. How they do not understand the computerized technology that they must use to carry out their role. How they do not have the credentials and education that THEIR people have.

It's the same in the physical security director's office. I have heard comments about how snobbish and condescending the IT director and his people are. How they do not understand the first thing about the role that physical security serves in the organization, and how ignorant and stupid their people are about physical security matters.

I have seen bitter, snide comments slung at each other across a conference table in meetings, and a complete lack of interest in cooperation between their respective units.

This is not healthy competition. This is useless and counter-productive competition. Because there is no IT security mission, there is no physical security mission. There is only the security mission. I have noticed for years that both people, who are responsible for the joint mission of securing the organization, blockade each other's roles, publically belittle each other's roles within the organization, fail to understand the importance of each other's roles, are largely ignorant of their opposite's role and would not, if ever invited, go to lunch with their opposite to learn anything about it.

It's not as simple as two units each paying attention to their own roles. These two roles are intimately intertwined, like interlacing fingers. The attributes of physical security are essential to IT Security. In the OSI network operating model, which has Level 1 through Level 5, Level 1 is the physical layer. The physical layer comprises digital switches, routers, cables, workstations, servers, and storage. It also intrinsically includes the physical environment housing (and protecting) the computerized elements.

Well THAT's interesting. IT has physical attributes! Go figure! Previouslyin this book, I related a story about a time earlier in my career I worked for a brilliant engineer, who taught me a high level of professionalism in design. I am forever grateful to him for that. However, brilliant though he was, he had a blind spot to security, and that was that the building was part of the security system. One day he was discussing the attributes of a security system that I had designed and he noticed that I had specified that a door be changed to a more robust door and frame, because it delineated a boundary from a controlled access space to a restricted access space. He criticized me for the note and said that the building architecture was not in any way important to security. I was dumbfounded! I said, "It most certainly is." He said if I could make a case for that then he would relent. I said: "Paul, imagine an important asset sitting in an open field. Imagine that we build a metal frame that defines the locations of corridors, door locations, offices, lobbies, etc., and we install on the metal frame card readers, locks, door position switches, motion alarms, video cameras, and security system servers and workstations sitting at a security console in one of the spaces defined by the metal structure. But there are no floors, no ceilings, no walls, no doors. Only a security system. The asset is sitting on a table in an open field, surrounded by a metal structure that supports the elements of a security system. But any person can walk across the field, through the metal structure and take the asset. Tell me, have we secured the asset with the security system?" He of course said no. But that was ridiculous. I agreed. "Of course it is. But now you see that architecture is as essential to security as is the security system." And he agreed.

So, understanding this, let us clearly understand now that physical security is essential to IT security. A server room that has no physical security is not secure no matter how many firewalls, IT intrusion detection systems, IT intrusion protection systems, no matter if it has antivirus, antimalware, no matter if it has ransomware defenses. It is not secure unless the servers, routers, switches, IDSs/IPSs, cabling, workstations, wireless access points and mobile devices are secure from unauthorized users accessing them. Physical security is part of the IT security mission. OSI Network Level One is the physical layer. Physical security is required to secure the Level One Network Layer. Thus, *physical security is essential* to the IT security mission.

It's no different for physical security. *It security is essential for the physical security mission.*

Virtually every electronic security system today comprises alarm sensors, card readers, electric door locks, and video cameras. And in today's systems these all reside on a TCP/IP infrastructure of CAT-6 cables, digital switches, routers, servers and storage. All of this IP infrastructure must be secured in order for the physical security system itself to be secure. Like the example of the metal structure in an open field, what good is a physical security system that is itself not secure against an IT security attack? How much protection does a physical security system provide if it, itself is not secure? I'll save you the time from figuring that out. The answer is "None." A physical security system that is not secure from IT intrusions offers no protection whatsoever to the organization. It is security theater.

Physical security relies on IT security. IT security relies on physical security. There is only one security mission. There are two units responsible for different parts of that mission. If they don't work together, if they don't understand each other's roles, there is no security for the organization whatsoever. Without both working together, there is only security theater. And threat actors know this.

WHAT VULNERABILITIES EXIST BETWEEN IT & PHYSICAL SECURITY?

What? Why would threat actors care about the squabbles of two managers in an organization? They care because it opens up vulnerabilities that can be exploited. Many organizations that I have seen seem almost entirely blind to the cross-discipline vulnerabilities between IT security and physical security. But those vulnerabilities can be deadly to an organization.

In 2015, at the DEF CON international hackers conference, presenters demonstrated how hackers can break into common variety, name brand alarm/access control systems and digital video systems (ones you might see at any major Fortune 500 corporation). They demonstrated how, once inside the system they could bypass alarms, remotely unlock doors to allow an intruder inside, and cover their tracks in the logs and by replacing video of the intruders moving through the building with previously recorded video showing no activity. So... No problem there!

It's no better on the IT security side. Many major recent IT security hacks have been conducted through exploits in the organization's physical security.

Two recent National Security Agency hacks were both carried out by allowing threat actors essentially unlimited access inside the agency itself. In both cases (Edward Snowden and the NSA Extra Bacon attacks) contractors were initially vetted and then never watched as they ran rampant with their access credentials and stole ever more important material. In both cases, highly sensitive material was either put up on www.WikiLeaks.com, or offered for sale on the DarkNet.

As I was conducting a survey at a government facility, I was in the parking structure and noticed a large white free-standing cabinet next to a parking space, near a vehicle entrance. The cabinet had two locked doors, one of which was ajar. Inspecting, I noticed that the lock was broken and it had been held together with duct tape at the top and bottom, which had failed over time. I opened the cabinet and saw a digital switch and several access control panels. This was the equipment supporting the parking structure security system. Several ports were unused on the digital switch. As I started asking questions, I discovered that the parking structure security system was not firewalled from the buildings' security system, and that the security system itself resided on the same VPN as the business network. It was a totally merged system with only one VPN. And it was possible for a car to pull into the space next to the cabinet, for the driver to tap into the digital switch and he/she was exactly inside the entire network!

It was exactly this kind of vulnerability that allowed hackers to attack target store point of sale devices in 2014. They first attacked the HVAC system company that serviced target stores, infecting their field service laptops, and through those into the improperly VPNed Target network to the point of sale devices when the HVAC contractor's field service techs remotely accessed target's improperly configured network (Krebs on Security—February 14, 2014—Brian Krebs—"Target Hackers Broke in Via HVAC Company").

And as I write this, there is a new trend of using lightly guarded "Internet of Things" (IoT) security devices including webcams, digital video cameras, digital video recorders, thermostats, IP enabled appliances, learning tools like the Raspberry Pi Zero, a $5.00 computer, to combine into a massive BotNet of as many as 145,000 different IP source devices from as many as 105 different countries that can deliver as much as 1.5 Tbs (TerraBits per second) against a Web or DNS server across the internet. Absolutely no server whatsoever can stand up against that. At this time, the maximum input port throughput of any server is 100 Gbps (GigaBits per second). That is delivering nearly 15 times the amount of data that even the most robust servers and routers can sustain. This simply crushes the ability of the router or server to operate and it simply crashes until the input load is gone. Why is this of importance to physical security? Because when an attack like this is directed at critical infrastructure, it can cripple an entire city! (NY Magazine—June 19, 2016—Reeves Wiedeman—"Envisioning the Hack That Could Take Down New York City").

When a medical center no longer has control over its IP infrastructure, the alarm system can be directed not to alarm, the access control system can be directed not to allow entry to anyone, or worse, lock people inside (with some types of systems), the security center can be blinded to ongoing physical intrusions, and patients can be killed by causing drug pumps to overdose (RT (Russia Today)—June 11, 2015—"Hackers can remotely kill hospital patients with drug pumps, IT expert discovers"). And, in the larger city attack, ambulances cannot be dispatched.

The author is not going to go into detail in this book on how such an attack can be carried out, but believe me. It is entirely possible. And the most terrifying part is that the example of a city attack is not yet to the full scale of what is possible. So... Is IT Security essential to physical security? You bet it is!

SOPHISTICATED THREAT ACTORS ARE EXPLOITING THOSE VULNERABILITIES

So... Vulnerabilities exist between IT and Physical Security that can be exploited. Is it happening? Yes, it is.

On December 23rd, 2015, at 3:30 p.m., at the Prykarpattyaoblenergo control center, a part of the Ukraine Power Grid, a power distribution switching facility in the Ivano-Frankivsk region of Western Ukraine it was just a short time before shift change when an operator was looking forward to ending his shift and going home for the day, when he noticed the cursor on his workstation drifting from right to left all by itself. It hovered over an icon for a substation and then clicked it from online to offline. A dialog box appeared asking if the operator was sure he wanted to turn off the substation. The cursor moved to "OK" and clicked. The cursor then moved on to another substation icon and clicked that one off too. He reached across the counter to the mouse and tried to move the mouse to reverse the clicks, but to his horror, and to the horror of others who were gathering around his workstation, he discovered that he had no control over his mouse. Someone else did. It was someone not in the room, someone not in the building, someone who wanted to turn off substations. Ukraine states with certainty that it was a hacker in Russia turning off a total of about 30 substations on that cold December day. A total of 230,000 homes and businesses were put out of electricity. The hackers not only turned off the substations, but they also overwrote the firmware with trash, requiring the physical replacement of equipment with new in order to restore power at many substations. Due to this, it took several months to get all of the substations back online (Wired Magazine—March 3, 2016—Kim Zetter—"Inside the Cunning Unprecedented Hack of Ukraine's Power Grid"; Wikipedia—Casualties of the U).

In related attacks, dams are being hacked to release their stored water. Demonstrations exist of medical devices being used to murder patients through drug overdoses.

LEARN HOW TO REDUCE AND MITIGATE THOSE VULNERABILITIES

So, knowing that vulnerabilities exist between IT and physical security, and knowing that attacks on either system can be conducted through vulnerabilities in the other, how do we approach securing both from the vulnerabilities of the other?

First, the "C Suite" and the Directors of both IT and physical security must understand that there is only one security mission, and that is to secure the organization itself, not the IT system or the physical facility. Every organization has the same four kinds of assets:

- People—Employees, Contractors, Vendors, Visitors and Customers
- Property—Real Property, Fixtures Furnishings and Equipment, and the IT System
- Proprietary Information—Trade Secrets, Customer Lists and Data, Vendor Lists and Data, Healthcare Data, Payment Card Data, in fact, all data that is unique to that organization, and essential to its sustainable operation
- The Business Reputation—The Brand itself.

Both IT and physical security are necessary to secure these. So is Human Resources (employee vetting and continuous vetting), and so is Safety (OSHA issues, etc.) All of these organizations share the common Security Mission.

The directors/managers of IT security and physical security should lunch together at least once every month, to get to know each other, to develop empathy for their counterpart, to understand his/her role, his/her needs, his/her challenges, and the lingering vulnerabilities in each system that only the other director/manager can see. They need to discuss these vulnerabilities and how they can work together to overcome them. Name them, prioritize them, attack them, overcome them together.

Look for the unlocked cabinet. Look for the shared VPNs that shouldn't be shared. Look for suspicious people asking questions about the IT system. Look for physical security system devices that are outside the skin of the building that are not firewalled. Anything can become vulnerability. And sophisticated attackers know where to look. So be alert to each other's needs, and look before the bad guy does.

CHAPTER 21A—CHAPTER SUMMARY

Early access control systems all used a proprietary wiring scheme, making true enterprise-class alarm/access control systems virtually impossible. The obvious answer was to change all these individual wiring schemes into a TCP/IP infrastructure. This occurred across the industry in the early and mid-2000s. Once one major manufacturer offered this, all the others had to follow to remain competitive. However, a new problem emerged. The systems, now on TCP/IP infrastructures, were not being secured from threats on the IP network. Like the computer networks before them, the threats to access control systems on the IP network was vastly underestimated. Today, skilled attackers are exploiting these vulnerabilities to reach into the IP network with the specific intent of exploiting the security system to bypass alarms, remotely unlock doors to allow unauthorized intrusions, and cover their tracks by overwriting video of the instrusion with video showing no intrusion.

Any organization secured by an unsecure security system is not secure. Such systems that are configured to permit remote web access facilitate intrusions without the threat actor even having to have the security system software. Thus, anyone who can hack in, and gain administrator access, can control the system. This is not an insurmountable barrier for most skilled hackers, certainly not those representing a foreign power.

An annual threat assessment is necessary to help keep systems secure. The threat assessment uncovers: (1) What threat actors are at work; (2) What threat scenarios they are using; and (3) What mitigation measures can work effectively against these scenarios.

The worst system vulnerability possible is an access control system with web access enabled that is on an unsecured or poorly secured network. Systems can reside on either a proprietary or converged network. Proprietary networks contain nothing but the security system. Converged networks typically cradle the business network and the security system; and they may also support other networks too such as the building automation system or other systems. Anytime two or more systems reside on the same network, each system puts all of the others at risk. Even the act of placing two or more security system elements on the same network can put the others at risk. So an access control system on the same network as the video system can put each other at risk. This is especially true for security systems that have some IP devices outside the skin of the building, such as on camera poles, or in a parking structure.

Security measures may include next-generation firewalls, intrusion protection and detection systems, real-time monitoring and management, and security information and event management systems.

Chapter 23, Securing the security system, presents a nine point plan for securing IP based security systems.

CHAPTER 21A SUMMARY—THE MERGING OF PHYSICAL AND IT SECURITY

Virtually every large organization has both a physical security unit and an IT security unit. These two units share a common mission (securing the organization), but rarely work together on that common mission. The two units often compete for budget and see their own mission as the only important one, taking little or no interest in the other unit's security mission. This is not healthy competition. This is useless and counterproductive competition.

The roles of these two units are intertwined, like interlacing fingers. The attributes of physical security are essential to IT security, and the attributes of IT security are essential to the proper functioning of the physical security system.

The IT system is not secure if there is no or inadequate physical security on the IT system, including servers, storage, network infrastructure, cabling, workstations, wireless access points and mobile devices.

It's the same for physical security. IT security is essential for the physical security mission. Virtually every electronic security system today comprises elements that reside on a TCP/IP infrastructure of CAT-6 cables, digital switches, routers, servers and storage. All of this IP infrastructure must be secured in order for the physical security system itself to be secure. If the physical security system sits on an IP infrastructure, and that infrastructure is subject to an IT security attack, then the physical security system itself can be attacked. How much protection does a physical security system offer, which itself is not secure? Little to none.

Additionally, vulnerabilities exist between the IT system and the physical security system. In 2015, at the DEF CON international hackers conferences, presenters demonstrated how hackers can break into common, name brand alarm/access control systems and digital video systems. They demonstrated how to bypass alarms, unlock doors and gates, and cover their tracks by replacing intrusion video with clean video, showing no intrusion. Hackers are using vulnerabilities in related IT systems (like the HVAC system) to break into servers, storage and point of sale systems. IoT devices are being used to conduct distributed denial of services attacks against individual businesses and even the internet itself. Power grids are being taken offline in the dead of winter. And dams are being hacked to release their stored water. Demonstrations exist of medical devices being used to murder patients through drug overdoses.

These vulnerabilities can be reduced. The "C Suite" and the directors of both it and physical security must understand that there is only one security mission, to secure the organization itself. Every organization has the same four assets:

- People—Employees, Contractors, Vendors, Visitors and Customers
- Property—Real Property, Fixtures Furnishings and Equipment, and the IT System
- Proprietary Information—Trade Secrets, Customer Lists and Data, Vendor Lists and Data, Healthcare Data, Payment Card Data, in fact, all data that is unique to that organization, and essential to its sustainable operation
- The Business Reputation—The Brand itself.

Both IT and physical security are necessary to secure these. So is human resources (employee vetting and continuous vetting), and so is safety (OSHA issues, etc.) All of these organizations share the common security mission.

The directors/managers of IT security and physical security should lunch together at least once every month, to get to know each other, to develop empathy for their counterpart, to understand his/her role, his/her needs, his/her challenges, and the lingering vulnerabilities in each system that only the other director/manager can see. They need to discuss these vulnerabilities and how they can work together to overcome them together.

Q&A

1. *What two security units do most large organizations have?*
   a. Guards and Dogs
   b. Posts and Patrols
   c. IT Security and Physical Security
   d. Physical Security and Point of Sale Security
2. *How many security missions are there between these two security units?*
   a. One
   b. Two
   c. Three
   d. Four
3. *These two units often:*
   a. Cooperate
   b. Compete
   c. Organize the Deck Chairs
   d. None of the above
4. *The attributes of _____ are essential to IT security.*
   a. Guards and Dogs
   b. Posts and Patrols
   c. Physical Security
   d. None of the above
5. *The attributes of _____ are essential to physical security.*
   a. HVAC Security
   b. IT Security
   c. Point of Sale Security
   d. None of the above
6. *_____ exist between the physical security system and the IT system.*
   a. Danger
   b. Opportunities
   c. Vulnerabilities
   d. None of the above

7. *At the 2015 _____ international hackers' conferences, presenters showed...*
   a. DEF-CON
   b. COMIC-CON
   c. Symantec
   d. Cisco
8. *(From 7 above), presenters showed how to break into*

   _____.
   a. Residences
   b. IT Systems
   c. Safes
   d. Alarm/Access Control Systems and Digital Video Systems
9. *(From 7 & 8 above), They showed how to:*
   a. Bypass alarms
   b. Unlock doors
   c. Cover intrusion video with clean video
   d. All of the above
10. Directors/Managers of _____ and _____ should lunch together at least once every month.
    a. Human Resources and Facilities Services
    b. Facilities Services and Physical Security
    c. IT Security and Physical Security
    d. Human Resources and IT Security

Answers: (1) c; (2) a; (3) b; (4) c; (5) b; (6) c; (7) a; (8) d; (9) d; (10) c.

Securing the Security System

CHAPTER OBJECTIVES

- Understand that the organization isn't secure if the Security System isn't secure
- What kinds of vulnerabilities can exist in the Security System itself?
- What can we do to secure the Security System?
- An 9 Point Plan for securing the Security System

CHAPTER OVERVIEW

I was an early advocate of using an IP infrastructure for electronic security systems. In the late 1990s and early 2000s every manufacturer had a unique proprietary wiring scheme. Even systems that used common protocols, such as RS-232 or RS-485 utilized their own in unique and proprietary ways so that once a customer was committed to their equipment, they were in fact completely committed. Wiring a security system was expensive and complicated. And for true enterprise-class organizations, with many sites spread across state and national borders, the dream of a true enterprise-class system was only that—a dream. It was virtually impossible.

Myself and many other forward thinking consultants realized that the logical answer was to replace all of this proprietary wiring with a TCP/IP Ethernet infrastructure. We pressured the manufacturers saying that whichever system converted first would begin getting all of our specifications. It worked. By 2002, in only a couple of years, with all this pressure from consultants and the US Government specifiers pretty much of the industry converted to an IP infrastructure.

As stated in Chapter 22, "Problem solved, right? Wrong. Sorry ... new problem emerges." This was an infrastructure that could easily be hacked. And manufacturers, consultants and system integrators were doing essentially nothing to secure the IP infrastructure underlying the electronic security system. I was horrified, as were many other consultants. And yet

Electronic Access Control. DOI: http://dx.doi.org/10.1016/B978-0-12-805465-9.00023-3

we were security system consultants, not IP consultants. We had created a true enterprise-class infrastructure, one that could easily work across state and national borders, and we had created an entirely new problem: a remotely hackable security system. And worse yet, we were incompetent to resolve the problem. Something had to change. We had to learn how to specify and secure IP infrastructures. The first was easy, the second, not so much. And so, specifiers and integrators would eventually live or die by their ability to do this.

UNDERSTAND THAT THE ORGANIZATION ISN'T SECURE IF THE SECURITY SYSTEM ISN'T SECURE

Even today in 2017, many electronic security systems are being designed on unsecure IP networks. This sad fact allows those security systems to be hacked.

So, let me ask a difficult question. What good is a security system that itself is not secure? If a hacker can gain entry to an alarm/access control system, and disable alarms and remotely unlock gates and doors for an intruder, and then cover the tracks of the intruder by replacing video showing the intrusion with previous video showing a quiet scene, What good is that security system? The answer is obvious. It is truly of no use whatsoever.

This is "security theater." It gives the feeling of security without providing real security. And all of the monies paid to the consultant and system integrator for all of the equipment and labor is money entirely wasted if threat actors can use the security system itself to provide them with illegal entry.

Many security systems are equipped with remote web access. This convenient function is entirely too convenient when the hackers can use that to control the security system without so much as even having to have a copy of the security system software!

An unsecure security system is in many ways worse than no security system.

WHAT KINDS OF THREATS PRESENT A PROBLEM TO SECURING THE SYSTEM DATA?

Threats evolve constantly to counter the measures that system owners implement to counter the last series of threats. Accordingly, what worked last year may not be good enough to counter the threats this year. An annual threat assessment is the only way to stay up with emerging threats. Smart organizations subscribe to an up-to-date threat report. The threat assessment should review:

- Active threat actors and their profiled motivation and preferred targeted systems and data
- Threat Scenarios: How do these threat actors attack systems? What tools do they use? What kinds of things do their threat scenarios leave as damage and what typical consequences to breached organizations face from these scenarios?
- Effective Mitigation Measures: What kinds of measures can be taken to mitigate the various types of threat scenarios?

WHAT KINDS OF VULNERABILITIES CAN EXIST IN THE SECURITY SYSTEM ITSELF?

The worst vulnerability is that just mentioned above—A security system with remote web access, connected to the Internet, with no firewall, or with an improperly configured firewall.

This allows a hacker with no real experience to have free reign inside the security system, with no software owners' manual, nor owning any of that software. A hacker with malicious intent in a system like that can result in very tragic consequences for the organization that owns that system because it allows an unskilled operator (the hacker) to "learn on the job," by watching how other operators work, and trying things themself on their own web connection to the system.

Other vulnerabilities can be created by the way the security system is architected onto the network. There are basically two major kinds of system architecture:

- Proprietary security system networks
- Converged networks

Proprietary security system networks sit on their own segregated IP networks that are not connected to any other network. Converged networks combine the business network with the security system network on the same IP infrastructure. There are advantages and disadvantages to both.

Proprietary security system networks assure that there can be no vulnerabilities from or to the business network. Vulnerabilities can include any connection to the Internet, and poor network design allowing for vulnerabilities in the proprietary network itself.

For example, if there are access control and digital video cameras located in an adjacent parking structure, and the digital switch(es) in the parking structure are not firewalled to the security system in the building, a hacker can tap directly into the security system from a digital switch in the parking structure, especially if the switch cabinets are themselves not physically well secured (in a secure room, and in a locked cabinet). Thus, a hacker can control the security system directly from the same parking garage digital switch that a camera and access control panel are plugged into.

Security systems that are converged onto the business network can present vulnerabilities to the business network from the security system, and to the security system from the business system. Additionally, the stresses of massive amounts of digital video on a business network that is poorly designed can crash the business network repeatedly. In the worst case, if the security system elements in the parking garage are not firewalled from the business/security system in the building, then the hacker in the garage can gain access not only to the security system but to the business system itself, right from the parking garage. Not a very good plan.

WHAT CAN WE DO TO SECURE THE SECURITY SYSTEM?

Network security changed since 2014 from something that we should do to something that we must do. In the past, network security could be relatively well done if the designer included a good firewall (properly configured), and antivirus and antimalware software, plus maintaining software patches. Or by designing a proprietary security system network that is "air gapped." That is, a network that is not connected in any way to the Internet. That has all changed. Sophisticated hackers and ransomware have overcome all of these defenses. "Air gapped" networks are no longer safe because advanced threat actors have figured out ways around and into air gapped systems. (There are several ways. I don't want to go into that because I don't want to educate hackers here). Ransomware is software that encrypts the organization's data (or compromises its accessibility in some other way) and holds the data for ransom until the organization pays the ransom. Ransomware can infect a system in a variety of ways, including ways that bridge the "air gap."

Yesterday's hackers were low to moderately skilled hackers. Today's hackers include exceptionally highly skilled and highly trained hackers from other countries, including those acting in behalf of other nation states. Yesterday's solutions do not stand a chance against ransomware and state-sponsored hackers. No chance at all.

Today's IT security solutions include:

- Next-generation firewalls and sophisticated configurations
- Intrusion Protection Systems (IPS) and Intrusion Detection Systems (IDS)
- Real-Time Monitoring of the firewall and the IPS/IDS
- Security Information and Event Management (SIEM) Systems with Real-Time Intervention on the Firewall/IPS/IDS systems.

A 9 POINT PLAN FOR SECURING THE SECURITY SYSTEM

IT customers have been coming to IT integrators for years with problems and asking them for pieces to solve the problems (sell me a firewall, sell me an IDS). This does not work. This is like a person needing transportation and going to the auto parts store where he is sold a tire, or a fuel pump. This does not solve the transportation problem. This is not a parts problem, it is a security problem. "If you think technology can fix security, you don't understand technology and you don't understand security. The root cause of (an IT) security incident is rarely about the technology and almost always about the implementation." (www.KrebsOnSecurity. com—https://krebsonsecurity.com/2014/02/target-hackers-broke-in-via-hvac-company/Comment 3.) What is needed is a whole car, not car parts. What is needed is a security solution, not just a piece of technology.

What is needed is a plan. So here is a plan.

1. Base the security solution on a major compliance standard
2. Conduct a current threat analysis
3. Understand the consequences of a data breach
4. Learn what vulnerabilities the network has
5. Prepare a gap analysis
6. Draft an IT security improvement plan in conformance with ISO-27001, the NIST Framework or Cobit, to resolve the vulnerabilities in the gap analysis
7. Implement the plan
8. Train and test
9. Rinse and repeat

Let's take a look at each of these in detail.

1. Base the security solution on a compliance standard

Security system data may fall under several different Government Compliance Standards including possibly HIPAA (Healthcare Insurance Protability and Accountability Act of 1996), and SOX (The Sarbanes-Oxley Act of 2002). Both of these US Federal statutes require information security professionals to protect data that falls under the data defined by the act. Security system data falls under Sarbanes-Oxley, if the company is a publically traded company, and under HIPAA if the security system data supports a healthcare facility. There are three compelling reasons to base the security plan on a compliance standard:

a. Data under a compliance standard *must* be properly protected. There is no choice available not to properly protect the data. The law requires it.

b. Budget must be made available to comply with the compliance standard. Again, no choice in this.

c. Penalties can be significant, and can reach past the corporate wallet directly into the pocket of the corporate executives themselves for failure to comply, if a post breach audit indicates both foreknowledge of the need and negligence of duty to protect.

2. Conduct a threat assessment

Threats are constantly evolving. A threat assessment should review current and emerging threats against networks and data. The current threats (2016) are so severe that most IT Security professionals put the probability breach for most organizations at 1. That is: every organization will be breached within the next 3 years. Don't think that is as severe as it gets. If, e.g., we are evaluating a major hospital and there is a string of ransomware attacks targeting major hospitals, that can elevate the threat above 1 (a probability of breach in less than 3 years).

3. Understand the consequences of a data breach

Data has quantifiable value to the organization. The loss of critical data also has a forecastable cost to the organization. Taken together, one can project the consequential costs of:

a. What is the potential cost of loss of access to critical data for business operations (for an hour, for a day, for a week, etc.). This is especially true for organizations whose business involves access to online services and data.

b. What is the potential cost of loss of data to nefarious individuals, organizations or states (Wikileaks, Ransomware, Loss of personal identifying data to state-sponsored hackers)

c. What is the potential cost of loss of business reputation resulting from a massive data breach

    d. Consequential costs such as compliance audit penalties, mitigation costs, etc.

    These potential costs can be used to help justify the expenditure to mitigate existing system vulnerabilities.

4. Learn what vulnerabilities the network has

    a. Perform a Network Vulnerability Assessment

    b. Perform a Wireless Vulnerability Assessment

    c. Perform a Social Engineering Vulnerability Assessment

    d. Review the servers and workstations for signs of a previous or existing breach or malware.

    When we use an "Ethical Hacker" to find how a potential nefarious hacker can get into and exploit the system, we can better identify the measures needed to resolve the vulnerability

5. Prepare a Gap Analysis

    A gap analysis identifies the difference between where we need to be with protection and where we are today. The compliance standard identifies where we need to be and the vulnerabilities assessments identify where we are today. The difference is the gap that must be mitigated with security equipment, software, services and configurations. The gap analysis helps identify how to structure the gap mitigation program.

6. Draft an IT security improvement plan to resolve the vulnerabilities in the gap analysis

    The network security provider can, using the gap analysis, put together a program of recommendations to mitigate the gap. The program could include such things as:

    a. Revisions to IT Security Policies and Procedures

    b. End-user anti-phishing training

    c. Next-generation firewall with proper configurations

    d. Intrusion protection system/intrusion detection system

    e. SIEM (Security Information and Event Management) system

        i. Real-time monitoring

        ii. Real-time incident management

    f. Managed security services

    g. Training and testing

    The exact mitigation program will be determined by the gap analysis which should stipulate the vulnerabilities that should be mitigated. The selection of hardware, software, services and training can be determined in response to the gap analysis.

7. Implement the Plan

    Once the plan is approved and funded, the implementation program can begin. Elements should include:

    a. Implement the plan in segmented phases—Those who implement
a plan in its entirety all at once often find that something crashes,
and they have no idea what caused the system to crash. Ouch! By
implementing the plan in phases, one can install and test each
segment and back out only that piece that doesn't work.

    b. Stress test each segment to look for any instabilities and
unintended consequences

8. Train and test

Training should be conducted for both the end user's IT support
team and the users.

    a. IT support team training on:

      i. New hardware

      ii. New software

      iii. New configurations

      iv. New managed services

    b. User Training on:

      i. Recognizing phishing emails

      ii. Good security hygiene

        – Don't click on unfamiliar URLs especially from
unfamiliar people

        – Don't open attachments from unfamiliar people

        – Always check the sending email address to make *sure* that
it is from the person it says it is from

9. Rinse and Repeat

Business data usage changes continuously. Those changes require
changes in hardware, software, and required user access levels. This
necessarily changes the usage pattern and thus the system
vulnerabilities.

Additionally, the threat landscape is constantly changing, evolving
and progressing. Useful measures last year may not work against
emerging threats.

Accordingly, this plan should be reviewed annually. An annual
threat assessment system vulnerability assessment is indicated, and an
updated gap analysis.

CHAPTER SUMMARY

1. Understand that the organization isn't secure if the security system
isn't secure

2. Approach the problem systematically, not haphazardly

3. Base the IT security program on compliance standards, if possible

4. Understand what data must be protected and where that data resides

5. Understand what kinds of threats exist against the system—conduct a threat assessment
6. Understand what vulnerabilities the system has against these kinds of threats
7. Create a "gap analysis" that identifies what should be done to mitigate the vulnerabilities.
8. Define a mitigation program from the gap analysis. Define what can be done to mitigate the vulnerabilities to defend against the kinds of threats defined in the threat assessment.
9. Implement the mitigation program systematically and in segmented phases
10. Train and test the IT support team and the users
11. Review the threat and vulnerability assessments annually, and conduct training annually.

CHAPTER 21B—CHAPTER SUMMARY

Early access control systems all used a proprietary wiring scheme, making true Enterprise-class alarm/access control systems virtually impossible. The obvious answer was to change all these individual wiring schemes into a TCP/IP infrastructure. This occurred across the industry in the early and mid-2000s. Once one major manufacturer offered this, all the others had to follow to remain competitive. However, a new problem emerged. The systems, now on TCP/IP infrastructures, were not being secured from threats on the IP network. Like the computer networks before them, the threats to access control systems on the IP network was vastly underestimated. Today, skilled attackers are exploiting these vulnerabilities to reach into the IP network with the specific intent of exploiting the security system to bypass alarms, remotely unlock doors to allow unauthorized intrusions, and cover their tracks by overwriting video of the intrusion with video showing no intrusion.

Any organization secured by an unsecure security system is not secure. Such systems that are configured to permit remote web access facilitate intrusions without the threat actor even having to have the security system software. Thus, anyone who can hack in, and gain administrator access, can control the system. This is not an insurmountable barrier for most skilled hackers, certainly not those representing a foreign power.

An annual threat assessment is necessary to help keep systems secure. The threat assessment uncovers: (1) What threat actors are at work; (2) what threat scenarios they are using; and (3) what mitigation measures can work effectively against these scenarios.

The worst system vulnerability possible is an access control system with web access enabled that is on an unsecured or poorly secured network. Systems can reside on either a proprietary or converged network. Proprietary networks contain nothing but the security system. Converged networks typically cradle the business network and the security system; and they may also support other networks too such as the building automation system or other systems. Anytime two or more systems reside on the same network, each system puts all of the others at risk. Even the act of placing two or more security system elements on the same network can put the others at risk. So an access control system on the same network as the video system can put each other at risk. This is especially true for security systems that have some IP devices outside the skin of the building, such as on camera poles, or in a parking structure.

Security measures may include next-generation firewalls, intrusion protection and detection systems, real-time monitoring and management, and security information and event management systems.

This chapter presents a nine point plan for securing IP based security systems.

Q&A

1. *An example of an early proprietary wiring scheme might be:*
   a. RS-232
   b. RS-422
   c. RS-485
   d. All of the above
2. *TCP/IP infrastructures provide the opportunity for true enterprise-class operation.*
   a. True
   b. False
   c. Neither True nor False
   d. Don't Know
3. *TCP/IP infrastructures can be hacked, allowing the hacker into the security system.*
   a. True
   b. False
   c. Neither True nor False
   d. Don't Know
4. *"The organization isn't secure if the security system isn't secure."*
   a. True
   b. False
   c. Neither True nor False
   d. Don't Know

5. *Which kinds of threats do security systems face?*
   a. Active Threat Actors (Hackers)
   b. Malware
   c. Users who don't follow policies and procedures for security
   d. All of the above
6. *Which are the two kinds of security architecture?*
   a. Proprietary and Dedicated
   b. Converged and Local
   c. Proprietary and Converged
   d. None of the above
7. *A connection to the Internet is/is not a vulnerability.*
   a. Is
   b. Is not
   c. Doesn't matter
   d. Is the least of your problems
8. *Some ways to protect the security system include:*
   a. Next-Generation Firewalls
   b. IPS and IDS Systems
   c. SIEM Systems
   d. All of the above
9. *One of the key points in the 9 point plan to secure the security system is:*
   a. Base the security solution on a Compliance Standard
   b. Learn what vulnerabilities the Network has
   c. Prepare a Gap Analysis
   d. All of the above
10. *Why is it important to conduct a current threat analysis before implementing a security solution?*
    a. Threats are constantly changing
    b. Threats are likely to change in the future
    c. Threats could be smaller than you think
    d. None of the above

Answers: (1) d; (2) a; (3) a; (4) a; (5) d; (6) c; (7) d; (8) a; (9) d; (10) a.

Chapter 24

Related Building/Facility Systems and REAPS Systems

CHAPTER OBJECTIVES

1. Learn the basics in the chapter overview
2. Understand how building/facility systems relate to security
3. Learn about how to control and automate building functions
4. Learn about advanced reactive electronic automated protection systems (REAPS) functions
5. Pass a quiz on related building/facility systems

CHAPTER OVERVIEW

This chapter is all about interfacing alarm/access control systems with the environment around them. We will discuss how to use the alarm/access control system and building management systems (BMS) to control heating, ventilation and air conditioning systems (HVAC), lighting, sprinklers, and many other systems. We will also introduce reactive electronic automated protection systems (REAPS), which are active defensive systems that are powerful tools to handle truly aggressive threat actors.

BUILDING/FACILITY SYSTEMS

Just about all modern commercial, governmental, judicial, military, and industrial buildings today are equipped with BMS. BMS systems control HVAC, lighting, lawn sprinklers, and other automated building systems. Related BMSs control only one or two specific functions, such as elevators.

Remember, the secret to an excellent security program is to understand that the organization uses its physical and operational environment to carry out the mission of the organization. Threat actors also use that same physical and operational environment to carry out their mission.

Electronic Access Control. DOI: http://dx.doi.org/10.1016/B978-0-12-805465-9.00024-5

The key to an excellent security program is to manipulate the environment to the advantage of legitimate users and to the disadvantage of illegitimate users. Let me say that again. *The key to an excellent security program is to manipulate the environment to the advantage of legitimate users and to the disadvantage of illegitimate users.* BMSs provide an excellent opportunity to do just that.

BMSs are good at controlling many things throughout the building and campus and can also perform complex logical functions to control those things. When coupled with the alarm/access control system, they can do wonderful things for security as well.

Let's look at a few examples.

Elevators

Elevators often interface access control systems to control access to specific floors for each user, but they can also be used to prevent intruders from accessing controlled areas of the building. Years ago I designed security for a very high end residential tower in Los Angeles that housed a large number of celebrities, families of foreign heads of state, a couple of CEOs of major firms, and others who wanted and expected privacy. One of the celebrities had been stalked by a violent person. Another concern was that of civil disorder (protests) against a couple of the CEOs, whose industries were the target of the Earth Liberation Front (ELF), Animal Liberation Front (ALF), and Earth First! All of these are on the FBI's watch list for radical activist groups. Both have used aggressive tactics against individuals who were related to organizations that these activist groups opposed. A typical tactic of these groups was to storm the lobby of the target and swarm all over security overwhelming the one or two people at the security desk and then charge upstairs and elevators to the floor of their target to confront them directly.

These residents wanted security and lots of it. But they also wanted it to be discrete. To counter the anticipated lobby protests I first suggested that stairwell exit doors exit to the outside of the building rather than to the lobby, and that those doors have no hardware on the outside, making it impossible to use the doors normally to gain entrance to the stairwell. I also designed a "duress switch" at the lobby security desk that could be activated by the security guard if a group of protestors stormed the lobby. When the duress switch was triggered, it automatically put the elevators into "fire recall mode." This brought all lifts to the ground floor lobby and kept them there, unable to respond to any command from within the lift to go up or respond to calls from upper floors. This effectively isolated the lobby (and thus the protestors) from the rest of the building. An alert was sent to building residents advising them of the protest and each unit was equipped with a video intercom that could see the lobby so they could know what was going on. Triggering the duress switch also automatically called the police.

Once triggered, a key from the building engineer was required to reset the alarm and thus the elevator to normal operating mode. The building engineer was not, of course, stationed in the lobby.

All of these functions were triggered by the alarm/access control system, which was interfaced by dry contacts to the elevator controller, phone dialer, and the recorded announcement to individual units and to the video intercom system.

Stairwell pressurization

In modern high-rise buildings stairwells are the primary means of exit during a fire emergency. It is very important to keep the stairwells free of smoke to aid in the exit. To do this, stairwell doors are operated by automatic closers, and all stairwell doors are equipped with latching locks to make sure that smoke cannot enter the stairwell. Additionally, the building automation system supervises the HVAC system to maintain a higher air pressure within the stairwell than exists in adjacent hallways and floor spaces so that smoke cannot enter the stairwell. Obviously, it is important to keep doors closed to maintain the stairwell pressurization.

One building I worked on had a pair of stairwell doors at the ground floor where one entered the stairwell from a corridor off the building lobby and the other was an exit to an employee parking lot. It was common for employees to take smoke breaks just outside this door. Although the door was equipped with an access card reader, lock, alarm, and electrified panic hardware, invariably one person in the group would forget their access card, so there was a habit of wedging the door open for reentry without a card.

This was not only a security violation, but also a fire code violation because it compromised the stairwell pressurization and presented the possibility that if a fire broke out, smoke could fill the stairwell.

The fix was to combine the alarm/access control system with the building automation system to create a function that was not (at that time) available on the alarm/access control system. Here is how the function worked:

- During working hours:
  - When someone propped the door open, the door position switch triggered a 20-second countdown timer on the BMS.

- If the door was still open after 20 seconds the BMS triggered a local alarm at the door to advise those nearby to close the door. It also triggered a new 45-second countdown timer.
- After 45 seconds, if the door was still open, it signaled the alarm/access control system that the door was propped open. The security console officer dispatched a guard to close the door and reprimand those nearby for leaving it open (and for forgetting their card).
- After working hours:
  - If the door was opened by any means other than access card or panic bar exit sense switch or if the door was propped open, the timers were bypassed completely and the alarm was sent immediately to the alarm/access control system where an appropriate response was mounted from the security console.

This is just one example of how BMS and alarm/access control systems can work together to do more than either could alone.

Lighting

Most commercial building lighting systems are controlled by BMS. These are used to turn overhead lights on and off on a schedule corresponding to the hours of occupancy.

However, when a computer programmer is called late at night to come fix a software problem, he/she needs a safe lighted path to and from the car to the office and the server room. It would also be nice to have snacks available if 1 hour turns into 10 (as it often does with these things).

Using the BMS and the alarm/access control system one can design an interface that looks at the parking structure entry card reader, and using a structured query language (SQL) database, determines the path from the car to the office and turns on lights between those points. It will also turn on lights for the workspace (including office and server room), light the restrooms and snack areas, and turn on power to the snack machines. These lights will stay on until the card is read exiting the parking structure (if still outside of working hours). If the HVAC system is needed, that will be turned on too.

The Access Control System feeds the SQL database with the data required to calculate the path (user name) and with a command to do so. The SQL database then sends a command to the BMS including turn on the following zones (along with the zones to turn on). This command is reversed if the user exits the parking structure after hours.

CONTROLLING AND AUTOMATING BUILDING FUNCTIONS

BMS and alarm/access control system interfaces can be conducted in a variety of ways, including direct action interfaces and proxy action interfaces.

Direct action interfaces

Direct action interfaces are good for simple commands and logic sequences such as the local alarm example regarding stairwell pressurization.

A direct action interface is a simple one- or two-way interface that performs a predictable, repeatable function. The function will include basic inputs and outputs and may include counters and timers. These are all controlled by a simple logic cell.

In a direct action interface, the alarm/access control system will typically provide one or more dry contacts to the BMS, which will perform a logical function that may control HVAC, lights, or other devices. Feedback may be given to the alarm/access control system if further action is needed. The function may be controlled by counters and timers and work on a time schedule.

Proxy action interfaces

Proxy action interfaces are those that use one system to operate another. The most common form of a proxy action interface is to use an SQL program and database to operate the BMS from the alarm/access control system.

The use of SQL database and functions vastly expands what can be done and the number of variables available to control the functions. All of the examples below use proxy action interfaces.

Feedback interfaces

When systems are integrated one is the "initiating" system and the other(s) are the "commanded" system(s). With simple interfaces the initiating system commands the commanded system to perform a task, and that task is performed by the commanded system. Feedback interfaces are those that provide information back to the initiating system about the status of or completion of the task commanded by the original interface trigger. Feedback interfaces can be of either the direct action type or the proxy action type.

REAPS SYSTEMS

REAPS are reactive security elements that "reach out and touch the offender." These are rarely used inside the United States, but see a wider use elsewhere around the world. REAPS systems are especially powerful at stopping aggressive threat actors.

There are three common types of REAPS:

- Communications Elements: The most basic form of REAPS is the security intercom. When an alarm occurs a security console officer can observe the alarm scene with video and respond on a nearby security field intercom station. "Hey, stop that!" is a lot more useful than no response. This often stops minor offenders and has been known to stop armed gunmen when a disembodied voice in a convenience store identifies the offender by sex, race, and clothing and commands him to stop. It does little good to shoot a loudspeaker, and when the voice informs that police are seconds away, the gunman often runs without harming the cashier.
- Deployable Barriers: More sophisticated REAPS applications utilize deployable barriers, including rising bollards and wedges to stop vehicles and electrified locks, roll-down doors, and deployable operable walls to delay pedestrians. Environment disruption devices can be used to delay an attack until a more formidable response force can arrive and take control of the threat actor.
- Attack Disruption: Ultimately, in high-security environments it may be necessary to actually disrupt the attack. This can include deployable smoke, fast setting and sticky foam dispensing systems, drop chains, explosive air bags, automated weaponry, deluge water systems, acoustic weapons, and other effective, if rarely used systems. All of these have the common element of making it much more difficult for an attack to continue and can result in the capture of the attacker. There are two types of attack disruption systems: nonlethal and lethal. Even some nonlethal systems can cause injury. In either case, it is important to implement safety measures in the activation mechanism to ensure that accidental activation does not occur and possibly injure innocents.

Irrigation systems

Irrigation systems are most often controlled by BMS. These can be used to deter and delay Threat Actors who have breached a fence and are moving over terrain. They also help assess the determination of the threat

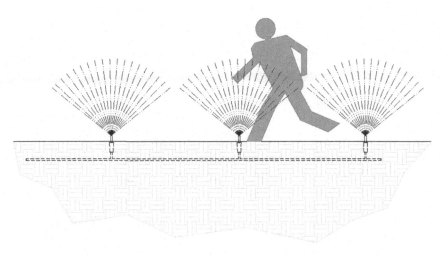

■ **FIGURE 24.1** Irrigation defensive system.

actors. Petty actors will usually run away from sprinklers, determined threat actors will not (Fig. 24.1).

Deluge fire sprinkler control

For any environment such as an outdoor shopping mall, where competing gangs of criminal youths could engage in gunfire, nothing can immediately disperse a crowd quite like 10,000 gallons of water being dispensed in 60 seconds. Afterward, although people are most likely upset and drenched, the gunfire will probably have stopped and the perpetrators will likely have fled, thus dissipating the danger and protecting the innocent public from escalating gunfire between gang members (who after all are so well known for their concern for innocent bystanders).

Coupled with gunshot recognition software, pan/tilt/zoom (PZT) cameras can instantly zoom in on the offender, providing visual verification of the condition requiring action.

Deluge activation systems should only be installed after careful consideration and advice from legal counsel, but they may be appropriate for any venue where there is a potential or history of gunfire in public spaces.

Acoustic weapons

Acoustic weapons deter inappropriate users by making the environment uncomfortable to the offender (Fig. 24.2).

■ **FIGURE 24.2** Acoustic weapons.

I once successfully addressed gang activity near convenience stores by implementing outdoor speakers playing symphonic pop elevator music on satellite radio. This drives gangbangers totally nuts. The designer is advised to use bullhorn speakers enclosed in a vandal-deterrent cage. Bullhorns are recommended because they sound very bad. The music is not offensive to legitimate customers, who may be very mildly annoyed by the tinny sound quality, but it creates an environment that does not support the business model of intimidators and drug dealers.

A Welsh firm makes a near ultrasonic tone generator, amplifier, and piezo speaker that creates a sound at the very top of the human hearing range that sounds like fingernails on a chalkboard. This also drives young people crazy. Older people cannot hear it due to natural age-related, high-frequency hearing loss. This will not bother most young people who are shopping, but it becomes more obvious and annoying to those who simply loiter outside.

The long range acoustic device (LRAD) uses two ultrasonic frequencies separated by an audio signal to propel the audio signal at painful sound pressure levels in a straight line out to incredible distances. LRAD has successfully been used to deter pirate attacks on large sea vessels.

High-voltage weaponry

Essentially a Tesla coil on steroids—high-voltage weapons are a Taser for protecting buildings. Useful to disperse crowds, they emit an attention-getting amount of high-voltage at very low amperage all around the vicinity of the spike transducers.

Another version of this weapon is the spark-gap transmitter, which can be used in very high security facilities that are under severe risk of intrusion by very high risk threat actors. (Don't be too skeptical; these facilities exist all over the world.) Similar to its portable cousin the stun gun, the spark-gap transducer can be mounted on top of a protective perimeter fence at approximately 10-ft intervals. When an intruder suspected of carrying satchel chargers or improvised explosive devices is attempting to make entry through the fence, and after detection and a video assessment of the situation and a polite verbal warning through a camera-mounted 30 Wt bullhorn, if this intruder continues to persist, the console officer and their supervisor can throw a pair of safety switches after selecting the appropriate fence zone, and all the spark gaps along that fence section ignite in a bright array of high-voltage arcs. If the intruder is carrying explosives and electrically activated detonators (the most common type), this display of high-voltage can also be accompanied by an explosion in the vicinity of the intruder, followed afterward by a repair of the fence. This approach is most appropriate in serious threat environments and in countries in which such methods are approved by the government. This approach is much less expensive than a helicopter gunship and requires less maintenance than automated guns.

Remotely operated weaponry

Except for high-voltage weaponry and some deployable barriers, most of the defeat technologies discussed above are nonlethal. When the asset value is exceptional or the risk of a successful attack to the public is high, and nothing else will get the job done, one can always rely on guns. At least two manufacturers make remotely operated weaponry. From 7.62 mm to twin 50 caliber guns and a range in between, remotely operated weapons are the final answer to intruders that just will not take no for an answer. These are products that couple a PTZ video camera

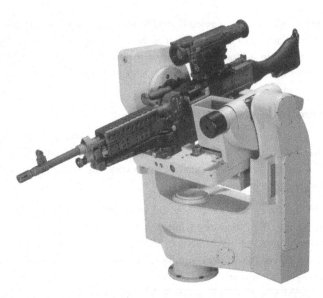

■ **FIGURE 24.3** Remotely operated weaponry. *Photo courtesy of MacKinzie River Partners, Inc.*

with an appropriate weapon (Fig. 24.3). They are available in either temporary or permanent installation versions, and are deployed in the United States at critical installations and also abroad.

Appropriateness

It is wholly inappropriate to apply lethal or even less than lethal countermeasures in anything but a serious threat environment in which the asset under protection could cause very great harm (munitions, nuclear reactors, etc.) and in which adequate public safeguards are not possible. The use of any of these systems in an environment in which an innocent could be harmed is completely inappropriate. I have refused to employ them where the government requesting the countermeasure was of questionable moral ethics. They should also be used in close consultation with the manufacturer and under a waiver of liability.

Operationally

The operation of such barriers and weapons should be limited to circumstances in which multiple layers of security are in place and the intruder has demonstrated firm determination to proceed toward a goal of harming or destroying a critical asset. Adequate evidence can be achieved by either the amount of force used or the determination with which the intruder continues undeterred by instructions toward the asset. If the attacker has or uses lethal force, it is arguably appropriate to defend with lethal force. If the intruder continues through other deployed barriers and refuses verbal instructions, it is reasonable to assume that they will be undeterred by mild measures.

In any case, the system should be configured with adequate operational safeguards to absolutely prevent accidental deployment. This typically requires a deployment action by both an operator and a supervisor or some action by the intruder that is unmistakably hostile and will result in the death of innocents, such as driving a large truck full of explosives toward a Marine barracks compound at 5:00 a.m. at high speed past a guard post.

Safety systems

For any REAPS technology that could cause bodily harm, it is imperative to build in deployment safety systems. These can be affected electronically, mechanically, and procedurally.

- *Electronic Safety Systems*
- Design REAPS systems to include at least two separate electronic circuits to trigger the system activation. For potentially lethal systems, it is advisable to use at least three triggering systems. These should be activated by totally separate electronic circuits such that no single point of failure could cause any more than one electronic triggering system to activate unintentionally. By separate systems, we do not mean separate dry contact points on the same circuit board. We mean separate circuit boards on separate power systems with their own battery or UPS back-up and with separate communications lines to each individual board. It is not inadvisable to locate the boards in different areas on separate circuit breakers, on separate electrical phases, and ideally on separate electrical transformers. The dry contacts should be triggered in series, not in parallel. It is also advisable to protect the conduit within which the triggering circuit runs with a conduit intrusion alarm system. Use only metallic electrical conduit, preferably rigid, not flexible, and waterproof where appropriate. For fast vehicle capture systems, e.g., at least two sets of loops should be used, coupled with optical sensors if possible, to prevent accidental triggering by any unusual positioning of slow-moving vehicles.
- *Mechanical Safety Systems*
- Mechanical safety systems are a back-up to electrical safety systems. For example, on drop-chain systems, the explosive bolts should be on separate electrical circuits and the bolts should be wired such that all bolts must break in order for the chains to drop. Four electrical circuits explode bolts set into a pattern where the chains will only drop if all bolts blow.

- *Procedural Safety Systems*
- Any lethal or less than lethal REAPS system should also have procedural safety systems. These can include the requirement for both a console officer and a supervisor to trigger lethal systems after verifying intent and capabilities to cause great harm to the protected asset or the organization's people or the public.

For nonlethal systems such as deployable doors, gates, grilles, and so forth, these measures are not necessary, since the deployment of a door is not an unusual or life-threatening event.

CHAPTER SUMMARY

1. Just about all modern commercial, governmental, judicial, military, and industrial buildings today are equipped with BMS.
2. The key to an excellent security program is to manipulate the environment to the advantage of legitimate users and to the disadvantage of illegitimate users.
3. Elevators often interface access control systems to control access to specific floors for each user.
4. In modern high-rise buildings stairwells are the primary means of exit during a fire emergency. It is very important to keep the stairwells free of smoke to aid in the exit.
5. Most commercial building lighting systems are controlled by BMS.
6. Lighting control interfaces can assist in controlling lighting to assist security functions.
7. BMS and alarm/access control system interfaces can be conducted in a variety of ways, including direct action interfaces and proxy action interfaces.
8. Direct Action Interfaces are good for simple commands and logic sequences such as the local alarm example regarding stairwell pressurization.
9. Proxy Action Interfaces use one system to operate another.
10. Feedback interfaces provide information back to the initiating system about the status of or completion of the task commanded by the original interface trigger.
11. REAPS are systems of reactive security elements that "reach out and touch the offender."
12. There are three common types of REAPS:
    a. Communications elements
    b. Deployable barriers
    c. Attack disruption

13. Irrigation systems can be used to deter and delay Threat actors who have breached a fence and are moving over terrain.
14. For any environment such as an outdoor shopping mall, where competing gangs of criminal youths could engage in gunfire, nothing can immediately disperse a crowd like a deluge fire sprinkler system.
15. Coupled with gunshot recognition software, PTZ cameras can instantly zoom in on the offender, providing visual verification of the condition requiring action.
16. Acoustic weapons deter inappropriate users by making the environment uncomfortable to the offender.
17. High-voltage weaponry is essentially a Tesla coil on steroids—high-voltage weapons are a Taser for protecting buildings.
18. Remotely operated weapons are the final answer to intruders that just will not take no for an answer. These are products that couple a PTZ video camera with an appropriate automated weapon.
19. It is wholly inappropriate to apply lethal or even less than lethal countermeasures in anything but a serious threat environment in which the asset under protection could cause very great harm (munitions, nuclear reactors, etc.) and in which adequate public safeguards are not possible.
20. The operation of such barriers and weapons should be limited to circumstances in which multiple layers of security are in place and the intruder has demonstrated firm determination to proceed toward a goal of harming or destroying a critical asset.
21. In any case, the system should be configured with adequate operational safeguards to absolutely prevent accidental deployment.
22. Design REAPS systems to include at least two separate electronic circuits to trigger the system activation.
23. Mechanical safety systems should be used as a back-up to electrical safety systems.
24. Any lethal or less than lethal REAPS system should also have procedural safety systems.

Q&A

1. *Just about all modern commercial, governmental, judicial, military, and industrial buildings today are equipped with:*
   a. White Noise Systems
   b. Pink Noise Systems
   c. Building Management Systems (BMS)
   d. None of the above

2. *The key to an excellent security program is to manipulate the*
_____ *to the advantage of legitimate users and to the*
disadvantage of illegitimate users.
  a. Environment
  b. Building Doors and Windows
  c. Elevators
  d. None of the above
3. *It is very important to keep the* _____ *free of smoke to aid in*
the exit.
  a. Elevators
  b. Stairwells
  c. Building
  d. None of the above
4. *Most commercial building lighting systems are controlled by*
  a. Light Switches
  b. Building Management Systems
  c. Automatic Motion Sensors
  d. None of the above
5. *BMS and alarm/access control system interfaces can be conducted in a*
variety of ways including:
  a. Indirect action interfaces
  b. Direct action interfaces and proxy action interfaces
  c. Proxy action interfaces
  d. None of the above
6. *REAPS stands for:*
  a. Reflective Electronic Automotive Powered Systems
  b. Reactive Electronic Automated Protection Systems
  c. Reactive Electronic Annunciator Powered Systems
  d. None of the above
7. *REAPS systems may include:*
  a. Intercoms
  b. Irrigation Systems
  c. Acoustic Weapons
  d. All of the above
8. *It is wholly inappropriate to apply lethal or even less than lethal*
countermeasures in anything but a serious threat environment in which
the asset under protection could cause very great harm (munitions,
nuclear reactors, etc.) and in which adequate public safeguards are
not possible.
  a. True
  b. False
9. *REAPS systems should be configured with adequate operational*
safeguards to absolutely prevent:
  a. Deployment
  b. Accidental deployment
  c. Intentional deployment
  d. None of the above

10. *REAPS safety systems may include:*
   a. Electronic Safety Systems
   b. Mechanical Safety Systems
   c. Procedural Safety Systems
   d. All of the above

Answers: (1) c; (2) a; (3) b; (4) b; (5) b; (6) b; (7) d; (8) a; (9) b; (10) d.

Chapter **25**

Cabling Considerations

CHAPTER OBJECTIVES

1. Learn the basics in the chapter overview
2. Learn about types of security cables
3. Understand when to use conduit or no conduit
4. Discover secrets to cable handling
5. Learn how to dress cables properly
6. Learn the importance of cable documentation
7. Pass a quiz on cabling considerations

CHAPTER OVERVIEW

Most new security system designers, installers, and maintenance technicians that I have met have little understanding about the importance of cabling. This understanding comes more often to maintenance technicians than designers and installers because they have to deal with the system troubleshooting problems that result from improper cable selection and installation.

In this chapter, types of cables, when to use conduit and when it is appropriate not to use it, and secrets of cable handling during installation will be discussed. Also discussed will be why good looking cable installations are more problem-free than those with jumbles of wiring, and how to document the cabling and why this is important to the system owner, the maintenance technician, and the next designer and installers who will expand the system in the future.

CABLE TYPES

A cable comprises its core (conductor or conductors), individual insulation around each core, and often an overall sheath. Each of these has characteristics that make the cable appropriate or inappropriate for its

Electronic Access Control. DOI: http://dx.doi.org/10.1016/B978-0-12-805465-9.00025-7

application. Understanding each of these will protect you from mistakes that can cost time, money, and even lives.

Copper/fiber

The top differentiator of cables is what they are made of. The three most common elements are copper, plastic, and glass. All cables that conduct electricity are made of copper. Cables that conduct light are called "fiber-optic cables" and made of either plastic or glass. Plastic fiber-optic cables are used for short distances (under a mile or so) and longer distances are handled by "single-mode" fiber-optic cables, which are made of glass.

Cable voltage and power classes

The National Electric Code (NEC) Article 725.21(B) has established three primary classes of wiring for cables inside buildings: Class 1, Class 2, and Class 3. Wiring Classes are determined according to the safety requirements of the circuits they serve.

Class 1 wiring is for powered devices requiring normal (mains) power or 120 V AC, actually, all wiring above 70.7 and up to 600 V. Class 1 wiring also includes low-voltage, low-current wiring that is essential for the safety of the building, such as boiler control wiring. For commercial buildings, Class 1 circuits must be installed within raceways, conduits, and enclosures for splices and terminations.

Class 2 wiring is low voltage and low power, typically not more than 30 V and less than 100 volt/amps (VA) power on the circuit. These are specifically for communication and signaling circuits. Class 2 circuits are limited power circuits to protect against fire by limiting the power on the circuit to less than 100 VA. Class 2 wiring also includes voltages from 30 to 150 V where the circuits are limited to 0.5 VA, which also includes Ethernet and low-voltage alarm wiring as well as microphone and line level audio wiring.

Class 3 wiring includes circuits over 30 V and exceeding 0.5 VA, but not more than 100 VA, which includes background music system loudspeaker wiring such as 70.7 V wiring schemes. It also includes nurse call systems, intercom systems, and most security systems wiring. Higher levels of voltage and current are permitted for Class 3 systems than for Class 2 systems.

It is all about voltage and power. Anything that powers directly from 120 V systems is Class 1 wiring. Low-voltage systems can be either

Class 2 or Class 3 wiring. Regarding power, a 12VDC circuit is considered Class 2 wiring if it is supplying below 0.5 VA and is Class 1 wiring if it is supplying more than 100 VA.

Wire gauge

Wires are available in a variety of thicknesses called gauges (Fig. 25.1). They are classified depending on how much power they must carry; the larger the gauge number, the smaller the cross-sectional area of the wire. Standard wire gauges include:

- Class 2 and 3 low power applications:
  - 26 gauge
  - 24 gauge
  - 22 gauge
  - 20 gauge
  - 18 gauge
- Class 1 power applications:
  - 18 gauge
  - 16 gauge
  - 14 gauge
  - 12 gauge
  - 10 gauge
  - 8 gauge
  - 4 gauge
  - 0 gauge

Insulation types

Cable insulation is rated for voltage and application.

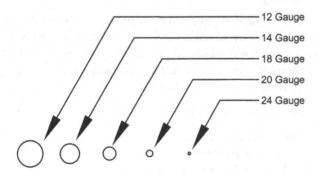

■ **FIGURE 25.1** Cable gauge.

Voltage Concerns: Class 2 cables should have insulation rated up to 300 V and Class 3 cables should have insulation rated up to 600 V (All code references come from the 2002 NEC.).

Applications: There are also different types of insulation according to the application. Insulation applications for cables to be installed:

- Within electrical conduits
- Above plenum ceilings without conduit
- Below ground within electrical conduits
- Below ground without conduits

Stranded versus solid core wires

Cables are also available as either solid core or stranded core. Solid core wire consists of a single piece of metal wire, whereas stranded wire is composed of a bundle of small gauge wires that are all twisted together to form a single, larger conductor. Solid core cables are used to power most devices that are powered from line power (120 or 220 V AC). Solid core wires can also be used in Class 2 and Class 3 circuits.

Stranded wire is more flexible than solid core wire; however, solid core wire is more likely to maintain its shape when bent. It is better to use where there may be repeated bending or where more flexibility is needed. For example, stranded wire would be used in an elevator traveling cable, whereas you could use solid core wire inside a conduit.

Stranded wire has a higher resistance than a solid core wire of the same wire gauge because less of the cross-sectional area is copper than solid core wire and there are unavoidable gaps between the strands of wire. Stranded wire is also more expensive to manufacture than solid core wire.

Cable colors

It is common to use cable colors to designate the purpose of the cable in installations. There are no color standards for communications and security wiring, but color standards do exist for fire, network, and fiber-optic cables. Fire cables are always red. For construction projects red should be reserved exclusively for fire cables.

For network cables the following colors usually apply:

- Gray: Standard Ethernet connection
- Green: Crossover Ethernet Connection
- Yellow: Wireless Power Over Ethernet (PoE)
- Orange: Analog Non-Ethernet

- Purple: Digital Non-Ethernet (RS-232/422/485)
- Blue: Terminal Server Connection
- Red: T1 Connection

Other colors may be selected to denote a difference between 100BaseT, 1,000BaseT, and 10,000BaseT wiring.

Fiber-optic cable colors:

- Orange: 62.5/125 μ multimode fiber
- Aqua: 50/125 μ multimode fiber
- Yellow: 9/125 μ single-mode fiber

Basic Cable Colors:

- Most low-voltage cables will be delivered as black unless specifically ordered otherwise
- Plenum rated cables (cables with insulation that can withstand direct exposure to fire) are typically white

Security system cables may be colored to denote system (alarm/access control, digital video, or intercom) and function (power supply cable vs analog or data cable). There are no standards for security wiring, so it is often helpful to designate specific cable colors to help delineate security from other cables in the building.

Cable brands

Cables are a bit like tires. There is an endless list of manufacturers. There are a few trusted brands that have withstood the test of time and countless others that you may have never heard of before. I recommend against buying cable from "Joe's Pool Hall, Used Cars, Yard Work, and Cables." Early in my career I was supervising a project where the contractor had used off-brand cable "to save money." There was one 3″ conduit with a very long cable run between two buildings. This conduit was packed to its legal capacity with wiring. Just to make things worse, the electrical contractor decided to coordinate this last remaining conduit between buildings between several low-voltage services, so there were three different contractors all sharing the same conduit. One of the security system cables did not work.

There was no continuity on this one wire between the buildings. There was also no more room to pull another cable. After 2 weeks of fruitless troubleshooting and struggling to find another way, the electrical contractor pulled all the cables out of the conduit. We isolated the bad cable and pulled all the cables back in, replacing the bad cable for a new one.

When the bad cable was examined, it was found that there was a strange lump in the middle of the cable run. We opened up the outer sheath and were astonished to find that the ends of two pairs of cable were tied together and these were run through the sheathing machine. No wonder there was no continuity from end to end. This one sheath held two separate cable runs, completely unconnected together. From that time until now, I have always specified either Belden or WestPenn wiring, both of which have always proven to be very reliable. More money has been lost in the name of saving money than any other thing.

CONDUIT OR NO CONDUIT
Why use conduit?

Conduit has only one purpose: to protect the wiring that is within it. Without conduit, wiring is exposed directly to the elements and environment. This means that every time some yahoo sticks his head above the ceiling and pokes around, tugging and pulling on wires to see what's up there, he is tugging and pulling on your cable unless it is enclosed within conduit. When he tugs, he may dislodge a connection. He will never know; he will never care. You will get a call the next day to come fix a problem and after 10 hours of troubleshooting, you will find a loose connection in a junction box. You ask yourself, "How did that happen?" You will never know. Just one of those things, you say.

Conduits make for more reliable security systems.

Types of conduit

There are seemingly endless types of conduit available. These include (National Electrical Code.):

- Rigid Metal Conduit (RMC): A thick threaded metal tubing that can be made from steel, stainless steel, or aluminum.
- Galvanized Rigid Conduit (GRC): Made of galvanized steel, and is thick enough to be threaded.
- Rigid Nonmetallic Conduit (RNC): Nonmetallic tubing that is always unthreaded.
- Electrical Metallic Tubing (EMT): Sometimes called "thin-wall tubing," it is a thinner, less costly, and lighter tubing than GRC. Although EMT is too thin to be threaded, it usually has threaded fittings that clamp to it. EMT uses clamps to make its connections. It may be either coated steel or aluminum.

- Intermediate Metal Conduit (IMC): Steel tubing that is heavier than EMT but lighter than RMC. IMC may or may not be threaded.
- Electrical Nonmetallic Tubing (ENT): Thin-walled tubing that is moisture-resistant and flame retardant. ENT is flexible enough to be bent by hand. It has a corrugated appearance and uses fittings that may be threaded.
- Flexible Metallic Conduit (FMC): Made of a self-interlocked ribbed strip of aluminum or steel that forms a hollow tube through which cable can be pulled. FMC is only practical in dry areas because it is not waterproof. It is forever flexible so it cannot be formed to a shape that it will maintain; it will always droop. FMC is commonly used in short "pigtail" segments to connect between electrical boxes. It is available from ½" to 4" diameters.
- Liquid-tight Flexible Metal Conduit (LFMC): This is FMC covered in a waterproof plastic coating. It is used to connect video cameras to their electrical box in outdoor installations.
- Flexible Metallic Tubing (FMT): It not considered a conduit, but a raceway. It is available in ½" and ¾" inside diameters.
- Liquid-tight Flexible Nonmetallic Conduit (LFNC): A class of conduits that refers to several types of flame-resistant, nonmetallic tubing. Interior surfaces may be either corrugated or smooth.
- Aluminum Conduit: Similar to galvanized conduit but is used where a corrosion-resistant form of conduit is needed, such as at food processing plants, chemical plants, and so forth. It cannot be directly embedded into cement, due to the corrosion that occurs as the aluminum contacts the alkalis in the cement.
- PVC Conduit: PVC is a polyvinyl chloride plastic tubing that is used as an electrical conduit. It is usually available in three wall thicknesses. The thinnest is used for embedding into concrete, the middle thickness is used for exposed applications, and the thickest is used for direct burial into Earth. PVC resists moisture and corrosive substances, but it must be used in accordance with codes because PVC creates toxic smoke when burned. Accordingly, it should not be used above ceilings.
- Other Metal Conduits: Conduits are also available in stainless steel, brass, and bronze to deal with extreme corrosion environments.
- Underground Conduit: Conduits for direct burial underground are usually made of PVC, polyethylene, or polystyrene to avoid corrosion. These may be buried directly or placed within a concrete "duct bank" that may contain many conduits. Older direct burial conduits may be made of metal, compressed Asbestos, fiber mixed cement, or fired clay.

Other wireways

- Surface Mounted Raceways (Panduit/Wiremold/Cache-Cable): Wiremold and its ilk are surface-mounted plastic "C-Channels" equipped with a snap-fit cover that provides a reasonably attractive alternative to loose cable. Wiremold is used to fit cables along a ceiling, wall, or floor from their source to an appliance such as a computer.
- Cable Trays: These are used to distribute quantities of cables in a back-of-house area. Cable trays are configured as a horizontal or vertical ladder rack into which the cables are laid. Cable trays can hold vastly more cable than equivalent conduit of the same cost. Use of Cable trays require Fire protective measures where the cable tray goes through a wall or ceiling. Cable trays are also useful where the installation is expected to undergo changes in wiring (adding, deleting, etc.) during the lifetime of the building, because it provides easy access to the wire way.

Indoor conduit applications

The question is not *when* conduits are recommended; they are always recommended.

- Use metallic conduits when the cables may be subjected to fire or physical damage. This includes virtually all cases of indoor use.
- Use PVC conduits when the conduit will be embedded into concrete or placed underground.

Outdoor conduit applications

Conduit should always be used outdoors. Outdoor conduit must be installed in a water-resistant or waterproof fashion; e.g., making all entries to boxes from below rather than on the sides or top where water could enter. Liquid-tight flexible conduits should be used wherever the conduit may come in direct contact with rain. All connections should be properly sealed from the weather.

When you can forget about conduit

You can use cable without conduit in any of the following conditions:

- Cables within enclosures
- Plenum rated cables (fire-resistant) above ceilings, where the cables are not likely to be subjected to physical damage, such as within a cable tray

Conduit fill

NEC 300.17 has standards for how much you can fill up a conduit with cabling. It is important to know what size conduit to use for a given assortment of cables. For this discussion, we will examine Class 2 and Class 3 cables (Article 725). Different rules apply for Fiber-optic cables. Here are the rules:

- One cable can fill only 53% of the conduit
- Two cables can fill only 31% of the conduit
- Three or more cables can fill only 40% of the conduit

This makes sense because one wire is easier to pull than two or more. And it is the total diameter of two cables, not their total cross-sectional area, that limits the amount you can pull. With three or more cables, it is the cross-sectional area that limits the amount of cables you can pull through the conduit.

Local authorities having jurisdiction may down-rate these figures, so be sure to check before ordering conduit.

There are numerous tables and calculators on the Internet to help with conduit fill calculations, but I am going to give you the actual formula (for a single size of cable).

- Take the outside diameter (OD) of the cable
- Square the cable OD
- Multiply by the number of cables with this OD
- Multiply by .7854
- This gives you the total cross-sectional area of the cables
- Select the appropriate size conduit from Table 25.1

Here is the formula for multiple cable sizes within the same conduit:

- Take the OD of the cable
- Square the cable OD
- Multiply by the number of cables with this OD
- Multiply by .7854
- This gives you the total cross-sectional area of the cables of this size (store the result)
- Repeat steps 1–5 for each additional cable size
- Add the total cross-sectional areas of all the cables
- Select the appropriate size conduit from Table 25.1

You can make a handy Microsoft Excel spreadsheet to calculate complex quantities of different cable sizes for each conduit by following these steps:

- Cell A1—"Conduit Size Calculator"
- Cell F2—0.7854

Table 25.1 Conduit Fill Sizes

| Conduit | | Total Cross-Sectional Area | | |
| | | 1 Cable | 2 Cables | 3 or More |
Size	I.D.	53%	31%	40%
1/2"	0.602	0.31906	0.18662	0.2408
3/4"	0.824	0.43672	0.25544	0.3296
1"	1.049	0.55597	0.32519	0.4196
1-1/4"	1.38	0.7314	0.4278	0.552
1-1/2"	1.61	0.8533	0.4991	0.644
2"	2.067	1.09551	0.64077	0.8268
2-1/2"	2.731	1.44743	0.84661	1.0924
3"	3.356	1.77868	1.04036	1.3424

- Cell A4 "OD"
- Cell B4 (insert OD of first cable in series)
- Cell C4 (insert OD of second cable in series)
- Cell D4−H4 (insert OD of additional cables in series)
- Cell A5 "OD Sq."
- Cell B5 (=B4*B4)
- Cell C5−H5 (square each OD above from row B)
- Cell A6 "Cable"
- Cell B6−H6 (insert shorthand for each cable such as CP for camera power, CR for card reader, A for alarm, AP for Powered Alarm, L for Lock, etc.)
- Cells B7−H7 (data entry fields—enter quantity for each type of cable for the conduit being calculated)
- Cell A8—"Total"
- Cells B8 (=B7*B5—total number of cables times the OD squared)
- Cells C8−H8 (same formula as B8)
- Cell A9—"Total CSA" (total cross-sectional area of all cables in this group)
- Cell B9 (B8*F2 or B8 *0.7854)
- Cells C9−H9 (same formula as B9)
- Cell I9 = sum(B9:H9—total CSA of all cables in this conduit)
- *Refer number in Cell I9 to Table 25.1 to select the correct size conduit*

Conduit bends

Do not exceed 180 degrees of total bends in any single run of conduit. Doing so is a sure way to create conduit pulling problems.

Conduit/cable fire protection

The jacket (sheath) of a cable is a source of fuel for fires within buildings. Fires can be limited by using plenum rated cables that are designed to withstand direct contact with heat and fire.

All cables that transit through a wall, floor, or ceiling from one fire space to another should do so through an entry-way that is packed with passive fire-resistant material (called "fire-stop"). Fire-stop is available in intumescents (which expand when exposed to heat), mortar, silicone foam, caulking, fibers, and pillows.

CABLE HANDLING
Cable handling nightmares

Cable pulling is one of the worst jobs in a project. Pulling cable through conduit is much easier than running cable without conduit because there are no obstructions in the way of the pull. Every installer has stories about nightmares pulling cables. Most of the horror stories revolve around unconduited cables. The others mostly revolve around older conduits that have a break somewhere midpoint that do not allow a new cable to be pulled, even though there is plenty of space in the conduit.

Before you have been in the security system industry very long you will certainly see what is commonly referred to as a "Rat's Nest" cable installation. These are installations where the installer just did not care. If it is your unenviable job to sort out this tangle of cables or add, modify, or troubleshoot a Rat's Nest, your troubles have just begun.

Good cable handling is a blessing. Bad cable handling is a curse on everyone.

Cable handling and system troubleshooting

Cables should be pulled neatly into conduits and never subjected to tension beyond their rated tensile strength. They should never be yanked, jerked, or connected to a 4-wheel drive truck with tires over 45 in. driven by a man named "Bubba."

All conduits should be equipped with a 200-pound test pull string that should be labeled with a number to easily identify which one to pull from both ends. Where there is no pull string you can use a fish tape.

All cables should be lubricated with an approved cable lubricant before pulling. Cables should be pulled through by the pull string, which should be attached firmly to the cables. The string should be bent back onto itself to create a loop and that loop should be tied off to itself. Cables should be fitted into the loop and twisted back. The twisted back portions should be tied back with phasing tape. This entire bundle should be coated with cable-pulling lubricant.

Make sure that all conduit ends are fitted with bushed chase nipples and deburred to prevent sharp ends from damaging the cable sheaths.

CABLE DRESSING PRACTICES
What is cable dressing?

Cable dressing is the art of installing cables into enclosures so that they are neatly organized. Good cable dressing promotes better operation due to reduced cable and circuit interference and better maintenance due to the ability to find and add, modify, and delete cables.

Good cable dressing is a beauty to behold (Fig. 25.2).

Cable dressing nightmares

Remember the Rat's Nest? When you encounter one, you will want to send a nasty message to the thoughtless fool who left you with the mess to clean up (Fig. 25.3). Encountering a poorly dressed cabinet or equipment rack can add many unnecessary hours to a project.

Cable dressing and system troubleshooting

Poor cable dressing virtually ensures system problems. With poorly dressed installations, technicians must tug and pull on cables to find out which one goes where. This invariably pulls connections loose that fail right then or much later. Poor cable dressing ensures that installers and technicians must spend countless additional hours to find circuits. It also adds many dozens of hours to a security system designer's work to figure out which old circuits to interface with and where those circuits are.

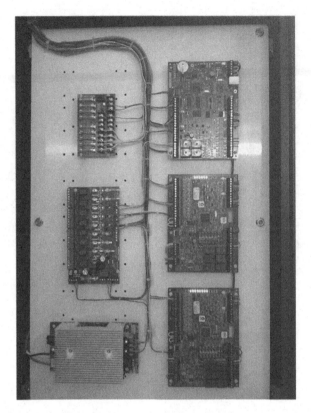

■ **FIGURE 25.2** Good cable dressing. *Image courtesy of Convergint Technologies.*

The proper way to dress cables

Cable dressing is the art of forming and wrapping cables into neat bundles. Here are some basic rules:

- Separate classes of cables and cables of different systems into their own bundles.
- Make certain that all Class 1 cables are well separated from Class 2/3 cables.
- On the front side of the rack, dedicate a minimum of one rack space (1.75 in.) for every two rack spaces of patch panels.
- Dress the cables by ensuring that all cables are parallel to each other as they are installed. Smooth them with your hand until they form a neat clean bundle.
- Cables may enter the rack from different angles and directions. This can result in cables of varying lengths. Cables should be cut to

■ **FIGURE 25.3** Rats Nest.

uniform lengths, which should be to the longest reach of the bundle. Label each cable with a wire number.

- Respect the minimum radius of each cable and do not create bends exceeding that radius.
- Use evenly spaced tie wraps or hook and loop straps to secure the cables. Tighten the tie wraps by hand only.
- Use cable management hardware to assist in creating good-looking cable installations.

Cable cross-dressing

Just as a "cross-dressing" man may look "a little odd," so too do "cross-dressed" cables. Alert designers, installers, and maintenance technicians may look into an equipment rack or cabinet that on first appearance seems like a work of art, only to discover that it holds its own nightmares.

The most common is the bundling of Class 1 and Class 2/3 cables together. This is a no-no. Class 1 cables create a significant magnetic current around them that can conduct power directly into nearby Class 2/3 cables, causing system instability and even equipment damage.

The second problem I have seen is fiber-optic cables that are installed with radiuses that are too tight for the fiber. This can cause bandwidth limitations (again poor system performance) and even total failure of fibers as they age.

Similarly, it is not uncommon to see cable dressing addressed as a complete afterthought. The result of this is that the installer must pull cables so tight that their connections are under constant tension, and may ultimately fail in the field in operation.

All of these poor cable practices are certain to bring the maintenance tech back again and again and make the system owner more unhappy with each trip to resolve the problems.

CABLE DOCUMENTATION
What is cable documentation?

Cable documentation is the recording of all cables in the system including:

- Their sources and destinations
- The path the cables follow to get from source to destination
- Their cable types
- Their cable numbers
- What colors go to which terminals for each individual conductor

Cables can be documented on drawings or on schedules (tables).

Who cares about cable documentation?

The short answer is virtually everyone who cares about the reliable operation of the system. This list includes:

- The original designer, who wants the system installed correctly
- The original installer, who wants to know how to install the system correctly

- The system owner, who wants the system to be well maintained
- The maintenance technician, who wants to maintain the system
- The next designer, who wants to know how to interface new equipment with old

When should cable documentation begin?

Cable documentation should begin with the original design. Every decision about how to run cables, terminate cables, which cables to use, what terminations to use, and which colors and wire numbers to use are all engineering decisions. These decisions are made by a qualified engineer or they are made by a guy with a screwdriver in his hand.

You cannot buy a more expensive engineer than a guy with a screwdriver, because he must make each decision independently, devoid of knowing what all the other guys with screwdrivers on the job are doing and how they are doing it. Will he use the same cable, the same wire numbering scheme, the same connectors, and the same power supplies? No. The installation will be a mishmash of helter-skelter decisions, none tied to any single design strategy. In short, it will be a mess.

What is the best way to document cabling?

There are two ways to document cabling: drawings and cable schedules (spreadsheets). Although both work, drawings are the best because they provide the most information. Cable schedules can "fill in the blanks" that drawings cannot, such as what color conductors to use on each type of terminal and so forth.

What is the best way to present cable documentation?

Cable documentation should be presented in the form of "As-Built" diagrams. Each cabinet or rack should have associated drawings in an envelope on the inside of its door. Each envelope should contain at a minimum:

- The floor plan showing device locations with device nomenclatures (CR-016 for Card Reader 016, etc.) and showing conduit paths and sizes and the equipment cabinets, racks, terminal cabinets, and junction boxes that are on that floor
- The cabinet or rack elevation and wiring diagram showing the equipment layout in the cabinet or rack and the cables that connect to

each termination of each device, including cable numbers and types for each connection

CHAPTER SUMMARY

1. A cable comprises its core (conductor or conductors), individual insulation around each core, and often an overall sheath.
2. Cable types include both copper and fiber-optic types.
3. The NEC Article 725.21(B) established three classes of cables for wiring inside buildings: Class 1, Class 2, and Class 3.
4. Class 1 wiring is for powered devices requiring normal (mains) power or 120 V AC, and actually, all wiring above 70.7 V and up to 600 V.
5. Class 2 wiring is low voltage and low power, typically not more than 30 V and less than 100 VA power on the circuit.
6. Class 3 wiring includes circuits over 30 V and exceeding 0.5 VA, but not more than 100 VA.
7. Wires are available in a variety of thicknesses called gauges depending on how much power they must carry.
8. Cable insulation is rated for voltage and application.
9. Cables are also available as either solid core or stranded core.
10. It is common to use cable colors to designate the purpose of the cable in installations.
11. Without conduit, wiring is exposed directly to the elements and environment.
12. Use of conduits is always recommended.
13. Conduit should always be used outdoors.
14. You can use cable without conduit in any of the following conditions:
    a. Cables within enclosures
    b. Plenum rated cables (fire-resistant) above ceilings, where the cables are not likely to be subjected to physical damage, such as within a cable tray
15. The NEC 300.17 has standards for how much you can fill up a conduit with cabling:
    a. One cable can fill only 53% of the conduit
    b. Two cables can fill only 31% of the conduit
    c. Three or more cables can fill only 40% of the conduit
16. Do not exceed 180 degrees of total bends in any single run of conduit.
17. The jacket (sheath) of a cable is a source of fuel for fires within buildings.

18. Good cable handling is a blessing. Bad cable handling is a curse on everyone.

19. Cables should be pulled neatly into conduits and never subjected to tension beyond their rated tensile strength.

20. Cable Dressing is the art of installing cables into enclosures so that they are neatly organized.

21. Cable Documentation is the recording of all cables in the system including:
    a. Their sources and destinations
    b. The path the cables follow to get from source to destination
    c. Their cable types
    d. Their cable numbers
    e. What colors go to which terminals for each individual conductor

22. Everyone who cares about the reliable operation of the system cares about cable documentation including:
    a. The original designer, who wants the system installed correctly
    b. The original installer, who wants to know how to install the system correctly
    c. The system owner, who wants the system to be well maintained
    d. The maintenance technician, who wants to maintain the system
    e. The next designer, who wants to know how to interface new equipment with old

23. Cable documentation should begin with the original design.

24. Cable documentation should be presented in the form of "As-Built" diagrams.

Q&A

1. *The three most common elements that cables are made from are:*
   a. Copper, aluminum, and plastic
   b. Copper, aluminum, and glass
   c. Copper, plastic, and glass
   d. None of the above

2. *The NEC Article 725.21(B) has established three primary classes of wiring for cables inside buildings:*
   a. Class 1, Class 2, and Class 3
   b. Class A, Class B, and Class C
   c. They can be called by either of the labels in a or b
   d. None of the above

3. *Class 1 wiring is for powered devices requiring normal (mains) power or:*
   a. 120 V AC
   b. Class 1 wiring also includes low-voltage, low-current wiring that is essential for the safety of the building, such as boiler control wiring

    c. Both a and b

    d. Neither a nor b

4. *Class 2 Wiring is specifically for:*

    a. Communication and signaling circuits

    b. Ethernet digital circuits

    c. Both a and b

    d. Neither a nor b

5. *Class 3 wiring includes:*

    a. Background music system loudspeaker wiring such as 70.7 V wiring schemes

    b. Nurse call systems

    c. Intercom systems and most security systems wiring

    d. All of the above

6. *In Class 2 or Class 3 wiring, 26 gauge cable _____.*

    a. Is bigger diameter than 12 gauge cable

    b. Is smaller diameter than 12 gauge cable

    c. Must be colored differently than 12 gauge cable

    d. None of the above

7. *Class 3 cables should have insulation rated up to:*

    a. 120 V

    b. 300 V

    c. 600 V

    d. None of the above

8. *Solid core cables are used to power most devices that are powered from line power (120 or 220 V AC).*

    a. Solid core wires can also be used in Class 2 and Class 3 circuits.

    b. Solid core wires cannot also be used in Class 2 and Class 3 circuits.

    c. Solid core wires are never used in Class 2 Circuits but may be used in Class 3 circuits.

    d. None of the above

9. *I recommend against buying cable from:*

    a. Any foreign manufacturer

    b. Cables that are more than 1 year old

    c. "Joe's Pool Hall, Used Cars, Yard Work, and Cables"

    d. None of the above

10. *Conduit has only one purpose:*

    a. To protect cable from water

    b. To protect the wiring that is within it

    c. To protect cable from wild animals in the ceiling

    d. None of the above

11. *Types of conduit include:*

    a. Rigid Metal Conduit (RMC)

    b. Galvanized Rigid Conduit (GRC)

    c. Rigid Nonmetallic Conduit (RNC)

    d. All of the above

12. *Types of Conduit include:*

    a. Electrical Metallic Tubing (EMT)

    b. Flexible Aluminum Conduit (FAC)

    c. Electrically Conducting Conduit (ECC)

    d. All of the above

13. *Panduit and Wiremold are trade names for:*

    a. Surface mounted raceway

    b. Molded wiring

    c. Paneled conduit

    d. None of the above

14. *Conduits are recommended whenever*

    a. Rain could affect the cable

    b. Animals could affect the cable

    c. Humidity could affect the cable

    d. Conduits are always recommended

15. *The one time you can forget about conduit is when*

    a. Cables are strung between buildings within a trench

    b. Cables are within an enclosure

    c. Cables are coated in PVC

    d. None of the above

16. *The NEC 300.17 has standards for how much you can fill up a conduit with cabling. Rules include:*

    a. One cable can fill only 53% of the conduit

    b. Two cables can fill only 31% of the conduit

    c. Three or more cables can fill only 40% of the conduit

    d. All of the above

17. *Do not exceed _____ degrees of total bends in any single run of conduit.*

    a. 45

    b. 90

    c. 180

    d. 360

18. *The jacket (sheath) of a cable is a source of fuel for fires within buildings.*

    a. True

    b. False

    c. Fires can be limited by using plenum rated cables that are designed to withstand direct contact with heat and fire.

    d. None of the above

19. *Cables should be pulled neatly into conduits and never subjected to tension beyond their rated _____ strength.*

    a. Textile

    b. Tensile

    c. Shear

    d. None of the above

20. *Cable dressing is the art of installing cables into enclosures so that they are:*

    a. Arranged by class

    b. Arranged by color

    c. Arranged by length

    d. Neatly organized

21. *Cable Documentation is the recording of all:*
   a. Devices in the system
   b. Conduits in the system
   c. Cables in the system
   d. None of the above

Answers: (1) c; (2) a; (3) c; (4) c; (5) d; (6) b; (7) c; (8) a; (9) c; (10) b; (11) d; (12) a; (13) a; (14) d; (15) b; (16) d; (17) c; (18) c; (19) b; (20) d; (21) c.

Chapter 26

Environmental Considerations

CHAPTER OBJECTIVES

1. Learn the basics in the chapter overview
2. Understand electronic circuitry sensitivities
3. Get to know environmental factors in system failures
4. Answer challenging questions about system environmental considerations

CHAPTER OVERVIEW

Environmental considerations are the forgotten information about designing, installing, and maintaining security systems. Many systems have failed early due to the designer and installer forgetting about or being unaware of the effects of the environment on the security system. When it comes to environmental effects on the performance of security systems, ignorance is not bliss.

This chapter explores all of the factors you need to know to protect alarm/access control systems from environmental harm, including electronic circuitry sensitivities and environmental factors in system failures. Today, the environment is expanded to include the IP Network on which the security system resides, and even the "Cloud," if the security system resides in the Cloud.

ELECTRONIC CIRCUITRY SENSITIVITIES

All electronic circuits are designed to operate within a limited range of temperatures and humidity. Operating circuitry outside these parameters can and will cause irreparable damage.

Electronic circuitry is also sensitive to other environmental factors. From rust to insects, to mice to snakes, to spilled Coca-Cola, to hackers, to loss of critical data, even the loss of the files of access control system itself; I have seen it all.

Electronic Access Control. DOI: http://dx.doi.org/10.1016/B978-0-12-805465-9.00026-9

Electronic circuits can be sensitive to a wide array of things, many of which you would never think of. For example, analog video circuits can be affected by the adjacency of coaxial wiring to fluorescent lights. Hum is a well-known phenomenon. When an A/C circuit is too near a high-impedance, high-gain audio circuit, you can hear the A/C signal changing polarity as hum in the audio circuit. Hum can also affect video circuits and data circuits. Very few people consider the effects of hum on a data circuit, but it can be devastatingly difficult to troubleshoot.

Grounding problems are also much more common than most designers, installers, and technicians think. They can induce hum into all types of circuits, which can show up in a variety of ways.

ENVIRONMENTAL FACTORS IN SYSTEM FAILURES

The environmental factors that can contribute to system malfunctions and unreliability are listed next.

Temperature extremes

All circuits have a minimum and maximum temperature range under which they can operate. Most security system circuits are designed to operate from −10°F to 0°F (−22°C to −17°C) on the low end to about 122°F (50°C) on the high end. Systems that are made to mil-spec specifications can handle temperatures up to 150°F (60°C).

Operating the electronics outside of these temperature ranges will result in malfunctions and ultimately damage to the electronics. This occurs most often in outdoor installations. Let me give you three examples:

1. Access control panel electronics are mounted within an outdoor rated steel enclosure in southern New Mexico. Afternoon temperatures inside the enclosure can easily reach 170°F (77°C), outside the operating range of 122°F (50°C) for which the circuitry was designed. What can be expected to happen?

   The electronics may exhibit unreliability and is certainly likely to fail entirely in half its rated life (typically 7 years). What can you do about this?

   Air-condition the enclosure or move the electronics inside an air-conditioned space. Wiegand wiring can run 550 ft (167.7 m) and dry contact wiring can span equal distances (use larger gauge wiring to assure good communications).

2. An access card reader is mounted next to the entrance door to the surface of a steel maintenance building in Houston, Texas. The reader

is mounted directly on the steel surface. The steel building acts as a giant solar collector and radiator, radiating heat into everything it touches. Afternoon temperatures measured by an IR thermometer on the steel reach 155°F (68.3°C). What can be expected to happen?

The card reader will almost certainly exhibit unreliability and will fail before its normally expected useful life (7 years). What can you do about this?

Mount the card reader onto a fiberglass junction box, which is then surface-mounted to the steel building surface. This will insulate the card reader from the high-temperature radiating steel surface.

3. An access card reader is mounted next to the entrance door to the surface of a steel maintenance building at Eielson Air Force Base, Alaska. The reader is mounted directly on the steel surface. The surface temperature of the steel building is −40°F (−40°C). With winds of just 5 miles per hour, wind chill at this temperature is −88°F (−66.6°C). This is well below the operating temperature of any alarm/access control system. What can be expected to happen?

At this temperature electronics may be expected to operate very poorly or to shut down completely. Depending on the construction, it may or may not operate again when restored to normal operating temperatures. What can you do?

Mount the card reader within a fiberglass enclosure that also includes a 7 W light bulb and a temperature sensor. The light bulb will maintain the enclosure within the normal operating temperature and the temperature sensor can be used to signal an alarm if the light bulb burns out, alerting maintenance to replace the bulb.

Humidity or condensation

Humidity is a long-term and silent killer of electronics. Condensation is its quick-acting cousin.

Condensation occurs when humidity reaches its dew point; i.e., when the air is so full of water vapor that it can no longer be supported as a vapor and it condenses back into liquid water. Condensation usually occurs when water vapor strikes a solid surface that is colder than the surrounding air. When this occurs, the water condenses onto whatever surface is near. As the water forms into droplets on the surface, it can conduct electricity, if that surface is an electrical or electronic circuit. When that occurs, the circuit can be short-circuited as electricity takes the path of least resistance (through the water droplets), bypassing resistors, capacitors, transistors, and integrated circuits. This forms unintended circuits that can apply unintended voltages and currents to electronics, damaging them permanently.

Humidity is the degree to which the air is filled with water vapor. As humidity percentages rise, so does the chance for long-term damage to electronics. As humidity forms near electronic circuits, it can trap dust, spores, bacteria, and other matter floating in the air, making them moist, and when they touch an electronic circuit they will settle there permanently. All of these things conduct electricity to a greater or lesser degree. Over time, they form a semiconductive layer on the electronic circuit that can cause the circuit to malfunction and become damaged. The effect is cumulative, and although this can take years to accumulate, the result is certain.

The answer is to protect electronic circuits from humidity and condensation.

Vibrations

If you have ever twisted a paper clip into a straight line and then twisted it back again into the shape of a paper clip you may have found that the metal broke while bending. If not on the first try, then pretty much certainly after a few more tries. This is because of metal fatigue. The molecular structure of the paper clip is broken by repeated bending. This phenomenon also occurs on circuit boards, connectors, wires, and paper clips.

If an electronic circuit is placed into an environment with continuous to infrequent vibrations (say near an HVAC fan unit or elevator machine room), that circuit can be placed under stress each time it vibrates.

Over time, things can fail. The most common failure points include:

- Circuit board to component connection solder joints
- Wiring connection points
- Ground connection points

When circuits fail due to vibrations, they may fail either catastrophically or by exhibiting infrequent anomalies. Troubleshooting intermittent circuits is always difficult. When dealing with circuits in a vibrating environment, be sure not to rule out vibration as a possible culprit.

Dirt

Any time that an electronic circuit is exposed to dirt accumulation, there will be problems. Any environment subject to the collection of dirt is also subject to humidity, and often condensation, insects, and possibly larger creatures. All of these are bad for electronics. Any of these issues can make electronics fail catastrophically or by an accumulation of intermittent problems over time.

Even if the electronics accumulated nothing but dirt, there would still be problems. Dirt itself is a semiconductor and can change the behavioral properties of any electronic circuit.

Insects, birds, snakes, and other creatures

It is surprising how many electronic circuits are exposed to insects, birds, snakes, and other creatures. Creatures like warmth in the winter (and sometimes other times too).

Electronic circuits make a wonderful place for nesting for small creatures and larger ones that can squeeze through a cabinet opening will go there for warmth. This is never good for the electrical or electronics circuits and is often also bad for the creature (see Fig. 26.1).

You can prevent this on indoor cabinets by filling all of the holes in boxes with vent plugs. A vent plug is a small plug that goes into an electrical cabinet conduit opening hole constructed of a small screen with a frame that pressure fits into the conduit hole.

Outdoor cabinets should not have any holes on the top or sides as these can facilitate entry of rain and moisture.

Lighting (at access control system portals)

One would not normally think that lighting could cause a problem for electronic circuits, but in some cases it can. Certain types of lighting (notably fluorescent and mercury vapor lights) operate by creating a plasma that reacts with a luminescent coating on the inside of the lamp. It is this reaction, rather than the plasma, that creates the light. But as a fluorescent lamp ages, the plasma it creates can become unstable. You may have noticed older fluorescent lamps that flicker like a flame licking the inside of the bulb. This condition can create variations in the plasma at twice the line frequency. This can create multipath reflections in adjacent circuits where the shielding of the circuit is poor (as is the case with most consumer and commercial quality circuits).

Most circuits designers assume that their circuits will be operating in a stable electrical environment. When they encounter flickering fluorescent

■ **FIGURE 26.1** Snake in a box. *Image courtesy of Ricky Ellis (www.safetyfirsthome.com).*

lamps nearby, the result can be unstable communications that can be a nightmare to troubleshoot.

It is best not to place circuits or run cabling close to fluorescent fixtures or any other light fixtures that create light by exciting a plasma.

Securing the IP network

Chapter 27, Access control design, contains an extensive description of the elements of securing the security system on the IP Network. The importance of this cannot be over-stated. As security systems continue to move from proprietary wiring schemes to all IP-Ethernet network wiring schemes, they intrinsically become much more vulnerable to many kinds of potential damage that did not exist in the old style of infrastructures.

These include, but may not be limited to:

- Network failures
- Network maintenance downtime
- Accidental or intentional loss or damage of security data
- Accidental or intentional loss of the security system programs themselves
- Hacking
- Disclosure of confidential security system information

Access control in the cloud

"Cloud Computing" is becoming a popular way to manage security systems for small and medium size organizations. Placing the security system in the "Cloud" (which means placing the security system software that is hosted in a large public datacenter that is available only on the internet) has numerous advantages for small and medium sized organizations, but it can also be frought with problems as well.

- The Good:
  - Cloud computing offers immediate availability from many types of devices. This frees the security staff from monitoring the system on a PC-based workstation and potentially allows for roving guards to be alerted of security incidents while they are on patrol, possibly even providing video of an ongoing incident right to the security officer's smartphone.
  - System monitoring and reports are available through a workstation's browser, allowing multiple authorized staff to access the system.
  - System maintenance is in the hands of data center technicians, who can keep software up to date, apply security protocols and back-up system data routinely.
- The Bad:
  - Some cloud data centers have been routinely hacked, so all security system data must be encrypted when in the Cloud and when it is in transit to and from the cloud.
  - If management staff are loose with their Cloud credentials, it is likely that unauthorized users could learn their username and password, allowing inappropriate users to access security system data and change security system settings.
  - Data center technicians are not likely to understand security system settings, if configuration changes must be made, so Cloud

security system users should keep their original installing vendor on hand to make such changes.

SECURITY-SYSTEMS AS-A-SERVICE

Some security system integrators are offering Security Systems as a Service. This concept combines Cloud computing with active security system management and monitoring, and it is growing in popularity. Security-systems as-a-service (SSaaS) answers a lot of problems for small and medium sized organizations, by allowing them to focus on their core business and outsource not only their security guards, but even their security system to a security vendor.

- The Good:
  - Eliminates on-premises employed security staffing costs
  - Reduces security system capital costs
  - Places liability for maintenance and monitoring onto the vendor
- The Bad:
  - Provides less visibility into security processes and success metrics for management to review the efficiency of their security and for their operating costs
  - Everything that applies to Cloud computing also applies to SSaaS

CHAPTER SUMMARY

1. All electronic circuits are designed to operate within a limited range of temperatures and humidity. Operating circuitry outside these parameters can and will cause irreparable damage.
2. Humidity is a long-term and silent killer of electronics. Condensation is its quick-acting cousin.
3. If an electronic circuit is placed into an environment with continuous to infrequent vibrations (say near a HVAC fan unit or elevator machine room), that circuit can be placed under stress each time it vibrates.
4. Dirt is a semiconductor and can change the behavioral properties of any electronic circuit.
5. Electronic circuits that are exposed to insects, birds, snakes, and other creatures can be damaged by them.
6. Certain types of lighting (notably fluorescent and mercury vapor lights) operate by creating a plasma that reacts with a luminescent coating on the inside of the lamp. But as a fluorescent lamp ages, the plasma it creates can become unstable. Older fluorescent lamps may

flicker like a flame licking the inside of the bulb. This condition can create variations in the plasma at twice the line frequency. This can create multipath reflections in adjacent circuits where the shielding of the circuit is poor (as is the case with most consumer and commercial quality circuits).

Q&A

1. *All electronic circuits are designed to operate:*
   a. Without failing
   b. Within a limited range of temperatures and humidity
   c. Without intervention on the part of maintenance personnel
   d. None of the above
2. *Operating the electronics outside of its temperature ranges will result in:*
   a. Damage to surrounding electronics
   b. Immediate failure of the electronics
   c. Malfunctions and ultimately damage to the electronics
   d. None of the above
3. *Humidity is:*
   a. A long-term and silent killer of electronics
   b. One of the most common ways to increase the life of electronics
   c. One of the most common ways to shorten the life of electronics
   d. None of the above
4. *Condensation is:*
   a. Humidity's quick-acting cousin
   b. Another good way to condense power into a power supply
   c. The result of using a condenser microphone
   d. None of the above
5. *Humidity is:*
   a. The degree to which the air is filled with rain
   b. The degree to which the air is filled with music
   c. The degree to which the air is filled with water vapor
   d. None of the above
6. *If an electronic circuit is placed into an environment with continuous to infrequent vibrations (say near a HVAC fan unit or elevator machine room):*
   a. That circuit can be placed under stress each time it vibrates
   b. That circuit will pass the vibrations on as hum
   c. That circuit will cause failures of other electronics
   d. None of the above
7. *The most common failure points include:*
   a. Circuit board to component connection solder joints
   b. Wiring connection points
   c. Ground connection points
   d. All of the above

8. *When circuits fail due to vibrations, they may fail:*
   a. Catastrophically
   b. By exhibiting infrequent anomalies
   c. Both a and b
   d. Either a or b

9. *Dirt buildup can make electronics fail:*
   a. Catastrophically
   b. By an accumulation of intermittent problems over time
   c. Both a and b
   d. Either a or b

10. *You can prevent creatures from entering indoor cabinets by:*
    a. Filling all holes in boxes with vent plugs
    b. Filling all holes in boxes with conduits
    c. Filling all holes in boxes with corks
    d. None of the above

11. *As a fluorescent lamp ages:*
    a. The plasma it creates can become green in color
    b. The plasma it creates can become unstable
    c. The lamp will turn yellow in color
    d. None of the above

12. *It is best not to place circuits or run cabling close to:*
    a. Dead creatures
    b. Dirty cabinets
    c. Angry geese
    d. Fluorescent fixtures or any other light fixtures that create light by exciting a plasma

Answers: (1) b; (2) c; (3) a; (4) a; (5) c; (6) a; (7) d; (8) d; (9) d; (10) a; (11) b; (12) d.

Chapter 27

Access Control Design

CHAPTER OBJECTIVES

1. Learn the basics in the chapter overview
2. Understand the different types of knowledge necessary for design versus installation versus maintenance
3. Understand all about design elements
4. Learn how to design robust portals—how criminals defeat common locks, doors, and frames
5. Learn access control system application concepts
6. Learn how to implement design ideas to paper
7. Understanding design elements of system installation
8. Learn to get commissioning right—in the specifications
9. Learn to avoid long-term problems by structuring the system acceptance correctly
10. Take a short test on access control system design

CHAPTER OVERVIEW

This chapter on alarm/access control system design draws significantly from my book Integrated Security Systems Design published in 2007.

In this chapter, we will begin by discussing the difference in skills necessary for design versus installation versus maintenance, and how being skilled in one discipline does not qualify a person in other disciplines.

The importance of designing to risk instead of to only a set of directions from the system owner will be discussed. The importance of designing not only for the problems faced today, but how to design to protect out to the future will be reviewed.

Each of the design elements will be discussed in detail including drawings, specifications, interdiscipline coordination, product selection, project management, and client management.

Electronic Access Control. DOI: http://dx.doi.org/10.1016/B978-0-12-805465-9.00027-0

How criminals defeat common locks, doors, and frames and how to design robust portals will be reviewed. Also reviewed will be the various application concepts necessary to understand to create a system that is robust, reliable, redundant, expandable and flexible, easy to use, and sustainable.

How to implement design ideas to paper and how to carry the installation to completion will be discussed at the end of this chapter.

DESIGN VERSUS INSTALLATION VERSUS MAINTENANCE (THE KNOWLEDGE GAP)

There are different common bodies of knowledge required for security system design, installation, and maintenance. The skills are similar but they are not duplicates. Few security system designers have also worked as both an installer and as a maintenance technician. Having worked at all three, I know that the skills necessary to be a great designer are completely different than the skills necessary to be a great Installer, and those skills are different than those required to be a great maintenance technician.

It is a shame that more designers do not realize this. I have found that many designers completely underestimate the value of installation and maintenance skills. It works the other way too. From my experience, installers and maintenance technicians vastly underestimate the requirements of security system design. In fact, installers and maintenance technicians are security system designers in the same way that auto mechanics are automobile designers.

This is not to denigrate installers and maintenance technicians. I have the highest respect for them. But in the same way that many designers do not understand the complexities of installation and maintenance, most installers and technicians rarely respect the complexities of security system design.

In this chapter, we will review some of those special skills.

THE IMPORTANCE OF DESIGNING TO RISK

The first thing to understand about security system design is that good designers design to risk. Security system design is not about electronic equipment. It is not about cameras, card readers, biometrics, locks, and alarms. It is not about gates and doors. It is only about reducing risk. It is that simple.

The vast majority of security systems that I have reviewed were not designed to risk. I know that because when I look at them I see blatant vulnerabilities that went completely unnoticed and, therefore, unaddressed by the designer. I know that because when I am called to review the facility, it is because of continuing security problems that went unaddressed by the first designer.

Good designs are predictable. You can trend-line the vulnerabilities to security events and you can trend-line security events back to vulnerabilities.

Risk comprises vulnerabilities, probability, and consequences. Any security system designed without a proper risk analysis *will be wrong*. I have talked to some designers who think that if they simply do a vulnerability assessment, they can design a good system. That is also wrong. Without understanding the entire risk equation for the facility you cannot perform a proper vulnerability assessment because you do not know what type of threat action it is vulnerable to.

A good risk analysis will tell the designer exactly what threat actions to be concerned about, what the potential consequences of the design basis threat are, and which vulnerabilities to be aware of.

Then, let the designing begin!

THE IMPORTANCE OF DESIGNING FOR THE FUTURE

After designing for risk, the most obvious difference between the way security consultants design and the way security integrators design is that security integrators usually design to solve today's problems.

The second important element of security system design is to understand that today is only a snapshot. Today is only one frame in a long movie. Designers who design the security system to solve today's problems are not solving anything but the problems on one frame in a movie.

The problems will return tomorrow ... with a vengeance. Nothing is more disappointing to clients than to spend literally hundreds of thousands and sometimes millions of dollars to solve security problems only to find that new problems arise quickly. I have seen cases where a new department was formed before the security system was complete only to find that the system does not address any of the security issues of the new department. This is poor planning.

A good designer asks a stream of questions of management all pointed toward where the organization has been and where it is going. He asks where it is going organizationally, structurally, geographically, and financially (new markets, etc.). A good designer implements all of this information into the design. Typically, there is no added project cost to do this. In its simplest form, you can design for the future by selecting a digital architecture that has capacity and redundancy by designing systems that are expandable and flexible in operations and by designing around operational procedures that are designed for growth and flexibility.

However, you can do a much better job after interviewing management about their growth plans over the next 5 years.

Anyone designing security systems should design to risk and design for tomorrow.

DESIGN ELEMENTS

The elements of a security system design include:

- Drawings
- Specifications
- Interdiscipline coordination
- Product selection
- Project management
- Client management

Drawings

Drawings are the heart of the design. They illustrate the designer's concepts about how the system should relate to the building, and they illustrate the relationship of devices to:

- Their physical environment (plans, elevations, and physical details)
- The conduit system and to power (plans and risers)
- Each other (single-line diagrams)
- The user (programming schedules)

Drawings must serve five distinct types of users:

The Bid Estimator: The bid estimator must determine what materials are needed. Helpful drawing tools include device schedules (spreadsheets, listing devices, and their attributes) and plans showing device locations and conduit lengths and sized and wire fills. Other drawings useful to the bid estimator include single-line diagrams, riser diagrams, and system interfacing diagrams.

The Installers: The installers need drawings that show both the big picture and the smallest details. Therefore, it is helpful if the drawings are formatted in a hierarchical fashion. Single-line Diagrams show the big picture. Plans show device locations and their relationship to the building and conduit system, and physical details and interface details show the smaller details.

The Installation Project Manager: The project manager needs to manage the progress of the installation, including coordinating the ordering and arrival of parts and supplies and coordinating manpower to the project at the correct time, in the correct place, and in coordination with other trades to get all devices mounted and all connections made. He/she will primarily rely on schedules for provisioning logistics, plans to measure installation progress, and single-line diagrams to gauge how close the system is to start-up.

The Maintenance Technician: After the system is installed, it is up to the maintenance technician to keep it running well. They will need single-line diagrams to determine how the system interconnects, plans to determine where devices are located and how they connect in the physical space, and risers and power schedules to know where to go from floor to floor and the source of power for each device. Drawings that illustrate how the equipment of that assembly interconnects with other equipment in the system and how the entire system operates should be located in a pocket in the door of each rack, console element, and panel.

The next Engineer Expanding the System: Virtually every system will be expanded in scope and/or function. This may happen a few months to many years after the original installation. The next engineer needs access to the original drawings to understand the context for his work.

Specifications

If drawings are the heart of the design, specifications are its head. Specifications generally take precedence in legal disputes. Drawings are there to illustrate the standards and practices that are required in the specifications. If you have been in your career for a long time, then you have seen some pretty bad specifications. We used to joke in our office about someday seeing a set of specifications that simply say: "Make it work real good." Some come pretty close to that. We have seen security system specifications that are only 5 pages long. There is a lot of room there for serious mistakes by a well-meaning contractor. Many security contracting problems are the result of incomplete or wrong specifications.

With very few exceptions, most integrators I have met sincerely want to do well for their clients. It is the designer's job to provide the integrator with enough information to do well. To the extent that drawings and specifications are incomplete, inaccurate, or misleading, the contractor can make unintended errors that will be costly and aggravating to the installer, the integrator for whom he/she works, and most certainly the system's owner.

Specifications should include a description of what the project entails; descriptions of the whole integrated system and each subsystem, a description of the services the contractor will provide; and a list of acceptable products and acceptable installation, testing, acceptance, training, and warranty practices. Different specification formats prevail in different areas of the world, and occasionally these may change as building code authorities evolve in their preferences.

Specifications should, to the extent possible, be based on "open sourced solutions" that allow for greater design flexibility, more bidding options and do not "lock" the client into a particular brand or model of hardware or software.

Interdiscipline coordination

Security systems are unique in that they relate to more building systems than any other building system. Security systems routinely relate to:

- Electrical
- Door and gate hardware
- Structural
- Elevators
- Parking
- Landscape
- Building automation systems
- Signage
- Concrete
- Lighting
- Traffic control systems
- Irrigation systems

Interdiscipline coordination makes or breaks the installation. It often determines whether a project works as intended or not, and whether the integrator makes a profit or a loss. There are many other trades working on a new construction project and the project manager must outline, communicate, and maintain coordination with all of these and with the main contractor at all times. Interdiscipline coordination will generally determine whether the integrator is on time or late. Every day an integrator spends on the jobsite after his designated completion date is money removed from his profit and irretrievable respect lost from his client.

Product selection

Specifying the correct products for the job can result in a wonderful system that can easily exceed the owner's expectations. The wrong products can leave the owner upset with the installer, the manufacturer, and the designer.

This is where the designer has free reign to do what is in the best interest of his client. If the designer is placed under pressure to specify one brand or another purely due to market forces, the owner will suffer. If the owner suffers, everyone suffers—the operator suffers, the maintenance tech suffers, and the integrator who has to listen to an unhappy client suffers.

Many designers express their product selection not only in their specification but also with a Bill of Quantities (BOQ). A BOQ is a spreadsheet listing all products to be used in the installation. At a minimum, the integrator's project manager should put together a BOQ and submit it to the designer for approval. This helps ensure that all necessary parts are included for the work.

Project management

The designer has to manage the design portion of the project. Design project management is all about delivering a design that meets the needs of the client, the integrator, and the client's project manager. The designer must do all this while working on other projects; they must provide the project deliverables on time and complete, and keep all parties happy. design project management has four phases:

- Initiating the project
- Planning the project
- Executing the project
- Controlling and closing the project

A number of things in each project need to be managed. The Project Management Institute (PMI) certifies project managers with both the Project Management Professional (PMP) and PMI certifications. The PMP certification is highly prized by everyone who knows anything much about project management. Get a PMP certification and your career as a project manager is assured. I highly recommend getting a PMP certification if you intend to have a career managing security projects. It will put you well out in front of others and help ensure that all your projects go well.

The PMP certification process includes the following:

- Establishing the framework for project management
- Managing the scope of the project
- Managing time
- Managing cost
- Managing quality
- Managing people
- Managing communications
- Managing risk
- Managing procurement
- Managing the project's integration aspects
- Maintaining a high level of professionalism throughout the project

Each project has three main aspects:

- The project scope of work
- The project schedule
- The project cost

Although it is beyond the scope of this book to teach project management, I will state that it is an essential skill for any engineer or designer and certainly for both design project managers and installation project managers. I strongly recommend that the reader should invest in several books on project management and spend several weeks getting familiar with the principles. Project management is all about providing structure and planning to what seems to many to be an intuitive process. However, without the necessary structure, project management can quickly descend into crisis management, and then further simply into project chaos and damage control. Hundreds of millions of dollars are lost each year by firms that entrust large projects to unqualified project managers. Your career will flourish if you have the requisite project management skills.

Client management

Client management is the process of managing the relationship between the firm you represent and the client's representative. *The most important aspect of good client management is to keep the project on scope, on schedule, and on budget.* Truly, whatever else you do is of no consequence if you do not do those three things.

Managing the relationship is about understanding who are the decision makers, the influencers, and simply the opinionated. Focus on decision makers; in particular, focus on those individuals who have the authority to approve work and issue checks. Spend some personal time before or after each meeting getting to know their interests, their personality, and their character. Find out what motivates them. Find out what they do not like. Use terminology they understand and resonate with. Be short in communications, be sincere, and always end with a "cookie"—a piece of good news that indicates the project is moving forward in a manner they can accept.

Everyone, and I mean *everyone*, brings three agendas to the project:

- Their role in the project: The interior architect has a different agenda than the shell and core architect, and the general contractor has a different agenda than the electrical contractor. The information technology contractor does not care about the agenda of the landscape contractor.
- Employer plans: The employer may have a technology bias or a business culture that is aggressive or conservative.
- Individual plans: They may be humble or pompous, technically competent, or covering up a feeling of inadequacy. They may be right- or left-brained, and may be kind or rude.

Besides the decision makers, others on the project also have particular interests and some of those may be influencers. Get on their wrong side of an influencer and you can have someone whispering bad things about your firm into the ear of the decision makers. This can make coordination very difficult. People who dislike you on a project routinely exhibit this by introducing delays, complications, and obstacles in the way of inter-discipline coordination.

Client management is by far the most complicated aspect of good project management, but it is far less important than managing scope, schedule, and budget. It does not matter how good you are at managing personal relationships if the project is wrong, late, and over budget. There is much more to good client management, but those are the basics.

DESIGNING ROBUST PORTALS—HOW CRIMINALS DEFEAT COMMON LOCKS, DOORS, AND FRAMES

I once wrote a popular article for a major Security Industry magazine titled "How to Plan, Design, and Install a Bad CCTV System." It outlined many things that can go wrong on an installation and provided a road map on how to avoid the most common pitfalls.

Understanding how to make systems fail is as equally important as understanding how to make systems work. If you do not understand how systems fail you cannot design or install a system that is robust against failure. Let's take a look at common methods used to defeat locks, doors, and frames.

Unlocking the door from the outside

I advanced the idea in the security industry of using what I call micro-controllers. These are small, single-door controllers located above the door on the secure side to which all the devices at the door connect. The microcontroller then connects by Ethernet to its host server. This approach has many advantages:

- Vastly reduces project wiring costs
- Increases system reliability because if a controller fails, it fails on only one door
- Microcontrollers can be configured with an integral digital switch to connect:
  - A local digital video camera
  - A local Intercom at the door
  - Another microcontroller down the hall

After my 2007 book *Integrated Security Systems Design*, Microcontrollers began to be introduced more widely into the market. (So far none have an integral digital switch.) However, there is one version of the microcontrol-ler that should be used carefully—the type that combines a microcontroller with a card reader all in one package. This type of product further simpli-fies the installation and reduces product costs, but it should only be used in relatively secure areas where users have already passed through at least one or two layers of access control. This is because if it is used on the exterior of a building (as I have seen it done), an intruder can simply smash the card reader and gain direct access to the lock power where they can open the circuit and simply enter the building. This will set off an alarm (usually only a communications alarm), but the intruder is now inside the building.

Lesson: Don't use these on exterior doors.

Double glass door exploit

Frameless glass doors are beautiful, especially when used in pairs on the front of a building. But because of settling foundations, they must be installed with a small opening between the doors through which an intruder can pass a yellow notepad, warmed by placing the notepad under his arm against his chest. The pad fits easily between the doors and its warmth and movement are seen by an infrared request-to-exit detector above the door. When this happens, the exit detectors interpret this as a legitimate request to exit and signals the access control panel to unlock the door, which it does, allowing the intruder to enter.

Lesson: Use a request-to-exit sensor that is adjusted to look out away from the door and place a plastic lip on one leaf of the doors to prevent insertion of a notepad.

Defeating electrified panic hardware

I was once called to evaluate how a burglar had gained entry to a gold and coin dealer's shop over a weekend. The burglar had disabled the alarm system from inside the store and made off with hundreds of thousands in gold and coins. Oh, and there was no evidence of forced entry. I got the call the Monday morning after the burglary. The shop was privately owned and the intruder used an access control system to enter through the back door. No one had access cards except the owner. There were no extra cards or keys out at all. So how did he get in?

Looking around, I noticed a bathroom next to the back door. I noticed that the access panel and alarm panel were inside the bathroom and the back door was equipped with electrified panic hardware. Outside the back door was a dumpster containing a small amount of trash and a broken umbrella. The inside ribs of the umbrella were all bent upward toward the top. Looking again at the back door I noticed some metal shavings on the ground just inside and outside of the door. Most uniquely, there were round metal plugs (in the form of large flat washers) in the door just below the panic bar on both the inside and outside of the door, held in place by a screw and nut.

I asked the owner when the plugs were installed and what it was for. He didn't remember ever seeing them before. As the police were there I directed the investigator to the umbrella and suggested he fingerprint the umbrella and the plugs, which they did. I suggested that the owner look through his video archive and look for anyone asking to use the bathroom in the last month. That is how they found the intruder, but how did he enter?

He drilled a 2″ diameter hole through the door from the outside just below the panic bar. He inserted the umbrella through the hole and pressed its button, opening the umbrella. Then he pulled the umbrella back through the door, which depressed the panic bar, opening the door and bypassing the security alarm. As there was no motion alarm at the back door on the inside and all the video was on the front of the building and in the shop and not in the back room or outside the back door, it was a simple effort to bypass the alarm from inside the bathroom and take all day Sunday to pilfer the store. He thought he had covered his tracks by plugging the hole. He had figured out this attack plan while using the bathroom at the shop and noticing all the details about how the physical and electronic security worked.

Lesson: Place a schedule operated motion detector focused on all exterior perimeter doors, and have a camera both inside and outside each exterior perimeter door.

Defeating door frames

Several criminal gangs have been using a device called a "frame spreader" for several years throughout the United States. I suspect this has happened in other countries as well. A frame spreader is a hydraulic ram that is inserted mid-height on the frame of a steel door (or pair of doors). The hydraulic ram is actuated and it expands by about 4–6 in. jamming the door frame into the adjacent walls. It achieves enough spread to release the door lock mechanism, allowing the damaged doors to be opened. This attack scenario is most common at remote warehouses with high-value assets such as computer CPU and memory chips, but it could be used on any facility.

Lesson: On high-value facilities, use multiple dead-bolt-equipped panic hardware (such as Securitech locks). This type of lock fits dead bolts into multiple points in the top, bottom, and sides of the frames, defeating this attack.

These are only a few innovative methods B&E (breaking and entering) criminals use to defeat security door hardware. Get together with local law enforcement and find out what has been used in your area. Be aware of the possibility of hard physical attacks on security systems. They happen all the time.

APPLICATION CONCEPTS

Alarm/access control systems use access control readers, electrified locks, request-to-exit sensors, door alarms, and volumetric and point alarms to defend facilities against inappropriate access by unauthorized users.

Use a layered security approach. This is a combination of detection and access layers placed over each other at geographical progression toward the most valuable assets (see Fig. 27.1).

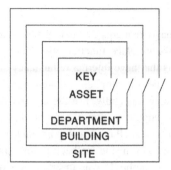

■ FIGURE 27.1 Layered security.

The basic goals of electronic security countermeasures include:

- Access control
- Deterrence
- Detection
- Assessment
- Coordinating response
- Evidence gathering

It is essential that all these goals should be met in the design of a comprehensive coordinated security program.

Security system designs should also be robust and redundant, expandable, flexible, and easy to use. And they should be sustainable.

Robust design

The quality of design and the quality of the installation work both have a strong bearing on how robust a system is. Poorly designed and installed systems have exposed wiring, exposed plug-in power supplies, fragile mounting of equipment, little or no shrouding of cameras, and exposed door position switches with loose wiring. The use of conduit alone instead of loose wiring creates a much more robust system. When you open an equipment cabinet, wiring should be neatly organized and well marked. There should be drawings in each cabinet to help the maintenance technician, or else he has to probe around in the wiring to figure out which wire goes to what equipment. All that probing can pull cables loose, creating another service call. All of these things create an unreliable system. If it doesn't look robust, trust me, it isn't.

Redundancy

The system should have redundancy such that if one component fails, another is there to take its place functionally. There are two ways to do this. First, use systems that have internal redundancy such as using equipment with redundant power supplies, redundant Ethernet connections, and redundant processing. Secondly, use the layered security approach so that if one component fails, detection occurs through another component. Remember the earlier umbrella intrusion example? A second motion detector facing the rear door inside the back room would have caught the intrusion the second the umbrella was inserted. A video camera on that area and on the outside of the back door might have helped identify the offender. This shop owner actually turned off his digital recorder after hours to preserve memory. He was using it to record activities in his customer area during open hours only, which was a very foolish procedure.

Expandable and flexible

A good designer designs systems so they are expandable and flexible. Even when everything about the project reeks of "this is a fixed design with no chance of ever changing," it is still best to incorporate expansion and flexibility into the design. For example, when designing a security system for my own home, I was *certain* of the design requirements, and those requirements could be filled by a very economical alarm system. I designed one with double that capacity. The alarm installer tried to sell me another alarm panel having just the needed capacity, which I declined. I wanted double. Within 1 year, things changed and I needed about half the available capacity for some unexpected changes. If I can't be sure about the future in my own home, how can any designer ever be certain about the future needs of a client. The short answer is you can't. Always design spare capacity and flexibility into the design. In almost every case, it can be done for little to nothing extra.

Easy to use

Please, please do not skip this section. I continue, after 35 years in this business, to find security systems that either require a PhD to operate or that are so confusingly configured that virtually no one knows what is going on in the system.

Please, do not get creative in system operation. Keep it simple. I actually have a design goal of 5 minutes of training to learn how to monitor a security console. I achieve that goal, and you can too. Still, there is plenty happening deeper in the system that you could find and operate, it is just that with only 5 minutes of training you can operate the basic monitoring functions. It is doable.

I witnessed part of the commissioning of a security system in Algeria where not a single one of the operators could read English, and the system was programmed entirely in English. That is just inexcusable. The contractor's excuse was that they should use operators that could read and speak English. Lots of luck on that in Algeria!

Sustainable

The idea of sustainability is relatively new to security system designers, but it should have been a primary part of design a very long time ago. All systems have a finite operating life, but you can extend that substantially through good design and good maintenance.

- Good design: First, understand that all devices have market life cycles.
  - Design idea
  - Prototype
  - Early market product used by early adopters
  - Commodity market (make lots of them and sell as many as you can)
  - Late market (old ho-hum technology)
  - End of life (sorry, we do not support that line anymore)

It is better never to design or install a system that is late market technology. I have sat in a conference room with product representatives who told me straight-faced that I should hurry to specify their old product line because it was going to be replaced very soon by a new line and then it would not be available anymore. Somehow they thought that was a selling point. Specify and install products that have considerable life left. Additionally, designing a robust system also helps make it sustainable.

- Good maintenance: Many systems die an early death due to poor maintenance.

You can sustain a system for many years by conducting good scheduled maintenance. I recommend daily checks of all field equipment. This can be done while on rounds by opening every access control door triggering every alarm and appearing on every camera. By conducting a routine guard tour and coupling it with a system operation checklist, you kill two birds with one stone. Reports each day result in either emergency maintenance needs or a device or two that must be put into the scheduled maintenance bucket. Have a certified maintenance technician visit the site at least once every month to take care of the scheduled maintenance work.

Use only technicians certified by the manufacturer of the equipment so they know what they are doing.

IMPLEMENTING DESIGN IDEAS TO PAPER

Designs should, oddly enough, start with a design concept. Surprisingly, very few designers seem to understand this. The first stage of the design is to write what I call a "basis for design" paper. This explains in simple language the goals and objectives of the design and how the designer plans to achieve those goals using an electronic security system.

The second step is to begin drawing the plan. But before that can be done, you need some information from the system owner. First, you need to understand how the access control systems will be zoned.

Creating access control zones

Every access control system uses card readers and electrified door locks to create access control zones. Users need an access card to enter a zone. Each zone may have one-to-many entries and an access control zone can include one or more other access control zones. For example, the perimeter of the building may comprise a "building zone," which encloses separate zones for each department within the building.

Most designers start by drawing card readers next to doors, but this approach almost certainly ensures the creating of "sneak paths" into access control zones. The better way is to use a colored pencil or pen and draw the boundary lines of each access area zone onto each drawing, starting with the overall site and going into each building and each floor and department. List each access area within its boundary. I like to use AutoCAD and create color-coded blocks corresponding to the access area zones. For nested zones, you can use an overall border around a group of individual access area zones.

Once you have all of the access control zones identified, then you can begin drawing locations for access control portals (access controlled doors and gates). Each access control zone may also have some doors or gates that could be used for exit, but not for entry, such as a rear fire stairwell door. Such doors may be equipped with only a door position switch and no card reader.

Door types

As you markup doors with card reader symbols, you should also note the type of door and frame and the number of leaves. From this information you can develop a list of door types. When I markup drawings, I use only two symbols: the door type symbol and a card reader. The door type symbol comprises one or two characters in a box next to the door and the card reader symbol shows where I want the card reader to be mounted (on the wall to the right or left, on the door mullion, etc.) You can do this in any way that makes sense to you. For consistency I use the following symbology (Fig. 27.2):

- Each Door Type symbol uses one or two characters, a number, and an optional letter.
- Odd numbers are used for single-leaf doors and even numbers are used for double-leaf doors.
- Door Type 1 is a single-leaf door with a surface-mounted door position switch.
- Door Type 2 is a double-leaf door with surface-mounted door position switches.
- Door Type 1A is a single-leaf door with a concealed door position switch.
- Door Type 2A is a double-leaf door with concealed door position switches.
- Door Type 3 is a single-leaf door with a card reader.
- Door Type 4 is a double-leaf door with a card reader.
- Door Type 3A/3B/3C, and so forth, is a single-leaf door with card reader where the letter designates the type of door, frame, swing, and lock combination.
- Door Type 4A/4B/4C, and so forth, is a double-leaf door with card reader where the letter designates the type of door, frame, swing, and lock combination.
- And so forth . . .

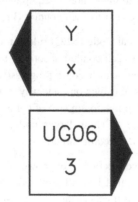

■ **FIGURE 27.2** Door type symbol.

Each door type symbol is illustrated in the physical details by its own detail (Fig. 27.3) comprising, e.g.:

- Detail 3B—A—A plan view
- Detail 3B—B—An elevation view
- Detail 3B—C—A schematic view showing devices at the door and the wiring to the Access Control Panel

Alarm devices

The next pass over the drawings will place alarm devices onto the drawings. Use a separate symbol for each type of alarm device and show its location on the drawing, including where it attaches to a wall or ceiling, under a desk, or other placement. I use a circle with a letter or letters within and a tag with a dot such that the dot shows the location on the plan where the device will attach to the wall. For my own symbols, ceiling-mounted devices have no tag (Fig. 27.4).

Racks, consoles, and panels

Next, show locations for equipment racks, security consoles, and equipment panels. Devise a symbol for each and show the location of each on the plan. I use a separate symbol for junction boxes, terminal boxes, and equipment panels so that the installer can tell which are used where (Fig. 27.5).

Conduits and boxes

Finally, connect all of the devices on the drawings with conduit and boxes. Use a "tree" approach so that conduits collect into ever larger boxes as they get closer to their "home" equipment cabinet, rack, or console. Show a conduit size next to each conduit (Fig. 27.6).

Observe National Electrical Code (NEC) rules on conduit fill so that you show the correct size for each conduit according to the type and quantity of cables within. Conduits can become progressively larger as they collect more and more device cables on their way to the equipment cabinet. Observe NEC rules on the maximum size of conduits for each junction or pull box. You will have to show larger pull boxes as more conduits collect and as the conduits become larger.

Finally, show power connections for powered devices. You should include a power schedule that lists all powered devices and for each, the electrical panel and circuit breaker that they are powered from.

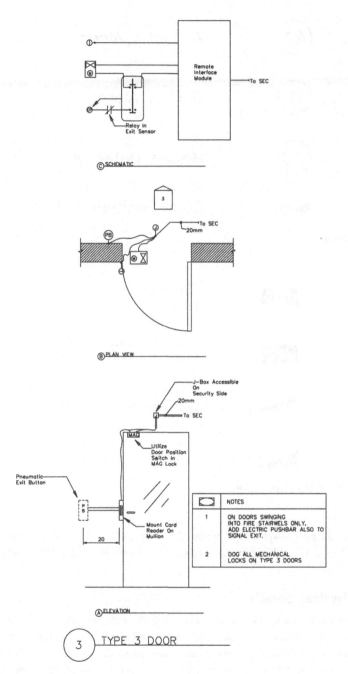

■ **FIGURE 27.3** Door type detail.

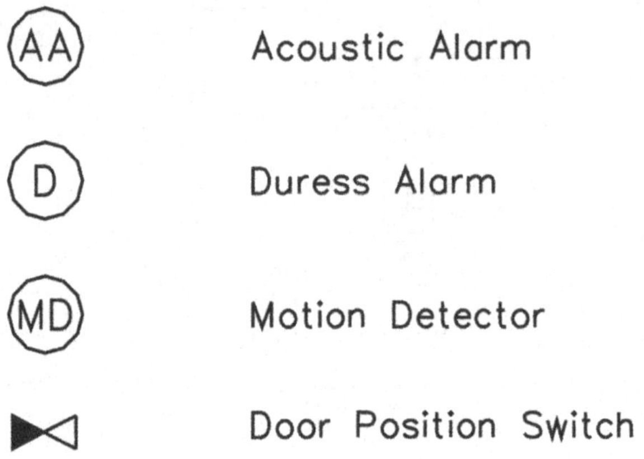

■ **FIGURE 27.4** Alarm device symbol.

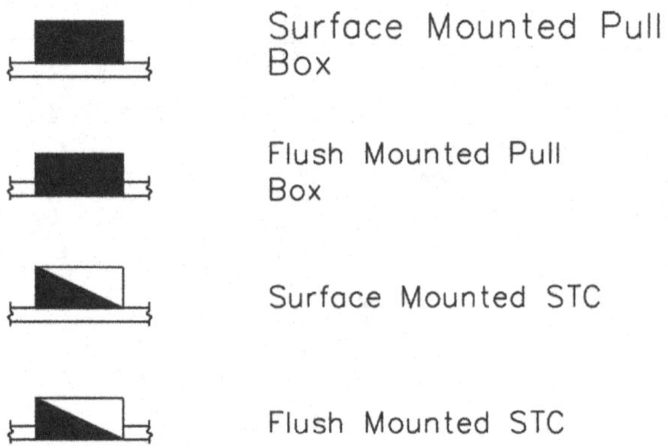

Acoustic Alarm

Duress Alarm

Motion Detector

Door Position Switch

Surface Mounted Pull Box

Flush Mounted Pull Box

Surface Mounted STC

Flush Mounted STC

■ **FIGURE 27.5** Equipment panel.

Do not mix power classes within the same conduit. Show Class 1 power cables in Class 1 conduits and Class 2/3 cables within Class 2/3 conduits.

Physical details

For each security device it is best to draw a physical detail that shows how you expect the installer to mount and configure the device in the field. If you want dome or enclosed cameras used, show that. If you want motion detectors to be either surface or flush wall-mounted, show that. Whatever you don't show, the installer may decide on his own what is best and neither you nor the owner may like that. Physical details should

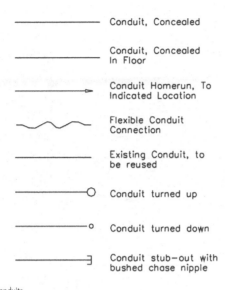

Conduit, Concealed

Conduit, Concealed
In Floor

Conduit Homerun, To
Indicated Location

Flexible Conduit
Connection

Existing Conduit, to
be reused

Conduit turned up

Conduit turned down

Conduit stub—out with
bushed chase nipple

■ **FIGURE 27.6** Conduits.

also have notes describing required dimensions, mounting methods, and so forth.

Riser diagrams

Many designers, myself included, like to show a drawing with all of the conduits and power points in the system. This is called a riser diagram. For a multistory building a riser diagram is normally divided into floors and on each floor is shown a main riser box and riser conduit (connecting the floors together), and from the main riser box conduits extend to the equipment panels and other terminal and junction boxes.

The riser diagram is also usually divided vertically to show the relationship between buildings on a campus or building zones within a building.

Finally, the riser diagram will show power drops, which are power connection points. All powered assemblies should be illustrated here. This is also a good place to show the panel and breaker to which each powered device connects.

Single-line diagrams

Single-line diagrams (Fig. 27.7) show all of the security equipment and their circuit relationships. Show a single-line diagram for each system.

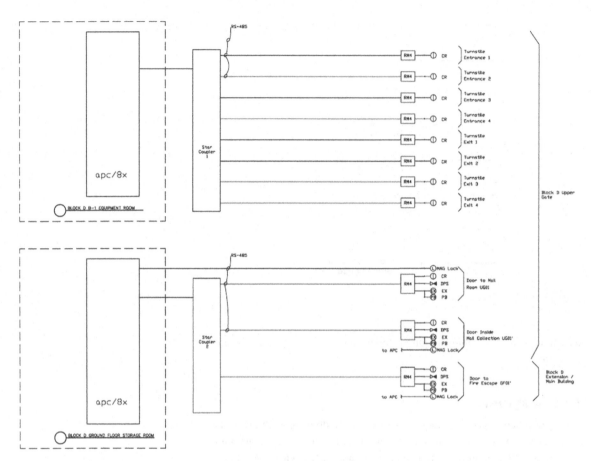

■ **FIGURE 27.7** Simple single-line diagram.

When you complete the single-line diagrams for each system you have a "biddable" set of drawings. The only additional type of drawing the installers need is wiring diagrams, which are not required for bidding.

Wiring diagrams

Wiring diagrams show the connections of each individual wire of each cable onto each terminal of each device. Many wiring diagrams can be typical; e.g., all RS-485 connections are wired the same throughout the project, so you only have to show that once, not once for every panel. Similarly, you only need one wiring diagram for card readers, door position switches, motion detectors (of a given type), and so forth.

						Security	Reader Type				Gates			Alarm		Lock Type				Exit Device								
CR#	Package	Floor	Location	Door #	Security Door Types	Mullion	Single Gang	Outdoor Reader	Elevator	Roll-Up Gate Operator	Roll-Up	Duress	DPS	EML	Mag	Delayed Egress	REX	Mortise	EX-PB	Delayed Egress	Elevator Access	SEC	To Switch	STC	R
GF01	Shell & Core	GF	Ground Floor Upper Gate Mail Sorting	GF01	3		1							1		1		1		1						
GF02	Shell & Core	GF	Ground Floor Upper Gate Mail Pick-Up	GF02	3	1								1		1		1		1						
GF03	Shell & Core	GF	Ground Floor Main Building Sheltered Service Court	GF03	3	1								1		1		1		1						
T1A	Interiors	GF	Turnstile 1 In	T1A	T								N/A													
T1B	Interiors	GF	Turnstile 1 Out	T1B	T								N/A													
T2A	Interiors	GF	Turnstile 2 In	T2A	T								N/A													
T2B	Interiors	GF	Turnstile 2 Out	T2B	T								N/A													
T3A	Interiors	GF	Turnstile 2 In	T3A	T								N/A													
T3B	Interiors	GF	Turnstile 2 Out	T3B	T								N/A													
T4A	Interiors	GF	Turnstile 2 In	T4A	T								N/A													
T4B	Interiors	GF	Turnstile 2 Out	T4B	T								N/A													

■ **FIGURE 27.8** Access control schedule.

You will need individual wiring diagrams for each access control panel, alarm panel, output control panel, digital switch, and so forth.

Schedules

In lieu of wiring diagrams, you may want to show those relationships in circuit or equipment schedules. This is a large spreadsheet that lists every field device and every attribute of that circuit. For example, an access control schedule (Fig. 27.8) may include the card reader number, door type, frame type, door handing, fire rating, type of door position switch, type of exit sensor, type of lock, what access control panel (and port) it wires to, what lock power supply, and so forth.

SYSTEM INSTALLATION
Project planning

It sounds obvious, but the success or failure of the project largely depends on the quality of project planning, which should include:

- Project schedule development
- Shop and field drawings
- Product acquisition and staging
- Permits
- Coordination with other trades
- Coordination for access to work areas
- Manpower planning

Each of these elements either fits into a well laid out plan or becomes a crisis to be solved. Every crisis delays the project and destroys the reputation and relationship with the owner, general contractor, and other trades on the project. Equally important, every day you spend on the project beyond the scheduled completion date is a day of labor costs removed from the project's bottom line. Planning is everything.

Project schedule

The project schedule is the single most important document in project planning. Simply put, any task conducted without being on the schedule is most likely to be late. The schedule should be built from the key milestone points and then down into greater and greater detail.

The most popular type of schedule format is the Gantt chart (Fig. 27.9). A Gantt chart shows the relationship between all the tasks, task groups and their durations, their start/finish dates, precursor tasks, and resources needed.

Start the schedule with the major task groups. Don't worry about dates just yet. The first step is just to get all the major task groups listed. Major task groups include all the points covered in this section.

After that, add a task line for each individual task under each task group. For example, the Installation task group might include a task for installing devices on each floor. You can also develop subtask groups, e.g., under the Installation task group for a small project, subgroups might include:

- Install conduit
- Install cabling

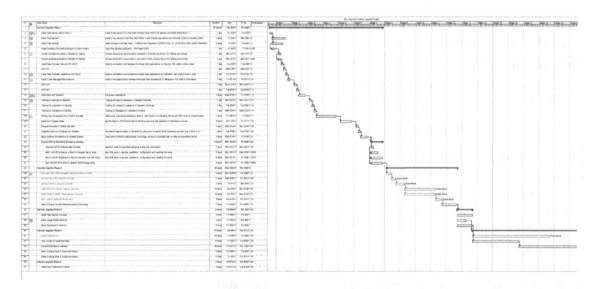

■ **FIGURE 27.9** Gantt chart.

- Install devices
- Install servers and workstations
- Test devices
- Commission system
- Acceptance testing

Each of these would then have a number of individual tasks within each subgroup.

After listing all of the tasks you can think of, go back to each task and list the time duration in hours or days that each task will take. Then list the order of precedence for the tasks. This is done by identifying for each individual task what other tasks must be completed before this task can begin. Finally, go to the top of the task list and list the start date. All the rest of the tasks will fill out their dates automatically from that based on their duration and order of sequence.

Shop and field drawings

Installers need a set of shop and field drawings to build from, and the integrator needs shop and field drawings on new construction projects to receive authorization from the consultant to begin the work. Shop and field drawings are also often required for building permits. They also show the consultant that the integrator understands the project and that the integrator is providing the necessary equipment to complete the work. For a complete list of drawings and how to put them together, read the section implementing design ideas to paper.

Product acquisition

It seems pretty obvious that you cannot install what you have not purchased, but a surprising number of installers don't seem to place much importance on product acquisition ... until they actually need the product on the jobsite. The development of shop and field drawings will help you identify every single product needed for each installation.

Early identification of products and their schedule of installation is important because some products take minutes or days to acquire while some take weeks and even months. Invariably, it seems that the product with the greatest lead time for purchase is the one that is identified last.

Most of the big-ticket items will already have been identified by the Installer's sales department during bidding. But there are many products that must be identified down to their last detail after the award. After identifying the products, the next step is to identify the suppliers and place orders.

Permits

In most municipalities, permits are required for installation of electronic security systems. Permits may be required by the building department and the fire department. It is usually the responsibility of the Integrator to obtain permits. Typically, these departments want to see a set of drawings and a work plan. Be sure to allow sufficient time in the schedule between the development of shop and field drawings and the beginning of construction work to obtain permits.

Coordination with other trades

Your installation can be stopped dead in its tracks over and over again by other trades that you must coordinate with if you do not get well out in front of the coordination needs and deadlines early in the project. I suggest having a meeting with each or all of the other trades with which you will coordinate. Go to the meeting prepared with a folder on each trade listing what the coordination will entail, what you need from them (schedule and resources), and what you will give to them (resources). Discuss any modifications that may be required to your procurement list and theirs. Discuss schedules so that the work of both trades goes smoothly. Record any decisions reached and action items for ether trade into a set of meeting minutes. Distribute the meeting minutes to everyone at the meeting.

Follow through on the discussion to ensure that all decision points are acted upon by you and the trades (let them know that you have fulfilled your list and ask about theirs). Then give them a heads-up before any work that requires the presence of both trades together on the jobsite is done, just to be sure that they "got the memo."

Access coordination

Many project delays are caused due to a jobsite that is unavailable to the installer when he is scheduled to be there. I strongly encourage each project manager to send out an e-mail 2 weeks ahead of schedule identifying the work to be accomplished in the next 2 weeks (day by day list) and the areas needing clearance for the installers. You should request written clearance for the next 2 weeks for all of the days and items on the list for the following reasons.

- List 2 weeks out because an e-mail on Friday may not be enough time to schedule installers for the following Monday or Tuesday. (So you are giving them an extra week.)
- List next week as a confirmation of work already scheduled and as a reminder.
- Get confirmation and have the installers carry that confirmation with them when they go to the area to show it to anyone who tries to deny them access.

Certain areas, such as bank vault areas and prison areas require accompaniment by authorized users for retrofit projects. So be certain to ask in all weekly coordination request e-mails if any escort is required, and ask that the escort be introduced to you ahead of time.

Preliminary checks and testing

Once conduits, cabling, and devices are installed and the servers and workstations, operating systems, and programs are installed, you can begin preliminary checks and testing.

Steps in preliminary checks and testing include:

- Test to verify that all devices are connected and show basic operation
- Test each device to show that all aspects of its operation are correct (doors unlock, alarms signal, in all four alarm states, etc.)
- Verify that all basic functions of the server and workstations, operating system, and security system programs are working

You are now ready to begin the final works.

Final works

Final works include system commissioning, completing punch list items, and system acceptance.

SYSTEM COMMISSIONING

System commissioning programs the security system servers and workstations with the software configurations necessary to carry out the planned tasks of the system. These include:

- Setting access control readers into access zones
- Setting alarm devices into alarm zones
- Setting system security schedules
- Setting users into access control groups

- Programming Graphical User Interface maps
- Programming automated actions and events

COMPLETING PUNCH LIST ITEMS

Following completion of all of the work and system commissioning, you will find a number of items that sort of "fell through the cracks" in the installation. These may include:

- A few items that arrived late and are left to be installed
- Replacing ceiling tiles and access panels
- Closing junction boxes
- Final wire and cable dressing and wire numbering
- Camera positioning
- Patching and painting
- Jobsite clean-up
- Repairing anything that does not work correctly

Before presenting the project for system acceptance, you should conduct your own punch list first. Walk the entire project and look for any remedial work items that need to be addressed before the owner's representative finds them. Complete all of these items. Let the owner and general contractor know in a memo that you are addressing preliminary punch list items before they see them. This will lay the groundwork for confidence in your work before they arrive for acceptance testing. It is better to hold up acceptance testing while you are finishing these items rather than to present a project that is not complete.

SYSTEM ACCEPTANCE

System acceptance is the time every integrator, project manager, and installer likes best, because the project is almost complete. For the system to be accepted, the project manager must show the owner that the project is complete in accordance with the project requirements.

The project manager should conduct a project tour to include the owner's representative, consultant, and others as required. The tour should include viewing of every equipment control panel, power supply, equipment rack, server rack, console, and workstation. It should also include a floor-by-floor tour of all installed devices. Finally, show all of the functions on the workstations so that they can see that everything works.

The owner's representative or consultant should note any discrepancies or remedial work items that still need attention. This will be issued by the owner's representative or consultant as an official punch list. Take your own notes of the punch list items they intend to list.

Begin working on the punch list immediately, even before receiving the official copy. Complete the punch list and resubmit for final acceptance. Don't submit bits and pieces; finish the whole punch list and then resubmit. This will finish the project much faster.

When the punch list is complete, submit a system acceptance form to include the warranty statement. Also include as-built drawings, manuals, cabinet/rack keys, and any portable items or spare parts called for in the contract. Receive a signature for all items. Have an initial beside each item delivered with the date of acceptance. You will need this in case there is a dispute over what has and has not been delivered.

Congratulations! Your project is complete!

CHAPTER SUMMARY

1. In the same way that many designers do not understand the complexities of installation and maintenance, most installers and technicians rarely respect the complexities of security system design.
2. Security system design is not about electronic equipment. It is not about cameras, card readers, biometrics, locks, and alarms. It is not about gates and doors. It is only about reducing risk.
3. Good designers design systems to address the problems of today and the problems of the future.
4. The elements of a security system design include:
   a. Drawings
   b. Specifications
   c. Interdiscipline coordination
   d. Product selection
   e. Project management
   f. Client management
5. Drawings are needed to illustrate the relationship of devices to:
   a. Their physical environment (plans, elevations, and physical details)
   b. The conduit system and to power (plans and risers)
   c. Each other (single-line diagrams)
   d. The user (programming schedules)
6. Drawings must serve five distinct types of users:
   a. The bid estimator
   b. Installers

      c. The installation project manager

      d. Maintenance technicians

      e. The next engineer expanding the system

  7. Specifications should include a description of what the project entails, descriptions of the whole integrated system and each subsystem, a description of the services the contractor will provide; and a list of acceptable products and acceptable installation, testing, acceptance, training, and warranty practices.

  8. Security systems are unique in that they relate to more building systems than any other building system. Security systems routinely relate to

      a. Electrical

      b. Door and gate hardware

      c. Structural

      d. Elevators

      e. Parking

      f. Landscape

      g. Building automation systems

      h. Signage

      i. Concrete

      j. Lighting

      k. Traffic control systems

      l. Irrigation systems

  9. Interdiscipline coordination makes or breaks the installation.

10. Many designers express their product selection not only in their specifications but also with a BOQ.

11. Project management has four phases:

      a. Initiating the project

      b. Planning the project

      c. Executing the project

      d. Controlling and closing the project

12. Each project has three main aspects:

      a. The project scope of work

      b. The project schedule

      c. The project cost

13. Client management is the process of managing the relationship between the firm you represent and the client's representative.

14. If you do not understand how systems fail you cannot design or install a system that is robust against failure.

15. Design using a layered security approach.

16. The basic goals of electronic security countermeasures include:

      a. Access control

      b. Deterrence

    c. Detection

    d. Assessment

    e. Coordinating response

    f. Evidence gathering

17. Security system designs should also be robust and redundant, expandable, flexible, and easy to use. And they should be sustainable.

18. The first stage of the design is to write a "basis for design" paper.

19. First, create access control zones.

20. Identify portals into each access control zone.

21. As you markup doors with card reader symbols, you should also note the type of door and frame and the number of leaves. From this information you can develop a list of door types.

22. Each door type symbol is illustrated in the physical details by its own detail comprising, e.g.:

    a. Detail 3B—A—A Plan View

    b. Detail 3B—B—An Elevation View

    c. Detail 3B—C—A Schematic View showing devices at the door and the wiring to the Access Control Panel

23. The next pass over the drawings will place alarm devices onto the drawings.

24. Next, show locations for equipment racks, security consoles, and equipment panels.

25. Finally, connect all the devices on the drawings with conduit and boxes. Use different line types of conduits in floors, walls and above ceilings so the installer knows where the conduits should be located.

26. For each security device it is best to draw a physical detail that shows how you expect the installer to mount and configure the device in the field.

27. Many designers like to show a drawing with all of the conduits and power points in the system. This is called a riser diagram.

28. Single-line diagrams show all of the security equipment and their circuit relationships. Show a single-line diagram for each system.

29. Wiring diagrams show the connections of each individual wire of each cable onto each terminal of each device.

30. In lieu of wiring diagrams, you may want to show those relationships in circuit or equipment schedules.

31. Project planning should include:

    a. Project schedule development

    b. Shop and field drawings

    c. Product acquisition and staging

    d. Permits

    e. Coordination with other trades

      f. Coordination for access to work areas

      g. Manpower planning

32. The project schedule is the single most important document in project planning.

33. Shop and field drawings show the consultant that the integrator understands the project and that the integrator is providing the necessary equipment to complete the work.

34. Early identification of products and their schedule of installation is important because some products take minutes or days to acquire, whereas some take weeks and even months.

35. In most municipalities, permits are required for installation of electronic security systems.

36. Your installation can be stopped dead in its tracks over and over again by other trades that you must coordinate with if you do not get well out in front of the coordination needs and deadlines early in the project.

37. Many project delays are caused because the jobsite is unavailable for the installer when he is scheduled to be there.

38. There are just a few steps in preliminary checks and testing:

      a. Test to verify that all devices are connected and show basic operation

      b. Test each device to show that all aspects of its operation are correct (doors unlock, alarms signal in all four alarm states, etc.)

      c. Verify that all basic functions of the server and workstations and operating system and security system programs are working.

39. Final works include system commissioning, completing punch list items, and system acceptance.

Q&A

1. *Security System Design is only about:*
    a. Locks, cameras, and card readers
    b. Alarms and output controls
    c. Cardholders
    d. Reducing risk

2. *Risk comprises:*
    a. Vulnerabilities
    b. Probability
    c. Consequences
    d. All of the above

3. *Any Security System designed without a proper Risk Analysis:*
    a. Will be more expensive
    b. Will be less expensive
    c. Will be wrong
    d. None of the above

4. *After designing for risk, the most obvious difference between the way security consultants design and the way security integrators design is that security integrators most commonly design:*
   a. To solve today's problems
   b. To solve tomorrow's problems
   c. To solve today's and tomorrow's problems
   d. None of the above
5. *The elements of a security system design include:*
   a. Drawings and Specifications
   b. Interdiscipline Coordination and Product Selection
   c. Project Management and Client Management
   d. All of the above
6. *Drawings illustrate the designer's concepts on how the system:*
   a. Should relate to the building
   b. Should relate to the conduits
   c. Should relate to the user
   d. None of the above
7. *Which of the following is not interested in the drawings?*
   a. The Bid Estimator
   b. The Installers
   c. The Maintenance Technician
   d. The Accounting Department
8. *Most integrators I have met:*
   a. Sincerely want to do well for their clients
   b. Are only interested in profits
   c. Are interested in profits more than the Owner's goals
   d. None of the above
9. *Security systems are unique in that they relate to:*
   a. Lighting and elevators
   b. Stairwells and elevators
   c. Doors and windows
   d. More building systems than any other building system
10. *Which of the following do security systems not routinely relate to?*
    a. Door and Gate Hardware
    b. Elevators
    c. Irrigation Systems
    d. Escalators
11. *Interdiscipline coordination:*
    a. Makes systems more difficult to use
    b. Makes or breaks the installation
    c. Both a and b
    d. Neither a nor b
12. *Specifying the wrong products can:*
    a. Result in more irrigation
    b. Result in less options for installation
    c. Leave the owner upset with the installer, the manufacturer, and the designer
    d. None of the above

13. *Each project comprises:*
    a. The Project Scope of Work
    b. The Project Schedule
    c. The Project Cost
    d. All of the above

14. *Understanding how to make systems fail is:*
    a. Unnecessary
    b. Equally important as understanding how to make systems work
    c. Neither a nor b
    d. This is a trick question

15. *The Layered Security Approach is a combination of detection and access layers:*
    a. Placed over each other at geographical progression toward the most valuable assets
    b. Placed together to ensure that all fences are protected
    c. Placed together to ensure that all gates are guarded by guards and dogs
    d. None of the above

16. *The system should have redundancy such that if one component fails:*
    a. Another will shift from its normal duties to cover
    b. Another is there to take its place functionally
    c. Both a and b
    d. Neither a nor b

17. *All systems have a finite operating life.*
    a. But you can extend that substantially if the system is operated properly.
    b. But you can extend that substantially through good design and good maintenance.
    c. But you can extend that substantially if the system is installed using high-quality cable.
    d. None of the above

18. *As you markup doors with card reader symbols, you should also note the type of door and frame and the number of leaves.*
    a. From this information you can develop a list of Door Types
    b. From this information, you can develop a list of cameras
    c. From this information, you can develop a list of users
    d. None of the above

19. *When marking up conduits and boxes:*
    a. Use a "ring" approach so that all conduits connect together
    b. Use a "tree" approach so that conduits collect into ever larger boxes as they get closer to their "home" equipment cabinet, rack, or console
    c. Use a "loop" approach so that all conduits connect together into a loop
    d. None of the above

20. *For each security device it is best to:*
    a. Draw a physical detail that shows how you expect the installer to mount and configure the device in the field
    b. Draw a schedule so that you know all devices are listed on a spreadsheet
    c. Draw a picture of the device from several angles
    d. None of the above

21. *Single-line diagrams show all of the security equipment:*
   a. And their circuit relationships
   b. And their conduit relationships
   c. And their electrical power supply relationships
   d. None of the above
22. *Wiring diagrams show:*
   a. The relationship between devices and their conduits
   b. The relationship between devices and the floor plan
   c. The connections of each individual wire of each cable onto each terminal of each device
   d. None of the above
23. *The _____is the single most important document in Project Planning.*
   a. Project Schedule
   b. Project Manager's Report
   c. Project Agenda
   d. Project Autopsy
24. *Shop and Field Drawings show the Consultant that the Integrator:*
   a. Understands what steps are in which order
   b. Understands that he can do what he wants
   c. Understands the project and that the Integrator is providing the necessary equipment to complete the work
   d. None of the above
25. *Early identification of products and their schedule of installation is important because:*
   a. Some products have to be ordered twice
   b. Some products are not received correctly the first time
   c. Some products take minutes or days to acquire while some take weeks and even months
   d. None of the above
26. *Permits may be required:*
   a. By the building department
   b. By the fire department
   c. Both a and b
   d. Neither a nor b

Answers: (1) d; (2) d; (3) c; (4) a; (5) d; (6) a; (7) d; (8) a; (9) d; (10) d; (11) b; (12) c; (13) d; (14) b; (15) a; (16) b; (17) b; (18) a; (19) b; (20) a; (21) a; (22) c; (23) a; (24) c; (25) c; (26) c.

Chapter 28

Access Control System Installation and Commissioning

CHAPTER OBJECTIVES

1. Learn the basics in the chapter overview
2. Understand jobsite considerations
3. Discover how conduit and cabling make or eliminate long-term problems
4. Discover device installation considerations
5. Learn how to do device setup correctly
6. Learn how to set up the access user database correctly
7. Learn how access schedules to make future system operation and maintenance easy or hard
8. Learn how using access groups can reduce work for everyone
9. Answer questions about access control system installation and commissioning

CHAPTER OVERVIEW

In this chapter a number of issues related to system installation and commissioning will be explored. System installation is pretty obvious, and commissioning is the process of making the system ready for use, including programming the servers and workstations to operate as desired.

Under installation, jobsite-specific considerations like safety rules and coordination will be discussed. Also looked at will be the issue of conduit and when to know if it makes sense to simply run cable without conduit. Device installation considerations related to making the system work reliably for many years will also be discussed.

Electronic Access Control. DOI: http://dx.doi.org/10.1016/B978-0-12-805465-9.00028-2

Finally, system commissioning will be reviewed. This will include setting up all devices and verifying their good operation, and programming the system databases for optimal operation.

JOBSITE CONSIDERATIONS

Surprisingly, the jobsite has a significant bearing on the conditions of installation. The main issues include:

- Jobsite safety rules
- Coordination with other trades
- Installation product storage arrangements
- Clean area rules

Most important of all are jobsite safety rules. Construction areas are accident prone. Hard hats, steel toed shoes, and careful behavior all contribute to a safe workplace. Maintenance of headroom in work areas, tying off belts when working above, and other safety considerations also help to prevent accidents. Each jobsite has its own rules. It is best to observe them all.

Operationally, it is astonishing how many points of coordination are required to successfully complete a security system project. No other system has as many points of coordination with other trades as an integrated security system. Trades to coordinate with may include:

- Electrical
- Door hardware
- Elevators
- Parking
- Signage
- Concrete
- Finish
- Hvac
- Fire alarm system
- And many more.

Additionally, as the project nears completion and it begins to be occupied, any work that requires access may need access permissions. It is important to place high consideration on coordination to complete a successful installation.

Many new installers forget that they must have a safe place to store their goods on site. You will have to provide one or more lockboxes to keep tools and materials stored safely. These in turn will need to be stored in a secure location.

Keeping work areas clean goes a long way in maintaining a safe and secure work area as well as enhancing the reputation of the integrator on the jobsite and with the client.

CONDUIT VERSUS OPEN CABLING

Conduit is expensive, so it would seem to make a lot of sense to eliminate it whenever possible, right? But that is not always the case. Generally, security system cabling should be contained within conduit. Why? Because in most cases, integrated security systems are mission-critical systems, and cables that are not within conduit are frequently damaged by workmen working on other systems. Conduit prevents damage to security system cables.

When running bare cables above a ceiling, in most cases these must be plenum rated cables in order to comply with fire codes. These cables are much more expensive than standard cables. So yes, you can save some money by not using conduit, but you run the risk of reducing system reliability in the process.

DEVICE INSTALLATION CONSIDERATIONS

We have already discussed the need to observe the way security devices interact with their environment, such as mounting card readers to a steel building in a way that will not cause them to overheat. This is one of many device installation considerations that are "not in the manual." Here are some other examples:

- Make sure that you are using the correct lock for the type of door. This is particularly important if the door is fire rated or is a fire-egress door.
- Be sure to use proper end-of-line resistors on all alarm devices.
- Be aware that when using a magnetic lock's integral door position switch function, it does not tell you that the door is closed, only that it is locked.
- When using pneumatic push button exit switches with magnetic locks, it is advisable to use two sets of contacts: one to notify the access control panel to bypass the alarm and the other to cut power to the door lock.

- When using pneumatic push button exit switches in the above manner, it is inadvisable to use the magnetic lock as a door position switch, because cutting power will cause a forced door nuisance alarm each time exit is requested.
- When installing video cameras be sure to verify correct line-of-sight no matter what the drawings show.
- Be certain that mounting heights conform to local codes.
- In earthquake-prone areas, be sure to mount all equipment in conformance with OSHPD (Office of Statewide Health Planning and Development—California standard for seismic compliance.) standards and with a minimum torsional load factor of five times the weight of the mounted device.
- Do not use custom circuitry. Virtually anything you want to do can be done with off-the-shelf circuitry or a programmable logic controller. This will be much easier to maintain than trying to find the installer who "jerry-rigged" a custom circuit out of bread-board and wire-wrap.

These are just a few cautions about device installations. As you talk with "old timers" who have been installing systems for many years, you will learn even more.

THE IMPORTANCE OF DOCUMENTATION

Proper documentation is essential to a good quality installation and to the installer's profitability. Every single decision about where and how to mount wire devices throughout the entire facility is an engineering decision, and they will be made either by a qualified engineer or by a guy with a screwdriver in his hand. You cannot buy a more expensive engineer than the guy with a screwdriver!

I have been to sites where the same circuit was wired four different ways on different floors of the building, because there were no design drawings and each installer used his own ideas about how to implement the security design concept. In that case they used different parts, which they each left the jobsite to go to buy on their own (four times the cost for acquisition), and different wiring, making the same circuit much more difficult to troubleshoot when failures occur.

Additionally, no matter what the situation, good "as-built" documentation is critical to be able to maintain a successful working system. Pity the poor technician who arrives at a complex facility to troubleshoot a problem with no system drawings available.

DEVICE SETUP AND INITIAL TESTING

After installing and wiring devices system testing begins. The first step is to test connectivity. Does every device show up on the computer or monitoring panel to which it is connected? Do all alarm-sensing devices show all four alarm states? Do all locks work? Do all request-to-exit devices unlock the doors? Do all card readers work with a sample card?

Additionally, the following work should be performed:

- Verify proper powering and grounding of all devices
- Verify the integrity of all insulation, shield terminations, and connections
- Verify the integrity of soldered connections
- Verify that all cables are properly dressed
- Verify all circuitry for continuity and operation
- Verify the mechanical integrity and eaesthetic acceptability of all mounted devices
- Verify that any devices that must be powered up or down in a specific sequence are set up to do so
- Adjust all devices for best operation and document the adjustments

To do this type of testing, send a worker around to every alarm sensor and door and have him exercise the device while observing its operation on the alarm/access control system workstation. Once the operation of every device in the system is confirmed, you can move on to database setup.

ALARM AND READER DEVICE DATABASE SETUP

Most alarm-sensing devices should be set up on a schedule such that they are armed at some times and bypassed at other times. The schedule should be set up in close coordination with the system owner/operator.

Many access control readers also operate on a schedule providing for the door to be under access control some times and unlocked at others.

USER ACCESS DATABASE SETUP

The primary database is the user access database. This includes a record on all access users in the system (everyone who will have an access credential). Each record in the user access database will include fields for Name (first/middle initial/last), Title (Mr./Ms./Mrs./Jr./PhD/MD, etc.), address, city, state, zip, organization the user is related to, department, access level, and other pertinent data.

The user access database also contains the information about the "clearance" that each user is assigned. The clearance comprises the permission granted to each user which will determine which gates and doors each user is allowed to enter with his/her access credential.

ACCESS SCHEDULES AND AREAS

Most users are not granted unlimited access everywhere 24/7/365. They are usually granted access to a limited number of doors and only during certain hours. There are two ways to program access privileges for users. The first way is to program each user individually to specific doors and to specific times. The second way is much more effective—assign individual users to one or more user groups. For example:

- Janitors
- Upper management
- Department X office workers

Then, assign individual doors to the access area groups corresponding to their areas and functions, e.g., Building 1, Department X. After that, you simply assign user groups to access area groups.

Access area groups comprise geographically related areas that are associated with an organizational component of the company or organization, e.g., a laboratory, clinic, research and development department, finance department, etc.

Access groups are a collection of doors or door groups that are related to a specific role within a department. door groups provide additional granularity to access groups, in order to tailor clearances more specifically.

Depending on the specific needs of each department, access groups may be assigned by:

- Geographical locations of portals, e.g.:
  - Clinics
  - Administrative
  - Research & development department
- Or by employee classification roles, such as:
  - Financial traders
  - Surgery personnel
  - Janitors

All clearances (user groups, door groups, access groups) should be audited at least annually to assure that their application is still relevant to any changes in the way doors and gates are actually used as the way the organization utilizes its facilities changes over time.

Special care must be taken with private user data in order to protect it from unauthorized access and improper use. Best practices include:

- Data encryption
- Restriction of personnel who can access private user data
- Elimination of sensitive information (such as social security numbers, home addresses, etc.) from reports where not absolutely essential.

CHAPTER SUMMARY

1. The jobsite has a significant bearing on the conditions of installation. The main issues include:
   a. Jobsite safety rules
   b. Coordination with other trades
   c. Installation product storage arrangements
   d. Clean area rules
2. Many new installers forget that they must have a safe place to store their goods on site.
3. Generally, security system cabling should be contained within conduit.
4. Proper documentation is essential to a good quality installation and to the installer's profitability.
5. After installing and wiring devices system testing begins.
6. Additionally, the following work should be performed:
   a. Verify proper powering and grounding of all devices
   b. Verify the integrity of all insulation, shield terminations, and connections
   c. Verify the integrity of soldered connections
   d. Verify that all cables are properly dressed
   e. Verify all circuitry for continuity and operation
   f. Verify the mechanical integrity and aesthetic acceptability of all mounted devices
   g. Verify that any devices that must be powered up or down in a specific sequence are set up to do so
   h. Adjust all devices for best operation and document the adjustments
7. Most alarm-sensing devices should be set up on a schedule such that they are armed at some times and bypassed at others. The schedule should be set up in close coordination with the system owner/operator.

8. The primary database is the user access database.
9. Assign portals to access area groups.
10. Assign individual users to one or more user groups. For example:
    a. Janitors
    b. Upper management
    c. Department X office workers
11. Then assign user groups to access area groups.

Q&A

1. *Which is not a main issue at a jobsite?*
   a. Jobsite Safety Rules
   b. Coordination with Other Trades
   c. Installation Product Storage Arrangements
   d. Keeping the area available for the Project Manager
2. *Which is the most important at a jobsite?*
   a. Jobsite Safety Rules
   b. Coordination with Other Trades
   c. Installation Product Storage Arrangements
   d. Keeping the area available for the Project Manager
3. *When running bare cables above a ceiling, in most cases these must be _____ cables in order to comply with fire codes.*
   a. PVC
   b. Fiber-optic
   c. Plenum rated
   d. None of the above
4. *Which is not an installation consideration?*
   a. Make sure you are using the correct lock for the type of door. This is particularly important if the door is fire rated or is a fire-egress door
   b. Be sure to use proper End-of-Line Resistors on all alarm devices
   c. Be aware that when using a magnetic lock's integral door position switch function, it does not tell you that the door is closed, only that it is locked
   d. Make sure you are using solid core cables
5. *Proper documentation is not a plus—*
   a. It is essential for users to know that the installation is on schedule
   b. It is essential to a good quality installation and to the installer's profitability
   c. Both a and b
   d. Neither a nor b
6. *After installing and wiring devices system testing begins. The first step is to test:*
   a. Connectivity
   b. If all locks work
   c. If all cards work

   d. If all users work
7. *Most alarm-sensing devices should be set up on a schedule such that:*
   a. They are armed at all times
   b. They are never armed unless dangerous
   c. They are armed at some times and bypassed at other times
   d. None of the above
8. *Most users are not granted unlimited access everywhere 24/7/365—*
   a. They are granted access to all doors at all hours
   b. They are granted access to some doors at all hours
   c. They are usually granted access to a limited number of doors and during only certain hours
   d. They are granted access to a limited number of doors at all hours

Answers: (1) d; (2) a; (3) c; (4) d; (5) b; (6) a; (7) c; (8) c.

Chapter 29

System Management, Maintenance, and Repair

CHAPTER OBJECTIVES

1. Learn the basics in the chapter overview
2. Learn the elements of system management
3. Learn how to minimize system maintenance and repair
4. Take a test on system management, maintenance, and repair

CHAPTER OVERVIEW

In this chapter, how to manage the alarm/access control system, including database management, merging databases of different systems on a common corporate campus, and alarm management and reports will all be discussed.

The difference between maintenance and repair, what resources are necessary for good system maintenance, and how to set up a sound system maintenance program will be reviewed. The need for system as-build drawings, what they should contain, and how wire run sheets can help the maintenance technician will also be discussed.

System infant mortality, how to provide well for emergency repairs, and how to keep that need as small as possible will be covered. Various maintenance contracting options including an in-house technician, an extended warranty, an annual maintenance agreement, or on-call maintenance and repairs will all be discussed.

MANAGEMENT

Alarm/access control systems are fundamentally business management systems; however, instead of managing profit-producing activities, they manage risk-reducing activities. The quality of systems management can

Electronic Access Control. DOI: http://dx.doi.org/10.1016/B978-0-12-805465-9.00029-4

vary greatly from one organization to another. What can you do to promote good system management?

GOVERNANCE, RISK MANAGEMENT AND COMPLIANCE

It's also important to understand that alarm/access control systems, indeed the entire security system, is an integral part of any corporate governance, risk management and compliance (GRC) program. The development and maintenance of a solid GRC program for the security business unit is essential to help assure the integrity and protection of the fundamental mission of the organization. Remember, the organization's assets are there to support its programs, which are developed to support the organization's mission. And the security business unit is the program whose job it is to protect the organization's assets. So the security business unit is fundamental to the organization's mission. The security business unit should work closely with the board of directors to develop and maintain a security governance framework (security policies). The security policies should be based on a solid risk assessment that considers threats, vulnerabilities and probabilities. The last piece of GRC is compliance. The security business unit must develop and maintain corporate compliance (procedures and metrics reports) that *prove* to the board that the security business unit is complying with corporate governance and with the compliance requirements of federal, state and local governments and regulating agencies, and more importantly that the security business unit is helping the board comply with the big picture on laws, codes and regulations.

Database management

The heart of the system is the data it produces and feeds to management. Those data are contained in and distributed from the system's database. The fundamental requirement of good system management is good database management.

Although most alarm/access control system databases are pretty well conceived, it is important to ensure that bad data do not creep into the system. Bad data can include:

- Access credential holders who no longer work for the company
- An accumulation of false or nuisance alarms
- An accumulation of system errors (communication errors, etc.)

It is imperative to keep access control system databases free of outdated card-holders. This can happen naturally as employees leave the company if the card-holders are not quickly marked as terminated. It can happen much more quickly if the database includes visitors whose unreturned cards are not immediately voided. I know of one system where this occurred and the company ended up ordering an additional 5000 visitor cards within a few months of its initial operation. As the database grows in size, each additional card can cause the system to operate more slowly. Over time, the system may become slower and slower, and eventually crash.

Similarly, an accumulation of false and nuisance alarms, or system error reports, when left unchecked can obscure real problems with the system. Most people know that the purpose of the alarm system is to notify the security console operators of an alarm, but fewer realize that it has a second purpose—to allow the operators to see system reliability trends that need attention. When false and nuisance alarms and system error reports go unchecked it is common for operators to ignore actual alarms and for the system to go into a general state of disrepair that affects its ability to do its primary job.

Merging databases of different systems on a common corporate campus

As systems grow many organizations discover that different buildings in their organization are operated by different access control systems. This can happen even on a common campus. At some time, the organization may want to consolidate its access control system operations. While this is simple when all buildings are on the same access control system, it is complicated when they are different systems. Typically, a common desire is to have the access cards of both systems useable in both, without having to manually enter and maintain both databases individually.

One effective way to do this is to use one of the systems as the enrollment and card management source for both. Then, a daily update from that one system to the other can ensure that both access databases are in sync. This requires that a script be written to adjust the data fields of the controlling system to write its data into a transfer database that is formatted in a compatible format to the second system. Then the second system can be updated daily, or with each record update of the controlling system's database.

Alarm management

After the access control database, the alarm database is the second important product of the alarm/access control system. It can be used to notify management of active alarms, and help management determine vulnerabilities in the facility.

A prevalence of alarm events at a particular location can mean that the system needs a redesign, or when such alarms occur unusually, that someone is testing your intrusion detection capabilities. In any event, false or nuisance alarms are a signal that something systematic should be done to remedy the problem.

Reports

The heart of alarm/access system usefulness lies in its ability to generate reports that can uncover trends and generate reductions in risks for the organization. Critical reports include:

- Alarm reports (see the section "Alarm Management")
- Access control system exception report (which valid cards in the system have not been used in the last 30 days?)
- System errors report (what's not working in the system?)
- Correlations between reports of losses and alarm/access control system reports (which employee, contractor, or visitor was present at each theft of laptops in a given department?)
- Correlations with other systems (facial recognition of unescorted visitors correlated to losses).

By using system reports, especially in conjunction with other systems, you can uncover potential risks that would otherwise go unnoticed.

MAINTENANCE AND REPAIR

Every system needs occasional maintenance and repair. The manner in which you do it can result in good or poor system performance or poor and in low or high costs. In this section we will begin by discussing system documentation, which is the basis of all good maintenance, and then we will discuss methods for performing maintenance and repairs in an effective and cost-effective manner.

System as-built drawings

Nothing beats a good set of security system drawings for clarity about how a system is designed and installed. A good set of as-built drawings begins with a good set of design drawings, which are updated by the field foreman with actual "as-built" conditions.

A good set of as-built drawings should include:

- A title sheet including a drawing list and information relevant on all other sheets
- Several sheets of schedules (spreadsheets) including:
  - Cable schedules
  - Circuit schedules for each system
  - Power circuit schedules
  - Conduit schedules
  - Equipment schedules
- Floor plans showing:
  - The background (floor plan)
  - Locations of all security devices including their device names
  - Detail callouts (references to other drawings that expand on information on this drawing)
  - Conduit paths with sizes and possibly fills (which wires are in what conduits)
  - Fields of view of video cameras
- Physical details
  - Mounting for all devices in all locations (typical mounting devices most common)
  - Notes on mounting, powering, etc.
  - Console and rack elevations
  - Equipment cabinet elevations
- Single-line diagrams of each security system element
  - Alarm/access control system
  - Digital video system
  - Security intercom system
  - Digital infrastructure
- Power infrastructure
  - Riser diagram
    - Conduit risers
    - Power riser
- Wiring diagrams
  - Typical wiring diagrams for:
    - Cameras
    - Card readers
    - Locks
    - Alarm devices, etc.
    - Workstations

- Custom wiring diagrams for:
  - Equipment cabinets
  - Equipment racks
  - Consoles
  - Servers, etc.

Each cabinet or rack should have a pouch on the inside of its door with relevant drawings to the equipment in that cabinet, which should typically include:

- Wire run sheets for the cabinet
- Floor plan(s) for devices served by the cabinet
- Appropriate single-line diagram
- Appropriate wiring diagram

Wire run sheets

Another way to document wiring is by using wire run sheets. A wire run sheet is a spreadsheet that includes each cable in the system and its component wire colors and connection points (source to destination). A wire run sheet will include:

- Location of source (floor plan drawing, building, department, floor, etc.)
- Type of circuit (video, card reader, lock power, etc.)
- Source equipment (device name like LPS-3 [lock power supply 3])
- Source equipment connector
- Destination equipment connector
- Destination equipment
- Location of destination device

This is done for every circuit in the system. Wire run sheets are usually assembled according to the devices to which their wires connect. For example, in a pouch inside the door of every equipment rack or cabinet there is a wire run sheet for the cables in that rack or cabinet.

System infant mortality

The most common time for electronic components to fail is when they are fresh from the box. In the electronics manufacturing business, this is called infant mortality. Every electronic product including camera phones to spacecraft suffers from varying degrees of infant mortality. The degree of quality control in manufacturing processes will determine the percentage of products that suffer from infant mortality, but all products suffer from it to some degree.

Installing contractors should be aware of infant mortality and have some spare products on hand to handle failures that occur soon after a product is installed.

Maintenance versus repairs

To people who are new to the installation or management of alarm/access control systems, the terms maintenance and repair are often used interchangeably; in fact, although maintenance and repair are similar, they are also different.

Maintenance is the act of monitoring and evaluating a system's operational condition and readiness and making such adjustments as may be required to keep it in good working order, whereas Repair is the fixing of the system when it malfunctions.

A surprising number of system owners wait for the system to malfunction before calling the installing contractor for a repair. This habit results in a system that is perceived as unreliable and, in fact, is unavailable for its duty until repairs are conducted. It also usually results in higher cost than is necessary to keep the system well maintained.

Scheduled maintenance

The best way to keep a system well maintained is to set up a scheduled maintenance program. Surprisingly, well-maintained systems need little, if any, repairs. There are a variety of ways to set up a good scheduled maintenance program, but the essentials include:

- The maintenance tech should visit the system at least once every 2 months.
- They should look at error reports and evaluate why they are happening.
- They should open cabinets and see that there is no accumulation of dust, moisture, insects, etc.
- They should note any changes in wiring such as other contractors interfering with the system's wires while they are working on their own systems.
- They should look at alarm reports and look for nuisance or false alarms.
- They should address any deficiencies found.

Emergency repairs

The maintenance agreement should also include provisions for emergency and nonemergency repairs, as well as scheduled maintenance. At some time, every system needs an emergency repair; however, not all repairs are emergencies. The maintenance agreement should define what is an emergency repair (cannot operate system), and what is a nonemergency repair (portion of system not working but its operation is not immediately critical).

The maintenance technician should be required to respond to emergency repairs within a suitable time frame (usually four hours after declaring the emergency), and should bring the system back into operation within a certain amount hours after arriving (typically within eight hours). It is important that either the system owner or maintenance service agency should stock critical spare parts for this purpose. For example, we recommend having a back-up workstation programmed and ready to go just by swapping it out with a bad workstation.

Spares for typical replacement parts should be kept in stock at the client's facility. This should include one spare card reader, door position switch, lock, request-to-exit device, and camera of each type.

Maintenance contracting options

The owner of the system should create a maintenance agreement to ensure that the system is well maintained. But how does he do that? What are the options open to him? The most common options are listed next.

In-house technician

The system owner can hire an in-house technician who is an employee to keep the system well maintained. This is often done on larger systems, where the cost to contract out to maintain the systems of a number of buildings on a campus or on multiple campuses could create an extraordinary budget.

Selecting the right in-house technician can be difficult. Ideal qualifications should include:

- Familiar with and ideally certified in the type of equipment at your site
- Best if the technician has experience in both installation and repair
- Someone with an active interest in the systems, not just looking for a paycheck
- At least 5 years experience (you want him to have seen and repaired just about every common problem out there)

The contractor or consultant may be able to help you find a good technician or you can advertise. It is important to check references carefully, looking for:

- Reliability
- Troubleshooting and repair skills
- Good work practices
- Good wiring practices
- Good documentation practices

The technician should be tasked with testing and maintaining the system daily, with keeping current records on the condition of all elements of the system, and setting up and maintaining a routine maintenance schedule. The technician should also set up a life cycle replacement schedule for all components of the systems.

Extended warranty

The installing contractor may be asked to offer an extended warranty. This should include all of the elements of the original warranty, and extend its time duration for up to 3 years.

Annual maintenance agreement

Similar to the extended warranty, the installing contractor should be asked to quote a renewable annual maintenance agreement that should cover all parts and labor. The requirements stated earlier are a good starting point for the terms of the agreement. This may be the most economical way to keep a system well maintained.

The advantage of the Renewable annual maintenance agreement is that the original installing contractor is the one servicing the system. He knows it well and has been through every start-up problem, so he knows what to expect and where to find everything.

On-call maintenance and repairs

The last option is to cross your fingers and pray that nothing goes wrong. When they do, you can call the installing contractor to hurry out to fix things. This always appeals to system owners because they somehow think that they are saving money. What they are doing is buying the most expensive repair per hour (remember, they will be the last on the priority list after those owners with contracts). Also, the repair technician will be unfamiliar with the system because he is not there regularly, so it will take longer to find and fix the problem.

SECURITY SYSTEM INTEGRITY MONITORING

1. *Goals and Objectives*
   a. The main goal of security system integrity monitoring is to help assure the health of the security system by identifying potential system problems long before they become system breakdown maintenance issues, or at least immediately upon incident of a component malfunction.
   b. The main objective of the security system integrity monitoring system is to provide the department that is responsible to maintain the security system with a tool to reliably forecast potential system maintenance issues well before a component breaks down in actual use, or to reliably notify the services department immediately upon a system malfunction, and to schedule system service in the timeliest possible manner.
   c. Notification: The security integrity monitoring system should provide system status of all security system subsystems, and malfunction incident notifications via dedicated software which will track individual systems (on a nested set of dashboards (all systems, each system), and individual elements within each system. Notifications should be prioritized as:
      i. Emergency: Must be addressed immediately—Top Urgent. Example, system shutdown.
      ii. Urgent—Must be addressed within 24 hours—Urgent. Example, camera or card reader off line.
      iii. Routine—Can be addressed within 48 hours to 30 days. Example, camera image shows some image aberrations, but is still useable.
2. *Scope of Integrity Monitoring*
   a. UPS, power, and battery integrity monitoring
   b. Power supply integrity monitoring
   c. Network infrastructure security at all 7 layers of the OSI network model
   d. Network infrastructure and integrity monitoring
   e. Alarm/access control system integrity monitoring
   f. Digital video system integrity monitoring
   g. Digital intercom and communication system monitoring
3. *System Descriptions*
   a. UPS, mains, and battery integrity monitoring:
      i. Monitored:
         1. All UPS status—dashboard
         2. UPS integrity at each UPS

        a. Input and output voltages

        b. Temperature

        c. Voltage

   3. Battery condition (for each individual battery) at each UPS

 ii. Method:

   1. Consider utilizing the building automation system to monitor UPS mains and output voltages, temperature and individual battery voltage.

   2. Alternatively, the contractor may utilize commercial off the shelf UPS monitoring hardware and software.

b. Power supply integrity monitoring:

 i. The power supply monitoring system should monitor the following on each low voltage power supply:

   1. AC low or lost voltage

   2. High/low battery voltage

   3. Internal over temperature

   4. Power supply internal failure

   5. Power supply blown fuse

   6. Battery presence

   7. Earth ground

   8. Individual load voltage monitoring

 ii. Method:

   1. Consider utilizing fully-monitored power supplies (LifeSafety power supplies or equivalent).

   2. Alternatively, the contractor may utilize the building automation system to monitor most of these conditions.

c. Network security monitoring:

 i. It is essential that the IP network that the security system resides upon must be itself a secure environment in order for the security system to help provide a secure environment for the organization's assets.

 ii. I have been appalled for many years at the industry-wide lack of regard towards securing the security system's IP network.

 iii. Problems abound. One of the most severe problems emerging today are the deployment of security system IP edge devices (card readers, intercoms video cameras) outside the skin of the building (parking structures for example), with no security on the network that these IP devices are connected to. Frankly speaking, one might as well place an open network drop on the side of the building! Even worse, are deployments of unsecured or poorly secured Wi-Fi networks for security systems whether indoors or outdoors.

      iv. Every security system IP Network must be secured at all 7 Layers of the OSI network model, or one can expect to have intrusions and compromises into the security system.

      v. However, after securing the network, it is incumbent on the designer to provide a means of monitoring the network security. There are a variety of highly regarded programs that can do this, available from major manufacturers. The monitoring software should be able to detect intrusion attempts into the system's firewall, look for any network security system edge devices or switch ports that are being unplugged (this action should disable the port until reset by a network administrator). The monitoring software should be monitored 24 hours per day by someone capable of understanding its alerts. This means that security console supervisors should be well trained on the software.

   d. Network integrity monitoring:

      i. Network integrity monitoring involves monitoring network throughput (overall and on each switch), looking out for rising ping times (too much traffic for switches to handle), switch power supply condition, switch performance conditions and network error incidents.

      ii. Better network integrity monitoring software is able to monitor the network right down to the condition of the data traffic through the switches right out to the edge devices. Like network security monitoring software, network integrity monitoring software should be monitored 24 hours per day by someone capable of understanding its alerts. This means that security console supervisors should be well trained on the software.

   e. Access control system integrity monitoring:

      i. Better alarm/access control systems have provisions for developing system logs of alarm/access control system anomalies. This can include devices going off-line (even momentarily), board overheat conditions, lost or intermittent connections, lost or poorly communicating edge devices, etc.

      ii. The alarm/access control system should be configured to present these conditions to a separate monitoring console and to prioritize repetitive or escalating conditions as alarm events, like any other alarm in the system. This should be monitored 24 hours per day.

f. Digital video system integrity monitoring:

    i. Much like alarm/access control systems, better digital video systems have provisions for developing system logs of system anomalies. This can include devices going off-line (even momentarily), lost or intermittent connections, lost or poorly communicating video cameras, etc.

    ii. The digital video system should be configured to present these conditions to a separate monitoring console and to prioritize repetitive or escalating conditions as alarm events, like any other alarm in the system. This should be monitored 24 hours per day.

g. IP security intercom system integrity monitoring:

    i. Few IP security intercom systems are well provisioned for system integrity monitoring. Regardless, a guard should call in from every intercom station at least once daily while on guard tour to check the end-to-end intercom function and audio quality, and a log should be made of these calls.

    ii. Additionally, the IP security intercom system should be integrated with the alarm/access control system such that each time a person makes a call from a field intercom station, an intercom call event is created in the alarm/access control system alarm incident database. Like every other alarm, an alarm disposition check box should be available with actions like:

        1. Grant access to the portal (this should open the gate or door and log the console operator as the user)

        2. Direct the intercom user to another location, etc.

        3. Additionally, there should be a checkbox for audio quality, which should be recorded in the alarm system database. This will allow the security director or manager to identify intercom problems as they develop in daily use.

h. Maintenance schedule monitoring:

    i. All security system elements have a finite life and should be repaired and replaced on a schedule that is appropriate for each element.

    ii. In its simplest form, a spreadsheet can track every device within the system for:

        1. Its installation date

        2. Warranty start/stop date (or service contract info)

        3. Device type, manufacturer, model and serial number

        4. Tcp/ip address

        5. Device location

        6. Service interval (bi-monthly checks, etc.)

        7. Estimated replacement date

        iii. A few rather elegant programs also exist for this purpose.

        iv. Maintenance schedule monitoring helps the organization get the most life out of its security system equipment at the lowest overall cost.

        v. It also provides essential device information for technicians to use for maintaining the system, and for engineers to use to expand or update the system.

4. Security system integrity monitoring provides the organization with the ability to identify arising security system issues before they become maintenance problems and this helps the organization maintain the highest quality of service and uninterrupted protection for the organization.

CHAPTER SUMMARY

1. Alarm/access control systems are fundamentally business management systems.

2. The fundamental requirement of good system management is good database management.

3. It is imperative to keep access control system databases free of old outdated card-holders.

4. Similarly, when left unchecked, an accumulation of false and nuisance alarms or system error reports can obscure real problems with the system.

5. After the access control database, the alarm database is the second important product of the alarm/access control system.

6. The heart of alarm/access system usefulness lies in its ability to generate reports that can uncover trends and generate reductions in risks for the organization.

7. Every system needs occasional maintenance and repair. The manner in which you do it can result in a good or poor system performance, and in low or high costs.

8. Nothing beats a good set of security system drawings to clarify how a system is designed and installed.

9. Each cabinet or rack should have a pouch on the inside of its door with relevant drawings to the equipment in that cabinet.

10. A wire run sheet is a spreadsheet that includes each cable in the system and its component wire colors and connection points (source to destination).

11. Installing contractors should be aware of infant mortality and have some spare products on hand to handle failures that occur soon after a product is installed.

12. Maintenance is the act of monitoring and evaluating a system's operational condition and readiness and making such adjustments as may be required to keep it in good working order, whereas repair is the fixing of the system when it malfunctions.

13. The best way to keep a system well maintained is to set up a scheduled maintenance program.

14. The maintenance agreement should also include provisions for emergency and nonemergency repairs, as well as scheduled maintenance.

15. Maintenance contracting options include:
    a. In-house technician
    b. Extended warranty
    c. Annual maintenance agreement
    d. On-call maintenance and repairs

16. Security System Integrity Monitoring
    a. Monitor each subsystem of the security system for its own security and proper operation.
    b. Keep track of warranty dates and maintenance and system update schedules.

Q&A

1. *Alarm/access control systems are fundamentally:*
   a. Business management systems
   b. Building management systems
   c. Digital Video Systems
   d. None of the above

2. *The heart of the system is the:*
   a. CPU
   b. Access Control Panel
   c. Scheduler
   d. Data it produces and feeds to management

3. *It is imperative to keep Access Control System databases free of:*
   a. Bad electrons
   b. Old power supplies
   c. Old outdated card-holders
   d. Old reports

4. *An accumulation of false and nuisance alarms, or system error reports, when left unchecked:*
   a. Can cause users to ignore card readers
   b. Can obscure real problems with the system
   c. Can obscure card-granting requests
   d. None of the above

5. *After the access control database, the _____ is the second important product of the alarm/access control system.*
   a. Alarm Database
   b. Schedule
   c. Access Zones
   d. None of the above

6. *The heart of alarm/access system usefulness lies in its ability to:*
   a. Generate alarms that guards can respond to
   b. Generate reports that can uncover trends and generate reductions in risks for the organization
   c. Generate Photo ID cards
   d. None of the above

7. *By using system reports, especially in conjunction with other systems, you can:*
   a. Uncover potential risks that would otherwise go unnoticed
   b. Uncover people trying to get access without using cards
   c. Uncover power supply problems
   d. None of the above

8. *Nothing beats a good set of security system drawings for:*
   a. Helping Project Managers to repair the system
   b. Helping Maintenance Technicians to Program the system
   c. Clarity about how a system is designed and installed
   d. None of the above

9. *A Wire Run Sheet is a spreadsheet that includes:*
   a. Every device in the system and the conduits that it connects to
   b. Each cable in the system and its component wire colors and connection points (source to destination)
   c. Both a and b
   d. Neither a nor b

10. *The most common time for electronic components to fail is when:*
    a. They are 7 years old
    b. They are fresh from the box
    c. They are over-used
    d. None of the above

11. *Maintenance is the act of monitoring and evaluating a system's operational condition and readiness and making such adjustments as may be required to keep it in good working order, whereas repair is:*
    a. The maintenance of the system when it is scheduled
    b. The fixing of the system when it malfunctions
    c. Both a and b
    d. Neither a nor b

12. *The best way to keep a system well maintained is to set up:*
    a. A schedule of devices
    b. A riser diagram
    c. A set of as-builts
    d. A scheduled maintenance program

13. *Security system integrity monitoring helps assure:*
    a. A reliable system
    b. Rapid response to system issues before they become reliability problems
    c. A secure security system network
    d. All of the above

Answers: (1) a; (2) d; (3) c; (4) b; (5) a; (6) b; (7) a; (8) c; (9) b; (10) b; (11) b; (12) d; (13) d.

Index

Note: Page numbers followed by "*f*" refer to figures.

A

Access card reader, 464–465
Access control panels
  attributes and components
    communications infrastructure, 275
    CPU, 273–274
    electronics panel, 275
    field elements, 274–275
    fourth generation systems, 274
    second and third generation systems, 273–274
    security system development theme, 274
    server(s), 275
    software, 275
    workstation(s), 275
  Communications Board
    CPU, 277–278
    digital protocol, 276
    EPROM, 278
    input/output interfaces, 278–279
    power supply and battery, 277
    RAM, 278
    TCP/IP, 276–277
  electronics, 464–465
  form factors
    2, 4, 8, or 16 door connections, 279
    cost reduction, 280
    functionality increase, 280
    Input and Output boards, 279
    microcontrollers, 281
    SQL, 281
    user experience improvement, 280
    virtually unlimited functionality, 280
  functions
    alarm status information, alarm inputs, 282
    antipassback event, 283
    authorized user, parking structure, 283–284
    building core lighting, 283
    building signage lighting, 283
    decision making, 279
    downloaded data, server(s), 282
    event data, 282
    fifth generation systems, 298
    night, 284
    output control, relay activation, 295
    weekend day/holiday, 284
    working day, 283
  local and network connections
    digital video and intercom systems, 288
    echelon systems, 286
    Ethernet (TCP/IP), 288
    multicast protocol, 288
    RS-232, 286–287
    RS-422, 287
    RS-485, 287
    UDP/IP and RTP/IP, 288
    VLAN, 288–289
  locations, 284–286
  networking options, 289–291
  and networks, 11
  redundancy and reliability factors
    A/C power, 294
    digital communications, 294
    good data infrastructure, 293–294
    good design, 292
    good power, 292–293
    good wiring and installation, 292
    heartbeat and watchdog timer, 294
    mission critical systems, 291
    servers, 294
Access control portals
  alarm monitoring, 50
  common type, 48
  credential readers, 49
  electrified locks, 49
  pedestrian and vehicle types, 48, 48f
  pedestrian portal types
    automated wall, 89
    automatic doors, 83–85, 84f, 86f
    man-trap, 88–89, 89f
    revolving doors, 85–87, 86f
    standard doors, 82–83, 83f
    turnstiles, 87–88, 88f
  portal passage concept
    anti-passback schemes, 81–82
    card entry/card exit, 78
    card entry/free exit, 78, 79f
    positive access control, 80, 80f
    scheduling, 81
    tailgate detection, 78–80, 79f
    2-Man Rule, 80–81
  request-to-exit sensors, 51
  safety systems, 49–50
  tailgating, 49
  vehicle portals
    automated roll-up vehicle gates, 92, 93f
    automated sliding vehicle gates, 91, 92f
    automated vehicle swing gates, 91, 91f
    high-security barrier gates, 92–94, 93f
    sally ports, 94, 94f
    standard barrier gates, 89–91, 90f
Access control system integrity monitoring, 530
Access credential and credential reader
  access cards, key fobs, and card readers
    capture card reader, 70
    magnetic stripe cards, 64
    mobile phone access control, 70
    multi-technology cards, 69
    125 K passive proximity cards, 68
    125 KHz active proximity cards, 68
    RFID wireless transmitter systems, 69
    13.56 MHz contactless smart cards, 68–69
    Wiegand wire cards, 66–68, 67f
  biometric readers
    authorized user, 72
    definition, 70–73
    fingerprint reader, 70, 71f
    hand geometry reader, 70, 71f
    identification readers, 72
    iris reader, 70, 72f
    photo identification, 73
    unique vs. individual attributes, 72
    verification readers, 72
  evidence of authority, 62
  identification card and bars, 63
  keypads, 63–64
"Air gapped" networks, 416
Alarm/access control system industry
  avoiding obsolescence
    automated password function, 268
    four-channel digital switch, 268
    microcontrollers, 268
    planned obsolescence, 267
    unplanned obsolescence, 267–268
  digital video industry, 256
  fifth generation
    BAS, 265–266
    CCTV, 266

Alarm/access control system industry
(*Continued*)
  detection system, 265
  DSP chips, 266
  edge devices, 266–267
  EPROM, 265
  Ethernet architectures, 266
  logical functions, 265
  microcontrollers, 266–267
  PLCs, 265–266
  reference databases, 266
  software-based functions, 266
 first generation
  Hotel Room Doors, 258–259, 258*f*
  McCulloh Loop system, 257*f*
  McCulloh Loop telegraph–type alarm
    system, 257–258
  Night Watchmen, 257
  police call boxes and fire pull
    stations, 257–258
  security technology, intercoms and
    CCTV, 259
 fourth generation
  CCTV, 262–263
  consumer videocassette recorders,
    262–263
  controller microcomputer, 261–262
  ENIAC computer, 261, 262*f*
  4-bit microprocessor, 261
  intercoms, 262–263
  mini-computer, 261–262
  6502 and 8088 8-bit microprocessors,
    261
  stalled progress, 263–264
  system interfaces, 263
 integrated security system, 256
 "me too" approach, 256
 second generation, 259–260, 260*f*
 third generation, 260–261
Alarm kit (ALK), 180
Alarm management, 522
Alternating current (AC), 157
Aluminum conduit, 447
Annual threat assessment, 409
Ashtray keypad, 63
Assets of organizations, 407–408
Astragal, 138
Authorities Having Jurisdiction (AJH), 101
Authorization, 8
Automatic doors
 bi-fold door, 85
 4-fold door, 85, 86*f*
 materials handling areas, 83
 one-/two-leaf swing door, 83
 sliding door, 84, 85*f*
 swing door operation, 84, 84*f*

B

Barcode cards, 65, 66*f*
Barium ferrite cards, 65
Bi- and 4-fold doors, 127–128
Bill of quantities (BOQ), 479
Bistatic microwave detection systems, 359
Blumcraft electrified door hardware, 221,
  223*f*
Building automation systems (BAS),
  265–266, 427
Building/facility systems
 direct action interfaces, 429
 elevators, 426–427
 feedback interfaces, 429
 lighting, 428
 proxy action interfaces, 429
 stairwell pressurization, 427–428
 Threat Actors, 425–426
Building management systems (BMS).
  See Building/facility systems

C

Cable Trays, 448
Cables
 brands, 445–446
 colors, 444–445
 conduit
  bends, 451
  conduit fill, 449–450
  fire protection, 451
  indoor applications, 448
  outdoor applications, 448
  types, 446–447
  wireways, 448
 copper/fiber, 442
 documentation, 455–457
 dressing
  cross-dressing, 454–455
  definition, 452
  nightmares, 452, 453*f*
  rules, 453–454
  system troubleshooting, 451–452
 handling
  nightmares, 451
  system troubleshooting, 452
 insulation, 443–444
 stranded vs. solid core wires, 444
 voltage and power classes, 442–443
 wire gauge, 443
Capture card reader, 70
Card readers, 259
Cardkey systems, 261
CCTV. *See* Closed-circuit television (CCTV)
Central processing unit (CPU)
 attributes and components, 273–276
 Communications Board, 276–279
Classroom lock, 171–172
Classroom security, 172
Client management, 480–481
Closed-circuit television (CCTV)
 fifth generation, 266
 first generation, 257–258
 fourth generation, 262–263
 second generation, 260
Cloud Computing, 469–470
Command, control, and communications
  (C3) Consoles, 394
Communication systems, 12
Consolidated communication systems
  (CCS), 392–393
Copper/fiber cables, 442
Corridor lock, 172
CPU. *See* Central processing unit (CPU)
Crash-rated gates, 246–247
Credential readers
 access control portals, 49
 credential authorization, 52

D

Database management, 520–521
Daughter boards, 279
Dead-bolt equipped electrified mortise
  lock, 203–204, 203*f*
Dead bolt monitoring switch, 184
DEF CON international hackers
  conferences, 410
Delayed egress locks
 electrified locks, 211–212, 211*f*
 fire ratings, 132–133
 magnetic locks, 163, 163*f*
Digital signal processing (DSP) chips, 266
Digital video recorders (DVRs), 382–383,
  383*f*
Digital video system integrity monitoring,
  531
Direct attached storage (DAS), 383–384,
  385*f*
Direct current (DC), 157
Distributed cluster management, 301

Door lock relay, 245
Door lockset selection
  door description
    automatic door, 219–220
    EPH with vertical rods, 220, 220*f*
    fire-rated doors, 219
    FM-200 fire suppression system, 218
    hollow–metal, 219
  double-egress doors, hospital corridor,
    228–229, 229*f*
  framed glass door, 221–222, 221*f*
  Herculite™ lobby doors, 222–223, 223*f*
  high-rise building stair-tower door
    dead bolt, 225
    hi-tower lock, 223–225, 224*f*
    hollow metal door, 225–226
    robust electrified lock, 224–225
    stair-tower door, 224
  inswinging office door, 229–231, 231*f*
    revolving door, emergency egress side
      door, 231–232, 232*f*
  office suite door, 226–228
  standard application rules, 217–218
  warehouse rear-exit door with hi-value
    equipment, 225–226
Door position switch (DPS), 332
Door types
  bi- and 4-fold doors, 127–128
  and frames, 9
  frames and mountings
    aluminum, 125
    hollow metal, 125
    mounting methods, 126
    wood, 126
  overhead doors, 126
  revolving doors, 126–127
  security, 115–116
  single- and double-leaf swinging doors
    balanced doors, 124–125, 124*f*
    framed glass doors, 119–120, 120*f*
    hollow metal doors, 117–118, 117*f*
    pivoting doors, 123, 123*f*
    solid core wood doors, 118–119, 119*f*
    total doors, 121–123, 122*f*
    unframed glass doors, 120–121, 121*f*
  sliding panel doors, 127
Dormitory lock, 173
Dot commands, 261
Duress Alarms, 362–364

E

Electric latch retraction (EL), 180
Electric mortise lock device, 180

Electric rim device, 180
Electric strikes, 150–151, 150*f*
Electrical metallic tubing (EMT), 446
Electrical nonmetallic tubing (ENT), 447
Electrified dead-bolt locks
  concealed direct-throw mortise dead-bolt
    lock, 202, 203*f*
  dead-bolt equipped electrified mortise
    lock, 203–204, 203*f*
  gate locks, 205–206, 205*f*
  safety provisions, 205–206
  surface-mounted, 201–202, 202*f*
  top-latch release bolt, 204–205, 204*f*
Electrified locks, 9. *See also* Electrified
    dead-bolt locks
  access control portals, 49
  circuit controls, 160
  CRL-Blumcraft panic hardware
    beautiful and egress door, 214
    C.R Laurence Co., Inc., 214
    panic bar mechanism, 214
    top-latching panic hardware montage,
      214
  dead-bolt equipped panic hardware,
    209–210
  delayed egress lock, 211–212, 211*f*
    Special school QID lock, 212
  electric strike pocket, 54
  electrified mortise and cylinder locks, 53
  electrified panic hardware, 53, 106, 106*f*
  essential attributes, 105–106
  evolution, 147
  fire ratings, 139–140
  hi-tower lock, 212–214, 213*f*
  infinite array, 105
  key distribution, 146
  lost keys, 147
  magnetic lock, 53
  mechanical, 105
  mortise lock, 106–107, 106*f*
  no special knowledge exit, 101
  operation
    cylindrical lock, 153
    electric strikes, 150–151, 150*f*
    electrified dead bolts, 154–155, 154*f*
    electrified panic hardware, 152, 152*f*
    electromagnetism, 149
    electromechanical locks, 149
    magnetic lock, 153–154, 153*f*
    mortise lock, 151, 151*f*
    paddle-operated electromechanical
      dead bolts, 155–156, 155*f*
  plate and shear locks, 53
  power supplies, 156–157, 156*f*

Securitech locks, 210–211
types, 148–149
  electrified cylinder locks, 162
  electrified dead bolts, 162
  electrified strikes, 162
  magnetic locks, 162–164
wiring considerations, 157–159
  constant voltage, 158
  Ohm's Law, 158
  total resistance, 158
  voltage drop, 158
Electrified panic hardware, 483–484
Electrified panic hardware with vertical
    rods (EPH-VR), 220
  on framed glass doors, 221*f*
Electronic access control systems
  access control panel, 46
  access groups, 47
  access zones, 46–47
  authorized users, 46
  credential authorization, 52
  data, data retention, and reports, 55–56
  digital networks, 45
  elements, 44
  locks, alarms, and exit devices, 52–55
  Manhattan Project physical security, 51
  portals
    alarm monitoring, 50
    common type, 48
    credential readers, 49
    electrified locks, 49
    pedestrian and vehicle
      types, 48, 48*f*
    request-to-exit sensors, 51
    safety systems, 49–50
    tailgating, 49
  scheduling, 47
  user groups, 46
Electronic circuitry sensitivities, 463–464
Electronic dead-bolt locks, 10
Electronic security system, 326
Electronic turnstile
  electronic beams and circuit embedded
    type, 87–88
  "paddles"/"glass wings,", 87, 88*f*
  positive access control, 80, 80*f*
  throughput and safety provision, 87–88
Emergency egress side door, 231–232,
    232*f*
Emergency generator, 285
End-of-line resistors, 511
Enterprise-class security systems,
    330–331, 334
Entry set, 173

Environmental factors, system failures
    access control in the cloud, 469–470
    dirt, 466–467
    humidity/condensation, 465–466
    insects, birds, snakes, and other
        creatures, 467
    lighting, 467–468
    securing the IP network, 468–469
    temperature extremes, 464–465
    vibrations, 466
Erasable programmable read-only memory
        (EPROMs), 263–264, 278
Ethernet architecture, 266
Exit push button, 108
Explosives detection methods and systems
    dogs, 364–365
    millimeter wave scanners, 366
    package X-ray scanners, 365
    personnel X-ray scanners, 365–366
    spectrographic detection systems, 366
    types, 364
    visual inspection approaches, 364

F

Facility Managers, 7
Fail-over host server, 303–304
Fail-Safe locks, 49
Fail-Secure locks, 49
Fiber-optic cables, 442, 455
Fingerprint reader, 70, 71f
Fire ratings
    basic fire egress concept, 132
    door assembly ratings
        glass doors, 135
        louvers, 135–136
        temperature rise doors, 135
        three-fourths rule, 135
    electrified locks, 139–140
    fire door frames and hardware
        fire exit hardware, 136–137
        latching devices, 136
    life safety code, 132
    pairs of doors
        astragal, 138
        double egress doors, 137–138
        inactive leaf, 137
        latching hardware, 137
        smoke and draft control, 138
    path of egress doors, 138–139
    penetration ratings
        time, 133
        wall ratings, 133

two broad exceptions, 132–133
Flexible metallic conduit (FMC), 447
Flexible metallic tubing (FMT), 447
Force Multiplier, 7
Foundational security
    access control portal, 21, 22f
    risk management
        credential programming, 38
        group and schedule
            programming, 38
        methods, 28
        portal programming, 38
        program elements, security, 29–30
        risk analysis, 30
        security and access control programs,
            29
        security countermeasures, 33
        security policies and procedures,
            30–31
        types of areas/groups, 37
        types of users, 37
        user schedules, 37
    understanding risk
        criticalities and consequences, 26–27
        organization asset types, 22–23, 23f
        probability, 27–28
        threat actors, 24–26
        types of users, 23–24
        vulnerability, 27
"Frame Actuator Controlled" strike plate,
    213
Frame spreader, 484
Framed glass doors, 119–120, 120f,
    221–222, 221f
Free egress electrified locks, 10
    electric strikes
        AC/DC version, 183–184
        fail-safe/fail-secure functions, 183
        latch keeper, 181–182
        mechanical lock, 181
        switches, 184
    electrified cylinder lock, 184–185
    electrified mortise locks
        additional lock switch fitting, 175
        advantages, 170
        dead bolts, 172–174
        door frame considerations, 174–175
        door handing, 175, 176f
        latch only — no lock, 171
        mechanical lock, 170, 171f
        no dead bolt, 171–172
        solenoid mechanism, 170
    electrified "panic" hardware

concealed vertical rod exit device,
    178, 178f
electrical options, 179–181
latch dogging, 179
mortise lock exit devices, 177, 177f
normal function, 179
popular double door applications, 181
purpose, 176
rim exit devices, 176–177, 177f
surface-mounted vertical rod exit
    devices, 177–178, 178f
three-point latching exit device, 179,
    179f
self-contained access control lock, 185,
    186f

G

Galvanized rigid conduit (GRC), 446
Gantt chart, 496, 496f
Glass-wing turnstiles, 243, 244f
"Global" system decisions/functions, 299
Governance, risk management and
        compliance (GRC) program,
        520–522

H

Hacking of security system, 414
Hand geometry reader, 70, 71f
Hard disk storage (HDS), 383
Herculite™ lobby doors, 222–223, 223f
Hi-Tech countermeasures, 325
Hi-tower lock, 212–214, 213f, 223–225,
    224f
HIPAA (Healthcare Insurance Portability
    and Accountability Act of 1996),
    417–418
Hirsch keypad, 63–64, 63f
Hollerith cards, 65, 66f
Hollow metal
    doors, 117–118, 117f
    frames, 125

I

Identification readers, 72
Institution lock, ANSI F30, 172
Integrated alarm system devices, 12
    alarm analysis, 369
    assessment, 351–353
    building perimeter detection systems
        balanced-biased door position
            switches, 359–360

glass break detectors, 360
photoelectric beam detectors, 360
seismic detectors, 360–361
communication and annunciation,
350–351
complex alarm sensing, 368–369
detection and initiation, 347–348
evidence, 353–354
filtering and alarm states, 348–350
intelligent video analytics sensors
conventional central processing
sensors, 368
edge sensors, 367–368
learning algorithm systems, 368
types, 367
interior point detection systems
Duress Alarms, 362–364
explosives detection methods and
systems. *See* Explosives detection
methods and systems
radiological detection systems,
366–367
interior volumetric sensors
dual-technology detection systems,
362
infrared detection systems, 361–362
microwave detection systems, 361
thermal imaging detection systems,
362
ultrasonic detection systems, 362
outer perimeter detection systems
capacitance detection systems,
355–356
fiber-optic detection systems, 355
gate breach detection system,
354–359
ground-based radar, 359
infrared and laser detection systems,
356–357
leaky-coax cable, 356
microwave detection systems, 358
outdoor passive infrared detectors, 357
pneumatic underground detection
systems, 357–358
seismic detection systems, 354–355
systems, 357–358
response, 353
trend analysis, 369
vulnerability analysis, 369
Intermediate metal conduit (IMC), 447
International Building Code (IBC), 101
Internet group management protocol
(IGMP), 293

"Internet of Things" (IoT) security devices,
405–406
IP infrastructure, 413–414
IP security intercom system integrity
monitoring, 531
Iris reader, 70, 72*f*
IT security, 409–412
mission, 402–404

K
K8 lift-arm barrier gate, 246*f*
K12 phalanx gate, 247*f*
K12 sliding gate, 248*f*

L
Latch-bolt monitoring (LX) switch, 180,
184
Layering approach, 6
Life safety, 9
codes and standards, 101–104
fire watch, 102–103
IBC, 101
legal action, 102
NFPA 72, 102
NFPA 101, 101
security contractor, 102–103
UL 294, 103–104
electrified locks
electrified panic hardware, 106, 106*f*
essential attributes, 105–106
infinite array, 105
mechanical, 105
mortise lock, 106*f*
no special knowledge exit, 101
and exit devices
access panel, 107
change of state, 108
intrusion, 107–108
mechanical switches, 108
motion detector, 108
request-to-exit sensor, 107–108
and fire alarm system interface,
109–111, 111*f*
first, 99–100
function, 49–50
vs. security, 101
Lift-arm barrier gates, 247
Liquid-tight flexible metal conduit
(LFMC), 447
Liquid-tight flexible non-metallic conduit
(LFNC), 447
Lo-Tech countermeasures, 325

Local area network (LAN), 266
Lock power supply, 285
Lock status monitoring switch, 184

M
Magnetic locks, 10
cautions
egress assurance, 195–196
operational and maintenance
warnings, 196–198
magnetic gate locks, 193–194, 194*f*
magnetic shear locks, 191–193, 192*f*,
194*f*
standard
basic forms, 191
direct current, 190
door configurations, 190
double door configuration, 192*f*
holding forces, 190–191
inswinging door configuration, 192*f*
large contact area, 189–190
outswinging door configuration, 191*f*
primary version, 189, 190*f*
Magnetic plate lock. *See* Standard
magnetic locks
Magnetic shear lock, 154, 154*f*
Maintenance concerns, 197–198
Maintenance schedule monitoring,
531–532
Man-trap, 88–89, 89*f*
Master host server, 308
Microcontrollers, 285–286, 482
Millimeter wave scanners, 366
Mobile phone access control, 70
Monostatic microwave detection systems,
358
Motherboard, 279
Mouse-operated graphical user interface
systems, 261
"Moving shooter" attack, 89
Multi-technology card readers, 70
Multitenant hi-rise buildings, 338, 341*f*
Municipal area network (MAN), 266

N
National Electrical Code (NEC), 490
National Fire Protection Association 101
(NFPA 101), 101
National Security Agency hacks, 405
Network attached storage (NAS),
383–384, 385*f*
Network integrity monitoring, 530

Network security monitoring, 529–530
No-Tech countermeasures, 325

O

Office lock, 171
"One-to-one" match reader.
  See Verification readers
OSI network operating model, 403

P

Package X-ray scanners, 365
Paddle-type electronic turnstiles, 243, 243*f*
Paneled overhead doors, 126
Parallel processing approach, 339–342
PDP-8/IBM Series 1 mini-computer,
  260–261
Pedestrian portal types
  automated wall, 89
  automatic doors, 83–85, 85*f*, 86*f*
  man-trap, 88–89, 89*f*
  revolving doors, 85–87, 86*f*
  standard doors, 82–83, 83*f*
  turnstiles, 87–88, 88*f*
Personnel X-ray scanners, 365–366
Phalanx gates, 247
Photo ID card
  combined functions, single credential,
    73
  printer, 73
Physical and IT security, 13
  merging of, 401
    mission, sharing, 402–404
    reduction and mitigation of
      vulnerabilities, 407–408
    sophisticated threat actors, 406–407
Pizza Guy, 63
Pneumatic controlled devices (PN), 181
Pneumatic push button exit switches, 511
Portal system, 9
Positive access control, 242
Power supply monitoring system, 529
Prison function lock, 173
Privacy set, 173
Programmable logic controllers (PLCs),
  265–266, 512
Project management, 479–480
Project Management Institute (PMI)
  certification, 479
Project Management Professional (PMP)
  certification, 479–480
Proprietary security system networks, 415
PVC conduit, 447

R

Random access memory (RAM),
  263–264, 278
Rare-Earth magnets, 65–66
Reactive electronic automated protection
    systems (REAPS)
  acoustic weapons, 431–433
  appropriateness, 434
  attack disruption, 430
  barriers and weapons operation,
    434–435
  communications elements, 430
  deluge fire sprinkler control, 431
  deployable barriers, 430
  electronic safety systems, 435
  high-voltage weaponry, 433
  irrigation systems, 430–431, 431*f*
  mechanical safety systems, 435
  operational safeguards, 435
  procedural safety systems, 436
  remotely operated
    weaponry, 433–434
REAPS. *See* Reactive electronic automated
    protection systems (REAPS)
Recommended Standard 232 (RS-232),
  286–287
Redundant host server, 304
Request-to-exit (RX) switch, 179–180
Request-to-exit sensors, 195
  access control panel, 195–196
  categorization, reliability, 195–196
  exit devices, 54–55
  magnetic lock, 196
  montage, 195*f*
  portals, 51
Revolving doors, 126–127
Rigid metal conduit (RMC), 446
Rigid nonmetallic conduit (RNC), 446
Rising bollards, 247
Risk management, 8
  credential programming, 38
  group and schedule programming, 38
  methods, 28–29
  portal programming, 38
  program elements, security, 29–30
  risk analysis, 30
  security and access control programs, 29
  security countermeasures, 33
  security policies and procedures, 30–31
  types of areas/groups, 37
  types of users, 37
  user schedules, 37
Roll-up overhead doors, 126

S

Safety concerns, 196–197
Securitech locks, 210–211
Security concerns, 197
Security countermeasures
  Hi-Tech systems, 31–32
  layering, 33
  Lo-Tech elements, 32
  mixing approach, 33
  No-Tech elements, 32
Security door controls (SDC), 204
Security mission, 402–404
Security program, 8
Security system integration
  advanced system integration, 335–336
  appropriate and inappropriate users, 324
  armed guards, 328
  assets, 323–324
  basic system integration, 335
  benefits of, 328–331
  challenge of, 326
  cost benefits, 331
  decision making, 326
  deterrence, 326
  digital video system, 327
  GUI Map, 327
  highly qualified guard at location,
    326–327
  operational benefits
    better communications, 330
    force multipliers, 329
    improved monitoring, 330
    improved system performance, 330
    multiple buildings, 329
    multiple business units, 329
    multiple sites, 329
    multiple systems, 329
    reduced training, 330
    uniform application of security
      policies, 328–329
  organization's mission, 324
  Police Emergency number, 328
  resources, 327
  security countermeasures, 325
  threat actors, 324–325
  types
    database integration, 334
    dry contact integration, 332, 332*f*
    serial data integration, 334, 334*f*
    TCP/IP integration, 334
    wet contact integration, 332–333
Security system integrity monitoring,
  528–532

goals and objectives, 528
scope of integrity monitoring, 528
system descriptions, 528–532
Security system, securing, 413
9 point plan for, 417–420
IT security solutions, 416–417
threats, 414–415
unsecure security system, 414
vulnerabilities, 415–416
Security systems
application
electronic security countermeasures,
485
expandable and flexible, 486
installation quality, 485
layered security approach, 484, 485*f*
monitoring functions, 486–487
redundancy, 486
sustainability, 487–488
architecture models, 393–394
C3 consoles, 394–395
cabling, 14
campuses and remote sites, 393–394
communications
CCS, 392–393
electronic security systems, 388–389
intercoms and bullhorns, 390–391
Nextel™ phones, 391–392
public address systems, 391
smart phones and tablets, 392
telephones, 389–390
two-way radio, 389
voice logger, 392
design, 4, 14
digital architecture, 475–476
elements
client management, 480–481
drawings, 476–477
interdiscipline coordination, 478–479
product selection, 479
project management, 479–480
specifications, 477–478
environmental considerations, 14
growth and flexibility, 475–476
implementation
access control zones, 488
alarm devices, 490, 492*f*
conduits and boxes, 490–492
door types, 488–490
physical details, 492–493
racks, consoles, and panels, 490
riser diagrams, 493
schedules, 495, 495*f*

single-line diagrams, 494, 494*f*
wiring diagrams, 494–495
integration, 12. *See also* Security system
integration
photo ID systems, 375–376
risk analysis, 475
robust portals
door frames, 484
double glass door exploit, 482–483
electrified panic hardware, 483–484
unlocking door from outside, 482
security video
assess events, 377
auto-white balance, 382
cameras and lenses, 379–380
display devices, 382
dynamic range, 382
history, 377–378
inappropriate/suspicious
activity, 377
lighting and light sources, 380–381
motion detectors, 386–387
movements of subjects, 377
recording devices. *See* Video
recording devices
remote investigations, 377
system interfaces, 387–388
video analytics, 387, 388*f*
visual evidence, 377
system installation
access coordination, 498–499
coordination with trades, 498
permits, 498
preliminary checks and testing, 499
product acquisition, 497–498
project planning, 495–496
project schedule, 496–497
punch list items completion, 500
shop and field drawings, 497
system acceptance, 499–501
system commissioning, 499–500
technicians, 4–5
visitor management systems, 376–377
vs. installation vs. maintenance, 474
vulnerabilities, probability, and
consequences, 475
Security-systems as-a-service (SSaaS), 470
Security technologies, 5
Security theater, 414
Semaphore arm barrier gate, 89–91, 90*f*
Servers and workstations
access control system networking
access control panel network, 311

business information technology
network, 312–313
core network, 309–310
integrated security system interfaces,
311–312
multisite network interfaces, 312
server network, 310
TCP/IP Ethernet networks, 309
VLANs, 312
workstation network, 310
decision process, 305–306
functions, 300–305
access control events, 302
access control module configurations,
301
access control panel configurations,
300
alarm events, 302
antipassback events, 302
cluster configurations, 301
communications management,
303–304
distributed cluster management, 301
operator logs, 302–303
output control events, 302
real-time data and reports, 304–305
scheduled events, 302
legacy access control systems, 313
system scalability, 306
master host, 308
super-host/sub-host, 308–309
unscalable systems, 306–309
basic scalability, 307
enterprise-wide system, 308
multisite systems, 307
system-wide card compatibility,
307–308
Signal switch (SS), 180
Silicon controlled rectifier (SCR),
332–333, 333*f*
Single- and double-leaf swinging doors
balanced doors, 124–125, 124*f*
framed glass doors, 119–120, 120*f*
hollow metal doors, 117–118, 117*f*
pivoting doors, 123, 123*f*
solid core wood doors, 118–119, 119*f*
total doors, 121–123, 122*f*
unframed glass doors, 120–121, 121*f*
Sliding gates, 247
Sliding panel doors, 127
Software developers kit, 279
Solenoid function, 150–151, 150*f*
Sophisticated threat actors, 401, 406–407

SOX (The Sarbanes-Oxley Act of 2002), 417–418
Special school QID lock, 212
Specialized portal control devices
  pedestrians, 238–245
    antitailgate alarm, 245
    automatic doors, 238, 239*f*
    electronic turnstiles, 242–244
    full-verification portals, 241–242, 242*f*
    man-traps, 239–241, 240*f*
  vehicles
    high-security barrier gates, 246–247
    sally ports, 247–249
SQL. *See* Structured Query Language (SQL)
Standard gamma ray detectors, 366–367
Standard magnetic locks
  basic forms, 191
  direct current, 190
  door configurations, 190
  double door configuration, 192*f*
  holding forces, 190–191
  inswinging door configuration, 192*f*
  large contact area, 189–190
  outswinging door configuration, 191*f*
  primary version, 189, 190*f*
Storage area network (SAN), 383, 385, 386*f*
Store/utility room lock, 172–173
Storefront doors. *See* Framed glass doors
Storeroom lock, 172
  dead bolt, 172
Storeroom lock, ANSI F07, 172
Structured Query Language (SQL)
  access control panel form factors, 280
  BMS, 428
  database integration, 334
  EPROM, 278
  fifth generation systems, 266
Surface-mounted magnetic lock, 153, 153*f*
Surface-mounted magnetic shear lock, 193*f*
Surface mounted raceways, 448
System installation and commissioning
  access schedules and areas, 514–515
  alarm and reader device database, 513
  conduit vs. open cabling, 511
  device installation considerations, 511–512
  device setup and initial testing, 513
  documentation, 512
  jobsite considerations, 510–511

user access database, 513–514
System maintenance and repair, 14–15
  as-built drawings
    floor plans, 523
    physical details, 523
    power infrastructure, 523
    schedule sheets, 523
    single-line diagrams, 523
    title sheet, 523
    wiring diagrams, 523–524
  contracting options
    annual maintenance agreement, 527
    extended warranty, 527
    in-house technician, 526–527
  definition, 525
  emergency, 525–526
  infant mortality, 524–525
  on-call maintenance and repairs, 527
  scheduled maintenance, 525
  wire run sheets, 524
System management
  alarm management, 522
  common corporate campus, 521
  database management, 520–521
  governance, risk management and compliance (GRC) program, 520–522
  maintenance and repair, 14–15
  reports, 522
  security system integrity monitoring, 528–532

T

Tailgate detector, 78–80, 79*f*
TCP/IP Ethernet communications, 309
TCP/IP infrastructures, 408
Temperature rise doors, 135
Thin-wall tubing, 446
Threat actors
  sophisticated, 406–407
Threat assessment, 409
Top jam mount surface magnetic shear lock, 194*f*
Triac, 332–333, 333*f*
Tripod turnstiles, 87–88

U

Underground conduit, 447
Understanding risk, 7

Underwriters Laboratories 294 (UL 294), 103–104
Uninterruptable power supply (UPS), 285
Universal Building Code (UBC). *See* International Building Code (IBC)
Unsecure security system, 414

V

Vehicle portals
  automated roll-up vehicle gates, 92, 93*f*
  automated sliding vehicle gates, 91, 92*f*
  automated vehicle swing gates, 91, 91*f*
  high-security barrier gates, 92–94, 93*f*
  sally ports, 94, 94*f*
  standard barrier gates, 89–91, 90*f*
Verification readers, 72
Video gamma ray detectors, 367
Video recording devices
  DAS, 383–384, 385*f*
  DVRs, 382–383, 383*f*
  fail-over servers, 383
  internal HDS, 383
  NAS, 383–384, 385*f*
  network video recording systems, 383
  open reel 2″ videotape recorders, 382
  RAID arrays, 386
  SAN, 383, 385, 386*f*
  server-based systems, 383
  time-lapse VHS recorders, 382
Virtual local area networks (VLANs)
  local and network connections, 288–289
  security architecture models, 393–394
  servers and workstations, 312
Virtual private networks (VPNs), 293–294
VLANs. *See* Virtual local area networks (VLANs)
Vulnerabilities between IT and physical security, reduction of, 407–408

W

Web fabric gates, 247
Web fabric K12 gate, 248*f*
Wide area network, 309
Wide area network (WAN), 266
Wiegand protocol, 67–68
Wiegand wire cards, 66–68, 67*f*
Wire run sheets, 524